ECONOMICS OF ELECTRONIC DESIGN, MANUFACTURE AND TEST

edited by

Magdy Abadir
Microelectronics and Computer Technology Corporation

Tony Ambler
Brunel University

A Special Issue of
Journal of Electronic Testing: Theory and Application
Vol. 5, Nos. 2 & 3
May/August 1994

SPRINGER SCIENCE+BUSINESS MEDIA, LLC

FRONTIERS IN ELECTRONIC TESTING

Consulting Editor
Vishwani D. Agrawal

Books in the series:

I_{DDQ} Testing of VLSI Circuits,
R. Gulati, C. Hawkins
ISBN: 0-7923-9315-5

Economics of Electronic Design, Manufacture and Test,
M. Abadir, A.P. Ambler
ISBN: 0-7923-9471-2

ECONOMICS OF ELECTRONIC DESIGN, MANUFACTURE AND TEST

A Special Issue of Journal of Electronic Testing

Vol. 5, Nos. 2 & 3, May/August 1994

Page numbers in this table of contents are the ones appearing at the bottom of the pages.

ISBN 978-1-4419-5142-7 ISBN 978-1-4757-5048-5 (eBook)
DOI 10.1007/978-1-4757-5048-5

Library of Congress Cataloging-in-Publication Data

A C.I.P. Catalogue record for this book is available
from the Library of Congress.

Printed on acid-free paper.

JOURNAL OF ELECTRONIC TESTING: Theory and Applications, 5, 127 (1994)

Foreword

The general understanding of *design* is that it should lead to a manufacturable product. Neither the design nor the process of manufacturing is perfect. As a result, the product will be faulty, will require *testing* and fixing. Where does *economics* enter this scenario?

Consider the cost of testing and fixing the product. If a manufactured product is grossly faulty, or too many of the products are faulty, the cost of testing and fixing will be high. Suppose we do not like that. We then ask what is the cause of the faulty product. There must be something wrong in the manufacturing process. We trace this cause and fix it. Suppose we fix all possible causes and have no defective products. We would have eliminated the need for testing.

Unfortunately, things are not so perfect. There is a cost involved with finding and eliminating the causes of faults. We thus have two costs: the cost of testing and fixing (we will call it cost-1), and the cost of finding and eliminating causes of faults (call it cost-2). Both costs, in some way, are included in the overall cost of the product. If we try to eliminate cost-1, cost-2 goes up, and vice-versa. An economic system of production will minimize the overall cost of the product.

Let us examine the roles of economists and engineers. As engineers, we design, manufacture and test airplanes, cars, computers, telephones, as well as components like printed circuit boards and VLSI chips. In this work, we deal with a variety of things like electricity, energy, heat, information, light and sound. Our designs process and produce these things. Our analyses deal with a variety of things through a system of standardized units. All quantities can be represented by units expressible in terms of the *fundamental* standard units of length, mass and time.

Economists, on the other hand, deal with systems where the end users are humans, societies or nations. Inputs to these systems, in addition to the engineering quantities I talked about, include labor, risk, etc. All quantities are converted into monetary units like dollar, mark, pound or yen. Engineering designs often form subsystems within a larger economic system.

The design of any subsystem should, therefore, examine the impact of engineering optimization on the overall system parameters like user cost and satisfaction. This is not always easy. Can we quantify user's satisfaction? Another problem is the standardization of the monetary unit. Can we express dollar in terms of the engineering units of length, mass and time? Or, if the monetary unit was to be included in the fundamental system of units, do we have a standard available? The present work does not provide a complete answer to any of the above questions, the last one of which, at some future time, may be addressed by our esteemed Standards Institutions.

The articles here add to the awareness of economics among electronic engineers and provide practical information related to the central theme. The material is derived from the *Second Workshop on Economics of Design, Manufacturing and Test*. This workshop series is a continuing forum for such awareness. I am thankful to Magdy Abadir and Tony Ambler, the organizers of the Workshop, for editing the material in the present form. I hope, its reading will make engineers' work more beneficial and satisfying to users of their products.

Vishwani D. Agrawal
Consulting Editor
AT&T Bell Laboratories
va@research.att.com

JOURNAL OF ELECTRONIC TESTING: Theory and Applications, 5, 129 (1994)

Preface

The issue of test economics is a long and tortuous one! It is a topic that will bring on high blood pressure, and reduce sane engineers and managers to tears. Why is this? What is it about this topic that causes such a state? Is it because of years of preconceived impact on empires and future empire building put a focus on the topic?

The answer is probably related to all of the above questions, and yet at the same time none of them. What is clear is that knowledge of test costs has not been as widespread as it could have been, and that had this knowledge been available, then different decisions would have been made in many situations. A prominent example of this has been the pro's and con's of area overhead associated with many DFT methods. It has consistently been stated that area overhead equals increased costs, therefore reject the method (we slightly simplify the issue here!). Although the fashion for DFT has greatly improved, there are still those that harbor feelings of antagonism towards the use of DFT based purely on factors such as the *perceived* cost of area overhead.

Fortunately the general opinion now is that DFT *pays* for itself, but there is still widespread unease about the financial details surrounding its application.

Back in the early 1980s Varma et al. (University of Manchester Institute of Science and Technology) began to develop cost models to more accurately predict the cost effects of DFT on circuits, but it has been a long path to follow to get to where we are now where discussion of test economic issues can be held in a fairly rational manner. Ben Bennetts in his 1992 ITC keynote rejected notion of "area overhead" for DFT stating that the term should be banned. Instead he introduced the term "QUIF—Quality Improvement Factor" for quantifying the benefits of DFT—the question is, of course, how should this be measured? One factor might seem to be a test method's cost effectiveness, i.e., the balance of increased costs through, for example, increased silicon area, vs. the savings in test generation costs, diagnosability etc. whilst assuming a minimum (stated) fault coverage.

In 1991, the first International Workshop on the Economics of Test was held in Austin, Texas, hosted by MCC with sponsorship by the ACM. This was a resounding success and led to the second workshop, now renamed The International Workshop on the Economics of Design, Test and Manufacturing, to reflect the wider issues. Again a success, articles from that Workshop have now been rewritten for presentation in this special issue of *JETTA*. An open call was issued both by *JETTA* and by the Workshop. Articles were subject to review prior to the Workshop, were rewritten after the Workshop and subject to further review.

This special issue contains 17 articles. The first by Vishwani Agrawal, the Workshop keynote speaker, introduces the topic by talking about the various issues surrounding engineering economics in general. This is followed by two articles by Dear et al., and Brendan Davis that review the topic area of test economics and present ways of analyzing and using the data in an engineering environment. Zarrinfar follows with a discussion of test economics with ASICs in mind with a focus on time-to-market savings that can be accrued through the use of DFT. Next, Varma looks at the economics of scan testing comparing full-scan with partial-scan.

Then there are two articles from MCC, Abadir et al., and Sandborn et al., that describe tools for multichip module design. The first article looks at comparing DFT methods through an economic analysis, whilst the second does a trade-off analysis of non-test related design parameters. In order to emphasize the non-test specific nature of the methods being described, the following two papers by Rao et al. and Alexander et al. look at packaging selection and pcb design.

The use of economics modeling as a design aid is addressed in the article written by Dick et al. In an industrial environment where data may not be available, or the accuracy might be in question, the sensitivity of the model to particular data will be of interest.

The next article by Millman discusses the relative cost merits of increasing test coverage vs. yield on improving quality levels.

The following group of five articles, Ziaja et al., Reventlow, Nicolaidas et al., Majumdar et al., Fares et al., and Kunzmann et al., address various aspects of test-related features that affect the ease of testability with the inevitable impact on costs.

We would like to thank the Program Committee of the Workshop and the reviewers for their efforts in helping to compile this special issue of *JETTA*.

Magdy Abadir
MCC
abadir@mcc.com

Tony Ambler
Brunel University
tony.ambler@brunel.ac.uk

JOURNAL OF ELECTRONIC TESTING: Theory and Applications, 5, 131-135 (1994)

A Tale of Two Designs: the Cheapest and the Most Economic

VISHWANI D. AGRAWAL
AT&T Bell Laboratories, Murray Hill, NJ 07974

> ...Economics is the study of how men choose to use scarce or limited productive resources (land, labor, capital goods such as machinery, and technical knowledge) to produce various commodities (such as wheat, overcoats, roads, concerts, and yachts) and to distribute them to various members of society for their consumption...
>
> — Paul Samuelson [1]

Abstract. Engineering economics is the study of how engineers choose to optimize their designs and construction methods to produce objects and systems that will optimize their efficiency and hence the satisfaction of their clients. We discuss some relevant concepts in engineering economics. We begin with the analysis of production and operational costs, and benefit-cost analysis and end with the economic arguments that justify design for testability.

Keywords: Benefit-cost trade-off, cost analysis, economics of design for testability, engineering economics.

1. Introduction

Engineers are often concerned with optimizing the *technological* efficiency. For example, in designing a heat engine, the prime consideration can be to have as much of the heat energy converted into mechanical work as possible. Economists, on the other hand, prefer to minimize the cost of obtaining mechanical work for the consumer. Apart from the technological (or energy conversion) efficiency, other factors like fixed and variable costs of material, equipment, labor, insurance, etc., will also enter into their consideration.

Lately, economics of test has received noticeable attention. The cost of testing has a complex relationship with the quality of products. For large electronic systems, testing accounts for a significant fraction of the total cost. Still, it has not been easy to justify the cost of design for testability at the component level. Perhaps, we will have better success in justifying testability if we looked at the impact of design for testability on the overall system cost rather than considering the component cost alone.

Study and practice of the principles of economics are as important for engineers as are the laws of physics. Even though we may, at times, find ourselves pursuing local goals like profitability, time to market, and beating the competition, we must not lose sight of the real goals such as usefulness to society and preservation of environment. Fortunately today, economics is a well-developed science. It is divided into two main streams. Microeconomics is the study of economic laws at a small scale as affecting a company. Macroeconomics deals with the wealth of society at the national or international scale. This article provides a brief introduction to microeconomics. We give basic ideas of costs, production, and benefit-cost analysis and study the case of built-in self-test in electronic products.

2. Defining Costs

Cost is a measurable quantity that plays a key role in economics. For example, we compare costs and benefits to select between alternatives or minimize costs to optimize a design. As we define below, there are several types of costs.

Fixed Costs (FC)

These are the costs of things that are necessary but do not change with use. For example, if we wish to produce cars, we require a factory building and machinery that contribute to the fixed costs. These costs do not change with the number of cars that are produced. The fixed costs are the same whether we build one car or one thousand cars. Even though the fixed costs remain unchanged, the fixed costs per car will reduce as the production is increased.

Variable Costs (VC)

These costs increase with the production output. For example, the variable costs of producing one thousand cars will be a thousand times greater than the variable costs of producing one car. Variable costs account for the labor, energy and consumable material.

Total Costs (TC)

Total costs are the sum of the fixed and variable costs. These also increase with production output.

Average Costs (AC)

Average costs are the total costs divided by the number of units produced.

To illustrate the role of costs analysis, let us consider the costs of transportation as incurred by an individual. Suppose the means of transportation in this case is a car. We take the purchase price of the car, say \$25,000, as the fixed cost. We estimate the variable costs that account for gasoline, maintenance, repairs, and insurance as 20¢ per mile. The total cost of traveling x miles is $(25{,}000 + 0.2x)$ dollars. Thus, the average cost is obtained as:

$$\text{Average cost} = \frac{25{,}000}{x} + 0.2 \text{ dollars}$$

This is a rather simple formula that illustrates the principle of computing average cost. Depending upon the situation, various types of fixed and variable costs should be included. For example, suppose the owner of the car drives 10,000 miles per year for five years and then sells it for \$7,500. The fixed cost then reduces to \$17,500 and the average cost will be 55¢ per mile. If the car is used for ten years and then sold for \$5,000, the average cost will be 40¢ per mile. Also consider the case where the car is used for twenty years and has no resale value. Then the average cost will be 32.5¢ per mile. These costs are shown in Figure 1. Notice that two average cost curves are given in the figure. The monotonically decreasing average cost is computed from the above formula. However, in reality, the repair costs may go up after the car has been driven over 100,000 miles. When the aging factor is taken into account, the average cost might be as shown by the rising curve. Thus, durability of equipment may be more important than the initial purchase price for realizing lower average operating cost.

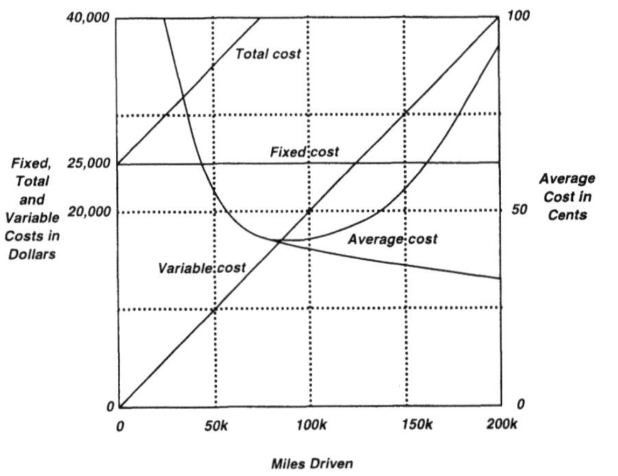

Fig. 1. Costs of transportation.

3. Production

Production is the process of making articles for which the society has needs. Inputs to production are labor, land, capital, energy, and enterprise [2]. Enterprise here can mean a variety of things like technical know-how, organiational skills, etc. The inputs account for the cost of production. Although inputs consist of widely different items, they can all be converted into dollar equivalent. Notice that both fixed and variable costs may be included.

A short-run production means that some of the inputs are fixed. An exmaple is the production in a factory over a period during which the size of the manufacturing facility remains fixed. Thus, in the short-run, output possibilities are limited. The company can hire more workers, order more raw material, perhaps add a shift, but that is about all. The long-run production is over a period during which the company can change all inputs, including the size of the manufacturing plant.

We first consider the short-run production. *Production output*, Q is a function of the inputs x:

$$Q = f(x)$$

Here x is the variable cost of production. We also define,

$$\textit{Average product} = \frac{Q}{x}$$

and

$$\textit{Marginal product} = \frac{dQ}{dx}$$

The average product, which is the product per unit of input, is also known as the *technological efficiency*. To find out how this efficiency can be maximized, let us set the derivative of Q/x with respect to x to 0. Thus,

$$\frac{d}{dx}\frac{Q}{x} = 0$$

This leads to the following relation:

$$\frac{Q}{x} = \frac{dQ}{dx}$$

Figure 2 illustrates the above relation. Maximum technological efficiency occurs when the average product equals the marginal product. The idea of maximizing this efficiency is to convert the variable cost into product as best as possible. This is like optimizing the energy conversion efficiency of a steam engine or an electric motor.

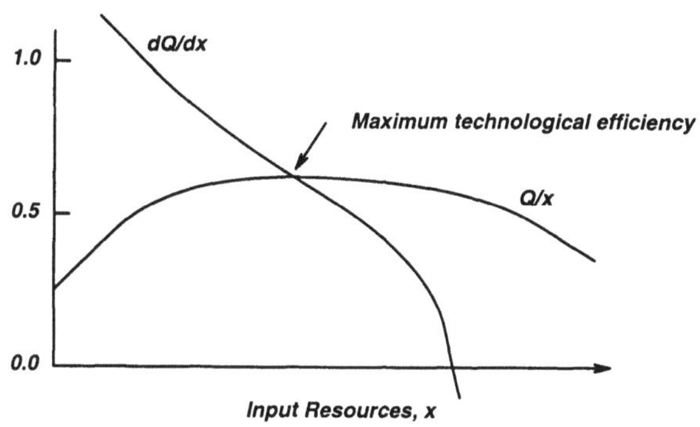

Fig. 2. Maximizing technological efficiency.

Engineers are good at optimizing the technological efficiency. What they often ignore is economic efficiency. The idea behind economic efficiency is related to the total cost of production. This cost includes both fixed and variable costs. Thus, economic efficiency may also apply to long-run production. Suppose, X is the total cost (TC). Then, the average cost (AC) is X/Q and marginal cost (MC) is dX/dQ. The economic efficiency minimizes AC and is achieved when MC = AC. This is illustrated in Figure 3. Consider the point shown on this graph as the "Max. Tech. Efficiency." If we neglect the fixed cost then this gives the lowest cost for the product. However, it is impossible to produce anything without fixed cost. For nonzero fixed cost, the average cost will be higher as illustrated by the other two average cost curves in Figure 3. Notice that if we were to produce at maximum technological efficiency, due to the actual nonzero fixed cost, the average cost of the product will be higher than the minimum cost given by the maximum economic efficiency.

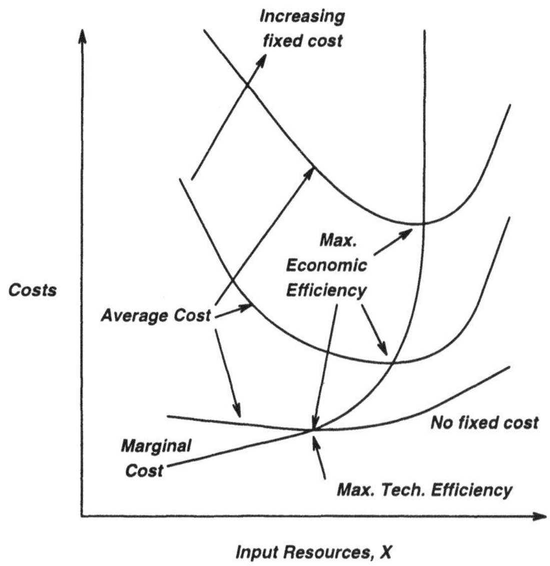

Fig. 3. Maximum economic efficiency.

This simple illustration shows that maximum economic efficiency may not be obtained at the same level of production that provides maximum technological efficiency. The difference here was in whether we minimize the TC or the VC. More detailed considerations in long-run production take factors like supply-demand and competition into account [2], [3].

Notice the shape of the production (Q/x) curve in Figure 2 which is typical of all production environments. As inputs are increased, first the production rises and then decreases. This is according to the fundamental principle known as the *law of diminishing returns*. This law is stated as follows: *If one input of production is increased keeping other inputs constant, then the output may increase, eventually reaching a point beyond which increasing the input will cause progressively less increase in output.* In practice, a variety of reasons cause diminishing returns. For example, when the input to a mechanical engine is increased, at some point its energy conversion efficiency begins to fall. This may be due to greater heat loss. If more workers are put on the same job, they usually come in each other's way and the rate at which the work is done does not increase proportionately. Similarly, the phenomenon of diminishing speed up in multiprocessor computers is well known.

The case of mass production is worth considering. It is known that production often increases faster than the increase of inputs. This is called *increasing returns to scale*. Some of the reasons are:

- Technological factors
- Specialization

For example, a large company can afford to have a research and development unit to help utilize new technology in their production. They can also justify hiring specialists that a small company may not be able to afford. In the long run, however, the law of diminishing returns prevails. A company that has grown beyond control tends to become inefficient due to reasons like lack of communication among employees and counterproductive rivalries.

4. Benefit-Cost Analysis

Benefits include income from sale of products or services, savings in cost and time, etc. *Costs* refer to the costs of labor, machinery, energy, finances, risks, etc. All items are normally quantified and expressed in the same units (e.g., dollars). We then define the *benefit-cost ratio* as follows:

$$B/C\ ratio = \frac{annual\ benefits}{annual\ costs}$$

Both benefits and costs should be estimated and averaged over the lifetime of the product. When there is a choice among products with similar utility, the B/C ratio provides a criterion for comparison. Product selection on the basis of just minimum cost or maximum benefits can lead to imperfect result.

5. Economics of DFT

In the computer and communication industry, testing of electronic components is an important part of the business. Customers want reliable product at a reasonable cost. A manufacturer, with a poor design, may try to improve reliability by increased testing with the added cost passed on to the customer. However, in a competitive market economy, the consumer benefits by selecting the best product. To stay in business, the manufacturer must find means to provide the best product at the lowest cost. This requires production at the maximum economic efficiency.

The new advances in technology have given the capability to quickly design and manufacture very complex circuits at reasonable cost. However, as the cost of these products has reduced, the percentage of the total cost attributed to testing has increased [4]. In order to keep the cost of testing under control, the designers must consider the complexity of test. In electronic parts, therefore, design for testability (DFT) is an important discipline. Techniques like scan, built-in self-test (BIST), and boundary scan simplify the test problem of electronic systems.

Consider the case of BIST. Test generation and response analysis circuitry is built into the VLSI chip or the printed wiring board. Test can be run at the rated clock speed, and the device can be efficiently tested even when it is embedded in a larger system. However, we must pay the price somewhere. The BIST circuitry will reduce the yield of the VLSI chip thus increasing the cost. Unless this cost increase can be offset by cost reduction elsewhere, the use of BIST is not justified.

Systems planners and designers often face a decision of choosing from several alternatives. A typical question is: Should one use BIST? Indeed, to answer the question, we must weigh costs against benefits. When such a trade-off is considered at the chip level, BIST offers some benefits in testing cost. However, if one considers the product life-cycle cost, because of the savings achieved, the decision is overwhelmingly in favor of BIST. Table 1 shows the impact of BIST on testing cost for chips, boards, and systems [5]. For simplicity, only qualitative impact (increase or decrease) of BIST on cost is shown in the table. Considering the cost of design and test development, we find that the additional cost of designing the BIST hardware somewhat balances the saving from test generation. Fabrication cost is increased at all levels due to the extra hardware. Production test cost is reduced due to more efficient tests and less expensive test equipment allowed by the self-test design. Also, at board and system levels, BIST allows improved troubleshooting during assembly and integration. Of course, the maintenance cost has the greatest impact on the system operation. Diagnosis and repair costs are reduced at board and system levels. In alternative strategies, the lengthy or improper diagnosis is often responsible for much loss in revenues due to service interruption.

The main point of Table 1 is the significant benefit that BIST provides at the system level. Testability decision at one level cannot be made independent of other levels. This is because the chips are produced for use on boards, which are used in the system. Benefits of BIST may be small at the chip level, but the chips with BIST reduce the cost of self-test at the board and system

Table 1. BIST Costs: + indicates cost increase, − cost saving, and ± increase almost balances saving.

	Design and Test	Fabrication	Production Test	Maintenance Test	Diagnosis and Repair	Service Interruption
Chips	±	+	−			
Boards	±	+	−		−	
System	±	+	−	−	−	−

levels where benefits are greater. Thus, even with relatively lower benefits at chip and board levels, BIST is still the best design for testability alternative. Published work on economics of BIST [6] must be critically examined on this point.

6. Conclusion

The economic laws of mass production and diminishing returns apply to all production environments. The basic ideas behind economic theories is to allow improved usage of the available resources for the good of the society. In the design of a product the overall benefit/cost ratio for design, test and manufacturing should be maximized. One must select the *most economic* design over the *cheapest design*. Every engineer should study economics because it teaches how to live and work intelligently.

Acknowledgment

This article is based on the keynote talk delivered by the author at the *Second International Workshop on the Economics of Design, Test and Manufacturing*, Austin, TX, May 1993. The author thanks Tony Ambler for illuminating discussions on the topic of this article.

References

1. P.A. Samuelson, *Economics, an Introductory Analysis*, McGraw-Hill, New York, 1976.
2. R.L. Mitchell, *Engineering Economics*, John Wiley & Sons, Chichester, UK, 1980.
3. W.P. Albrecht, Jr., *Microeconomic Principles*, Prentice-Hall, Englewood Cliffs, NJ, 1979.
4. I.D. Dear, C. Dislis, J. Dick, and A.P. Ambler, "Economic Effects in Design and Test," *IEEE Design & Test of Computers*, vol. 8, pp. 64–77, Dec. 1991.
5. V.D. Agrawal, C.R. Kime, and K.K. Saluja, "A Tutorial on Built-In Self-Test, Part 1: Principles," *IEEE Design & Test of Computers*, vol. 10, pp. 73–82, March 1993.
6. A.P. Ambler, M. Abadir, and S. Sastry, *Economics of Design and Test for Electronics Circuits and Systems*, Ellis Horwood Limited, Chichester, UK, 1992.

Vishwani D. Agrawal holds a BSc (1960) from the Allahabad University, India, BE (1964) from the University of Roorkee, India, ME (1966) from the Indian Institute of Science, Bangalore, India, and PhD (1971) from the University of Illinois at Urbana-Champaign. Since 1978, he has been at the AT&T Bell Laboratories, Murray Hill, New Jersey, where he is a Distinguished Member of Technical Staff in the Computing Science Research Center. He is also a Visiting Professor of electrical and computer engineering at Rutgers University, New Brunswick, New Jersey. Previously, he worked at the TRW Defense and Space Systems Group in California, EG&G, Inc. in New Mexico, and the Indian Institute of Technology in New Delhi, India.

Agrawal's intrests include VLSI testing and distributed computing. He has coauthored over 175 papers and four books, and holds five U.S. patents on electronic testing. He is the founding editor-in-chief of the *Journal of Electronic Testing: Theory and Applications (JETTA)* and consulting editor of the Kluwer International Book Series titled, *Frontiers in Electronic Testing*. From 1985 to 1987, he was the editor-in-chief of the *IEEE Design & Test of Computers* magazine. He chairs the Steering committee of the International Conference on VLSI Design that is held annually in India. He served on the board of governors of the IEEE Computer Society for the years 1989 and 1990.

Agrawal is a Fellow of the IEEE and the IETE (India) and a member of the ACM and the VLSI Society of India (VSI). He has received five best paper awards and one honorable mention award for his papers. For his contributions to the field of computers, the IEEE Computer Society honored him with their Outstanding Contribution Award (1988) and Meritorious Service Award (1989). In 1993, he received the Distinguished Alumnus Award from the University of Illinois at Urbana-Champaign.

JOURNAL OF ELECTRONIC TESTING: Theory and Applications, 5, 137-155 (1994)

Test Strategy Planning Using Economic Analysis

I.D. DEAR, C.D. DISLIS AND A.P. AMBLER
Brunel University, Uxbridge, Middlesex, UB8 3PH, UK

J. DICK
Siemens Nixdorf Informationssysteme AG (SNI), Munich, Germany

Abstract. This article will discuss the impact on testing of life-cycle costs and present an approach for minimizing the overall life-cycle costs of a product by selecting the most economic test strategy at each stage. The selection of test strategy is based on a detailed economic analysis of the different test techniques available.

Keywords: ECOtest system, life-cycle costs, test methods, test planning.

1. Introduction

The economics of testing electronic components, boards, and systems has long been ignored as a serious topic for discussion, other than to say, *Can we afford to test?* Fortunately, the predominant view today is that quality is what matters. From this we must conclude that test is/should be considered essential. There may still be a wide variety of views as how best to achieve this, or what level of test is considered sufficient, but at least the DFT practitioners can raise their heads above the parapet.

The following statements should put the issues into context:

i. 70% of life-cycle costs are determined at the design stage [1].
ii. >60% of a product's cost can be attributed to testing costs [2].
iii. ~50% profits of systems manufacturers comes from maintenance contracts [3].

Quotes like these, assuming that the test issue is taken seriously, emphasize that there is a lot of money to be attributed to test related issues that the designer(s) will have a direct impact upon.

Of course, there are other factors that have a major impact upon a product's commercial viability, e.g., time to market. If the product is too late to the market place, it doesn't matter what its testability or quality actually is—no money will be made from it.

It has certainly been the case that designers have not placed as great an emphasis on testability issues as perhaps they should have. The reasons for this can be many: the manager said, "No!"; there was no room for it; performance impact was too high; or they just don't know about test... Conversely, even for the test literate designer the profusion of DFT methods that can be found in the literature is quite large. Which one(s) to select? Add to this the fact that the parameters that are useful in helping to decide the optimum DFT methods, e.g., circuit size, design time, performance impact, test pattern generation effort, test length, ATE requirements etc., all vary with circuit style. Indeed, it must be said that *not* using a structured DFT method at all could be the most cost effective way of achieving a given level of fault coverage in some circuits.

Unfortunately, the choice of parameters is more than these already mentioned. When comparing two similar DFT methods, the problem is compounded somewhat. Each method will require an *overhead* in terms of extra gates, but to differing amounts; they will, hopefully, increase fault coverage over what might have been achieved without their use, but to differing amounts; they will, hopefully, reduce test pattern generation costs, but to differing amounts; one DFT method might not require fault simulation, but the other may need it to verify fault coverage; the number of pins required on the device package will increase due to test control signals, extra clocks etc., but to differing amounts (this may mean increasing to the next package size for example); there will be an impact on ATE requirements, but to differing amounts.

The average designer can, therefore, be faced with a very wide range of sometimes conflicting requirements that need to be juggled when trying to achieve not just a testable design, but one that (given the statements made earlier) is as cost optimal as possible: how to relate the effects of increased pin count with reduced test pattern generation costs, increased area overhead vs. reduced test time. A wide range of very disparate values need to be compared. One way to achieve this in a coherent way is to cost the impact of each parameter on the overall design. In this context, "cost" means £, $, ¥, etc.

Such a method would remove any doubts as to relative importance and, potentially, pacify all doubters/dissenters in design, manufacturing, test, and management provided that it is done in a rigorous way.

Cost data that refer to test-related issues are difficult to find. Even a widely known concept as the "rule of tens" is now under attack as not representing what is happening in the 1990s [4]. Other graphs, Figure 1, show a scenario around DFT cost impact upon total costs, but neglect to show how to achieve an optimum level of testability that will minimize overall costs.

Even if the process needed to achieve minimum cost were known, there would still be problems surrounding management culture in any given company. Implementing DFT may well increase the Design Department's costs in order to make even greater savings in the Test Department.

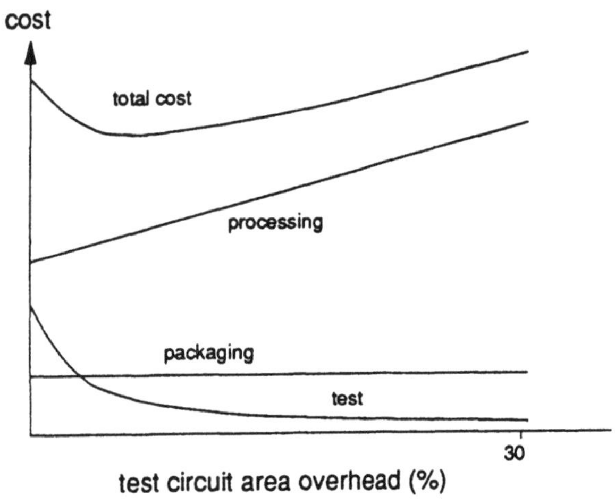

Figure 1. The effect of increasing area overhead.

This article discusses a set of software tools that have been developed at Brunel University with Siemens-Nixdorf Informationssysteme that address this issue. These tools are integrated around a suite of economic models that attach or calculate real costs for a wide range of parameters that cover all aspects of design, manufacture, and test so that any test-related decisions are placed in their correct contexts. The economic models then interface to a set of knowledge bases and algorithmic procedures to advise a designer as to the optimal mix of DFT strategies. The optimization process attempts to reduce the overall cost to the company and ensures that any specified engineering criteria is still met, for example, a minimum fault coverage, system operating speed.

2. Economics of IC Testing

2.1. Introduction

There are many different test strategies for testing integrated circuits. These vary from using no structured DFT approach, relying on good design practice and experts to generate an effective test, to adopting a single structured DFT approach [5]. These approaches have proven valuable and cost effective in many scenarios. However, whereas one method has proven to be economically viable for one company it is not guaranteed to be so for another. Indeed, what has proven to be economical for one design might not be economic for another within the same company.

As the complexity of devices and the variety of structured partitions of a chip increase there are more and more DFT approaches that could be used. For example, in many complex devices an effective test strategy can be obtained from a mixture of ad hoc methods, structured DFT and Built-in Self-Test (BIST). This adds to the complexity and permutations of what can be used.

For example, the number of BIST methods for an embedded PLA is greater than 20 [6]. Each BIST method will have a different effect on area overhead, test length/time, I/O requirement, fault cover, and performance. The relative importance of these characteristics for each BIST method will change depending on the PLA's functionality and size, number of devices to be tested, and also on the skill of the designer and/or effectiveness of the design tools. Thus it would be unreasonable to expect a designer to identify the "best" test method for each embedded PLA in the design. Furthermore, as we need to minimize the overall design, manufacture, and test costs for the device, our selection procedure for the "best" DFT method for the embedded PLA needs to consider what other DFT methods are being adopted for other partitions of the device.

2.2. *The ECOtest System*

The ECOtest System, a tool developed by Brunel University in collaboration with Siemens-Nixdorf Informationssysteme, is a test strategy planning tool utilizing economics modeling for cost optimal selection of DFT methods [7]–[10].

The ECOtest System was developed specifically for the design of ASICs. Chips are fabricated by external vendors, who also test them with test patterns supplied by the designers. This definition of the application area of the test strategy planner affects the design of the cost model. It thus highlights the specific nature of the economic results that will be presented later in this article. For example, the fact that an external vendor fabricates the chips means that the cost of production can be easily calculated from the vendors prices, if a good estimation of gate count can be achieved.

The structure of the ECOtest System can be seen in Figure 2.

The design description is acquired either from the user or from an existing netlist and is built in a hierarchical fashion in order to allow test strategy decisions to be made at several stages of the design process. It can read EDIF or Siemens specific hierarchical design descriptions.

The test strategy planner contains the programs for the evaluation of the cost model, as well as the evaluation of the effects of test methods on the design. It takes its input from the design description and cost model files, and uses stored knowledge on test methods to evaluate a variety of test strategies. Cell library data are linked to test method descriptions.

The test strategy control can be left entirely to the user, and a group of functions are provided to apply test methods and examine the cost model and the testability status of the circuit. Alternatively, the user has the option of employing one of a set of algorithms that perform automatic test strategy planning to accelerate the selection process [11].

ECOtest does not try to partition the circuit into testable units. It relies on the design hierarchical information to define testable units. Figure 3 shows an example circuit and the hierarchical level the system requires. This circuit will serve for demonstration purposes and be used in later analysis. A range of DFT methods, including the no DFT option, are considered for each block. It should be remembered that when applying a DFT method there will always be a set of constraints that have to be met before the economic issues can be considered, e.g., minimum overall fault coverage/area overhead. Thus ECOtest ensures that only applicable DFT methods can be applied to each block. In addition, after DFT has been considered, to ensure testability of the circuit, an analysis is performed to check that each block is 100% accessible from external pins.

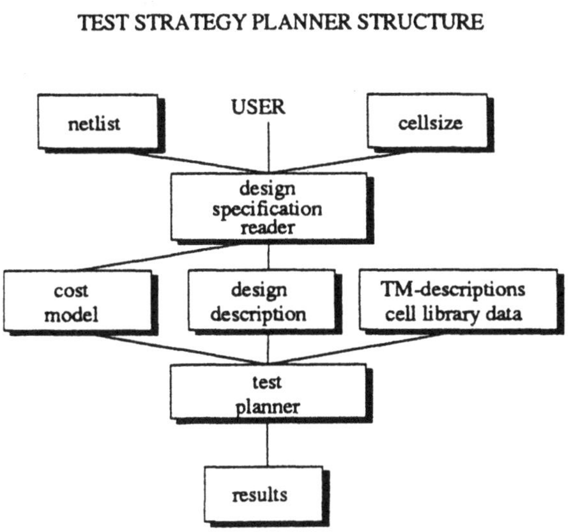

Figure 2. The ECOtest test strategy planner architecture.

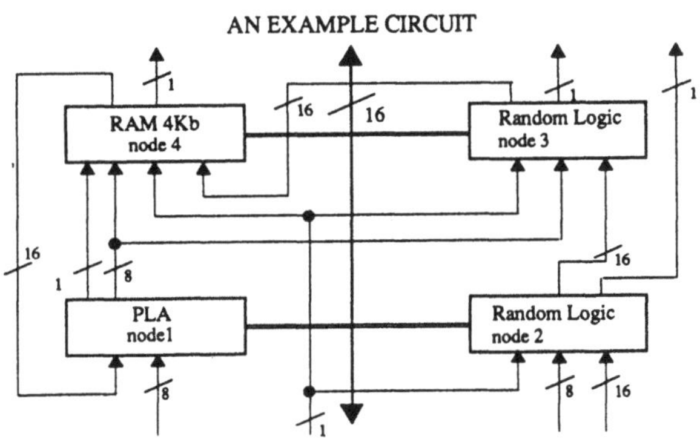

Figure 3. Demo circuit.

The description of test method contains information to enable an economic evaluation to be carried out and to ensure the test method is suitable for the specific design application. The description of test methods is partitioned into three types; cost model parameters, design implications, and design requirements. Parameters used by the cost model will include, but not be limited to, equivalent gate count, sequential depth, performance complexity, additional pins, originality number of extra function, number of test patterns, achievable fault cover, TPG method, and pin compatibility. Design implications consider such parameters as accessibility impact, test method type, and self-test capabilities to establish how the DFT method affects global considerations. Design requirements ensure the suitability of test methods to design styles and structures being adopted. Figure 4 shows an example test method description.

Test Method Name	Fujiwara
Test Method Type (ext./int.)	internal
Suitable Design Classes	PLA
Self Test	yes
Assures data-in accessibility	no
Assures control-in accessibility	no
Assures clock-in accessibility	no
Assures data-out accessibility	no
Assures bus accessibility	no
Performance implication	1.1
Test pattern generation method	none (universal test set)
Sequential depth	0
Achievable fault coverage	1
Number of test patterns	(3+2log$_2$ * number_of_product_lines)*(2* number_of_inputs + 3 * number_of_product_lines)
overhead_formula	(4*number_of_inputs+ 11* number_of_product_terms + 5*number_of_outputs + 6)/4
originality impact	0
inpin overhead	3
outpin overhead	2
bipin overhead	0
pin compatibility class	class 32
Design Style Requirements	asynchronous, synchronous latch or synchronous flip flop
Number of functions	0

Figure 4. Test method description.

The data for the cost model for a specific design are generated automatically from the design database. The cost model is stored in a text-based file that is easy to maintain, change, and add to as and when required. The primary parameters for the cost modeling exercise are partitioned into three types: design dependent, test plan dependent, and design independent.

Below is a list of some of the design independent cost parameters used by ECOtest.

cexp-	design complexity exponent
exptgs-	working hours per week
exper-	designer's experience measure
pcad-	productivity of the CAD system
desrate-	cost rate for external design center
equrate-	cost rate for cpu time for mainframes
costrate-	cost rate of designer
fpg-	average number of stuck-at faults per gate
sdlim-	sequential depth for 99% ATPG fault cover
glim-	gate count limit for 99% ATPG fault cover
tppf-	average number of test vectors to test a single fault
fcv-	verification vector fault coverage

The design dependent primary parameters are extracted from the design description automatically, or if not available, they can be entered. Below is a list of some of the design dependent cost model parameters.

nre-	non-recurring engineering cost
vol-	production volume
blocks-	number of testable units
fdblocks-	number of functionality described blocks
percuse-	percentage use of design center
cputime-	cpu time of accounted mainframe
fcreq-	required fault cover per TU
tpf-	time for manual tpg for a single fault in a TU
tpver-	number of verification vectors
pms-	test vectors in a single price band
pps-	price for additional test vector test
avs-	average sequential depth per TU
ppu-	selling price per IC
ndu-	non-delivered units, caused by delays
npcost-	cost for non-performance penalty

Test plan dependent primary parameter initial values are generated automatically as the Test Method description file is interrogated for different DFT methods. Below is a sample of some of the test plan dependent parameters.

sumtp-	number of externally applied test vectors
puc-	production unit cost
fun-	number of functions per TU

fun&-	number of function per chip
cells-	number of cells per TU
cells&-	number of cells per chip
in-	number of input pins
bi-	number of bidirectional pins
cperf-	performance cmplexity per TU
or-	originality per TU
cgate-	equivalent gate count per TU
cgate&-	equivalent gate count per chip

The list below shows examples of some of the secondary parameters that are calculated. All of these can be examined at each stage of the test strategy planning process.

pc-	production cost
plib-	productivity of the cell library
prod-	overall productivity
cpin-	pin complexity
orabs-	absolute originality per TU
orabs&-	absolute originality per chip
or-	relative originality of chip
cperfabs-	absolute performance complexity per TU
cperf-	relative performance complexity for chip
compl-	overall complexity
mp-	manpower in weeks
destime-	design time in weeks
numdes-	number of designers needed
engcost-	engineering cost of the design
descentcost-	design center cost
mainframecost-	cost accounted cpu time
descost-	overall design cost
faults-	number of possible stuck-at faults per TU
faults&-	number of possible stuck-at faults per chip
fcach-	achievable fault cover by ATPG per TU
remfaults-	number of ATPG aborted faults per TU
mtgtime-	time needed for manual TPG pr TU
mgtime&-	time for manual TPG per TU
mgtime&-	time for manual TPG for chip
mtc-	overall cost of manual TPG
faultsa-	faults per TU for which ATPG was successful
numtpa-	number of test vectors per TU generated manually
tpgen-	number of generated test vectors per TU
tsl-	test set length
tac-	test application cost
costtgs-	ATPG cost per TU
costsgs&-	ATPG cost per chip
tcost-	profit loss for non-delivered units
ttmcost-	time to market cost
ovcost-	final cost

The economics model equations are specific to the design environment and company. In the case of the examples given, it was specific for ASICs designed by Siemens-Nixdorf and then manufactured and tested by outside vendors. However, the models can easily be changed to represent a different environment. Because the models are based on Siemens-Nixdorf's own accounting system and generated within the company, they have proven to be accurate for their purpose.

It is also possible to obtain information from the system on the comparative risk of adopting different test methods based on the availability and accuracy of the input data. This has been extended to perform a sensitivity analysis of all the parameters in the model [12].

2.3. Cost Analysis Using ECOtest

Below are the results from performing an automatic test strategy planning run on the demonstration circuit shown in Figure 3 for two very different production volumes. The circuit contained 16,200 equivalent gates and 464 equivalent D-FFs. The production volume has proven to be a sensitive parameter.

For 5,000 ICs, the total design, manufacture, and test cost was 522,570 DM. The cost per IC is 104.51 The selected test strategy for each partition (node) is shown below. HFC and ILLMAN are both BIST methods [13], [14].

Node	Node Type	Test Method
node 1	PLA	hfc
node 2	random seq	int_scan
node 3	random seq	int_scan, ext-scan_set
node 4	RAM	illman

For 500,000 ICs the total design, manufacture, and test cost was 13554361 DM. The cost per IC is 27.11 DM. The selected test strategy for each partition is shown below. TREUER is a BIST method [15].

Node	Node Type	Test Method
node 1	PLA	treuer
node 2	random seq	int_scan
node 3	random seq	int_nodft, ext-d-latch
node 4	RAM	illman

It can be observed that changing the production quantity has resulted in a different selection of test strategies. Node 4, the RAM, is still tested with the Illman self-test method. However, the test method for node 1, the PLA, and node 3 has changed. In the case of node 3 it was the application cost impact that full scan has as production volume increases that caused the adoption of a non-DFT option to be selected. This is because test generation costs can be amortized over the larger production volume.

The increased gate count due to the additional test logic is often quoted to be a reason why DFT is not condensed for high production volumes. However in this case, some DFT is still cost effective over no DFT at all, although the amount of DFT is reduced as the production volumes are increased.

As an example of the effect gate overhead can have on the results, all the gate overheads associated with each test method was doubled in the Test Method Description file and the test strategy planning procedure repeated. The same test strategy plans were obtained for both 5,000 and 500,000 production volumes. For 5,000 IC the total cost increased to 573000 DM (114.6 DM per IC) a 10% rise. For 500,000 ICs the total cost was 14104181 (28.2 DM per IC) a rise of 4.4%. It should be pointed out that this result cannot be taken as a general conclusion, it is only valid for this circuit being produced in a specific way.

Figure 5 indicates the sensitivity gate count has to the production cost of an ASIC, using a specific ASIC vendor for a particular production volume range. Using such a pricing function and accurate gate overhead prediction methods for different DFT approaches results in highly accurate manufacturing cost calculations.

The effect of gate overhead on the total cost of design, manufacture, and test can be observed in Figure 6. Again this relation is only true for this particular costing environment. The sensitivity of very different production volumes can be observed.

The test throughput cost becomes an increasingly important factor as the production volumes increase. The calculation of ATE testing costs can be performed in detail by taking into account capital cost through to labor costs [4], [16], [17]. In this particular case all test application was performed by an outside vendor, and the application test cost per IC is based directly on the number of test vectors applied. Figure 7 shows a typical pricing function based on a vendor's pricing policy.

In this case the initial set of vectors is "free," corresponding to loading the ATE memory once. Subsequent test vector sets are charged for, the pricing being dependent on the memory available per pin. Due to this step function, the region between two pricing bands is often critical in selecting DFT methods, i.e., it is a region of high sensitivity.

Figure 8 shows the sensitivity such a pricing function can have on the total design, manufacture, and test cost for different production numbers. As is expected, the sensitivity of the overall cost to the test pattern application pricing function is most noticeable in the larger production volumes.

In some cases there will be different pricing functions depending on the type of ATE used. With such variations it can be important to consider the different

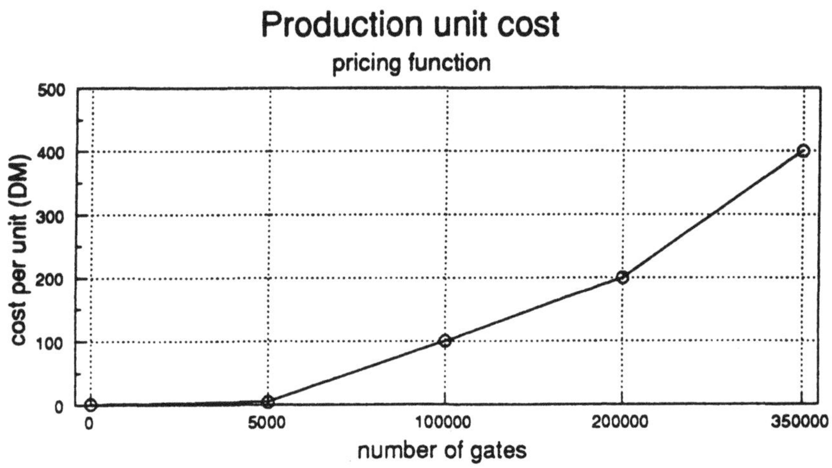

Figure 5. Manufacturing cost vs. gate count.

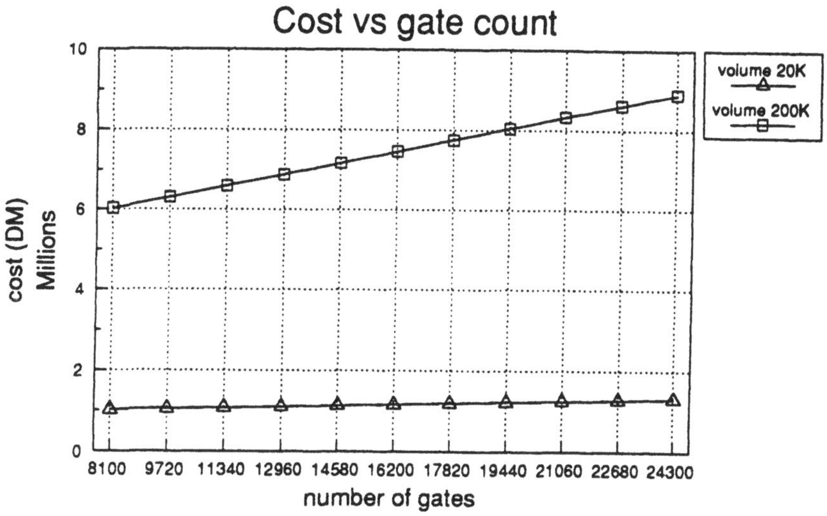

Figure 6. Overall cost vs. gate count.

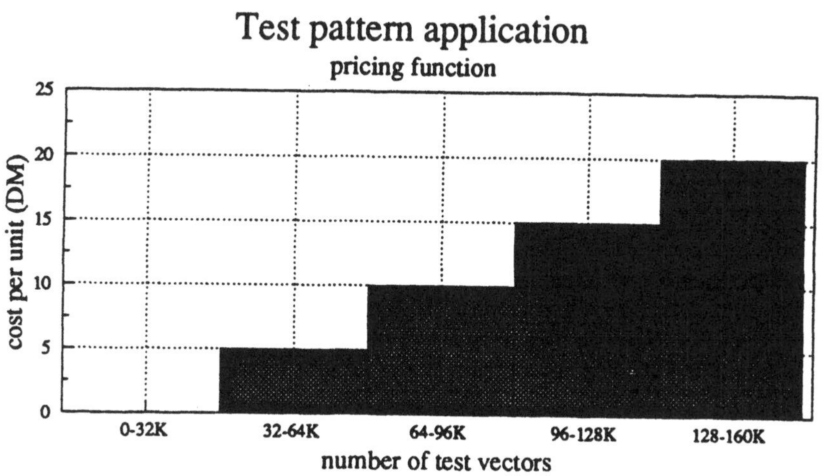

Figure 7. Test pattern application pricing function.

ATE available when DFT decisions are considered. Indeed for large production runs a new ATE could be considered for purchase if it reduced the overall cost. Below are the results from a DFT selection procedure for the demonstration circuit if a particular scan ATE is available.

For a production run of 5,000 ICs the total design, manufacture and test cost was 522,570 DM. Below shows the test strategy for each node.

Node	Node Type	Test Method
node 1	PLA	hfc
node 2	random seq	int_scan
node 3	random seq	int_scan, ext-scan_set
node 4	RAM	illman

For 500,000 ICs the total design, manufacture, and test cost was 13,565,278 DM. Below shows the test strategy for each node.

Node	Node Type	Test Method
node 1	PLA	hfc
node 2	random seq	int_scan
node 3	random seq	int_nodft, ext--d-latch
node 4	RAM	illman

However, if a scan ATE is not available and the same test strategies are applied using a non-scan tester then a penalty is incurred. For 5,000 ICs the new total cost is 547,570 DM, an increase of 25,000 DM (5 DM per IC). However, if the test strategy planning procedure

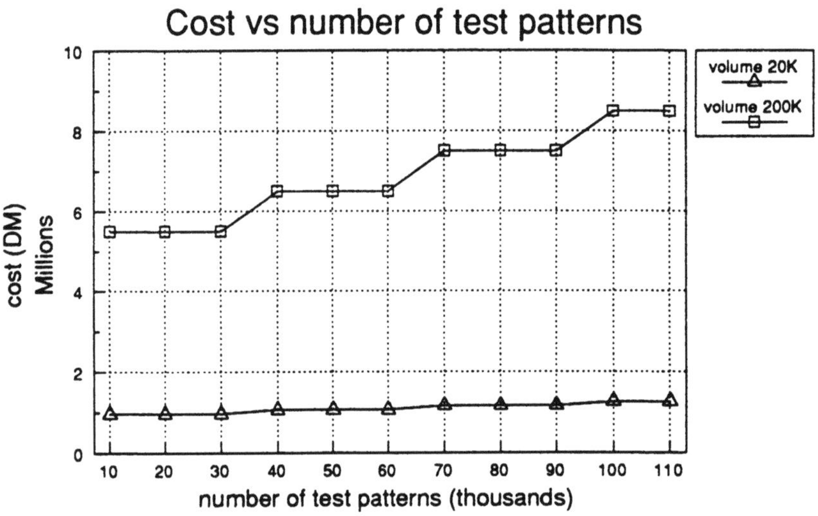

Figure 8. Overall cost vs. number of test patterns.

is performed again with the knowledge that no scan ATE is available, in this case no "better" test method can be selected. Thus the DFT method selected does not change but there is a cost incurred due to the lack of a scan ATE. The single cost incurred from this production run is not enough to consider purchasing a scan ATE. However, if all the production runs of a company over several years are considered the position might be different.

If a scan ATE is not available for the 500,000 production run of the demonstration circuit and the same DFT strategy is used, then the total costs are 16,054,360 DM, an increase of 2,489,082 DM (increase of 4.88 DM per IC). This is significant and the purchase of a scan ATE could be considered. If we considered the fact that no scan ATE is available when the DFT selection process is performed then a new test strategy plan is obtained. The cheapest test strategy for 500,000 ICs when no scan ATE is available is shown below.

Node	Node Type	Test Method
node 1	PLA	treuer
node 2	random seq	int_nodft
node 3	random seq	int_nodft, ext-d-latch
node 4	RAM	illman

The total cost is now 15,805,150 DM an increase of 2,239,872 DM. There is still a large cost penalty associated with not having a scan path tester but it has been minimized by reconsidering the test strategy plan. It can be observed that the new test strategy relies less on the use of scan paths and more on self-test and traditional nonintrusive test methods.

The above results are highly specific to the example given.

2.4. ECOtest Conclusion

In this section we have shown how test economics can be used in an integrated environment to aid in the selection of test strategies of ASICs. The results presented were highly specific and should not be treated as a general conclusion as there is a large set of design and company specific parameters to be considered.

The above test method selection criteria do not consider adding DFT methods onto an ASIC to aid board test, system test, and field maintenance. It only tries to minimize the total cost at the ASIC production level. It is often argued that adding DFT at the device level can greatly reduce the cost of testing at later stages. It should be remembered that not all device level DFT can be activated at later test stages. However, adding a boundary scan to an ASIC does not improve the testability of the ASIC or reduce its testing cost, but the cost penalty paid at the chip level could be covered at later stages of test and diagnosis. Thus, when total life-cycle cost is considered it might be the case that strategies that do not appear cost-optimal at the chip level should be considered.

The ECOtest system and its cost and quality models can be utilized to evaluate the cost and quality impact of adding, for example, boundary scan to an ASIC, and the information used to consider life-cycle cost implications.

3. Economics of Board Testing

3.1. Introduction

The testing of complex VLSI based systems is an activity where aspects of quality, time, cost, availability of equipment, and design choices are important. Traditionally, the choice of testing at board level (the test strategy) was simply a question of selecting one or more test stages to be used (for example, in-circuit test followed by functional test). However, if a certain quality level has to be met, if accessibility is hampered by the use of methods such as surface mounting, if a choice of ATE is available, or if different options are available for the components, it becomes clear that the traditional view is no longer adequate. It is important to define the term *test strategy*, in this case, as "the selection of design choices and test stages which will achieve the highest quality product at the lowest cost" [17], [18].

To continue our work on test strategy economic modeling, a board level test strategy planner that bases its decisions on economic factors as well as predefined specification criteria has been developed. Compared to chip-level test strategy planning, that for boards is not as straightforward.

At board level where several stages of test can occur each using different test equipment, test strategy planning refers to the optimum selection of multiple test stages. This in itself is not enough since the fault coverage obtainable by each test stage is imporant to the overall analysis. Any test strategy selection at the board level, therefore, must weigh between the straight financial calculations and those that determine overall fault coverage.

3.2. ECOvbs System

ECOvbs (economics of VLSI based systems) was also the result of a collaboration between Siemens-Nixdorf Informationssysteme and Brunel University [16]. This was a continuation of the the effort that resulted in the test strategy planning tool for ASICs (ECOtest), described above. The tool was intended for industrial use and needed to take into account industrial practices and needs.

Figure 9 shows the outline of ECOvbs.

3.3 Cost Modeling of Board Test Planning

As with the chip level system a test economics model for boards needs to include the all main phases of development and production as test decision affects all these areas. Figure 10 shows a typical flow diagram of the phases that need to be modeled.

The development phase includes the following subphases:

- The design phase comprises the initial design, design entry, and computer simulation. Any addition modification do the design for test purposes will affect the costing at this stage.
- The layout phase includes the placement and floor planning of the PCBs. The need for increased test access or the addition of a larger ASIC due to improved testability criteria will impact on the layout phase.

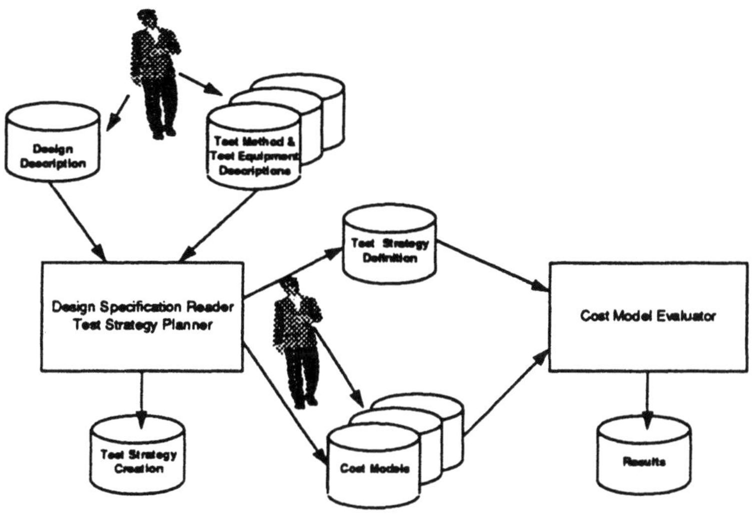

Figure 9. Architecture of ECOvbs.

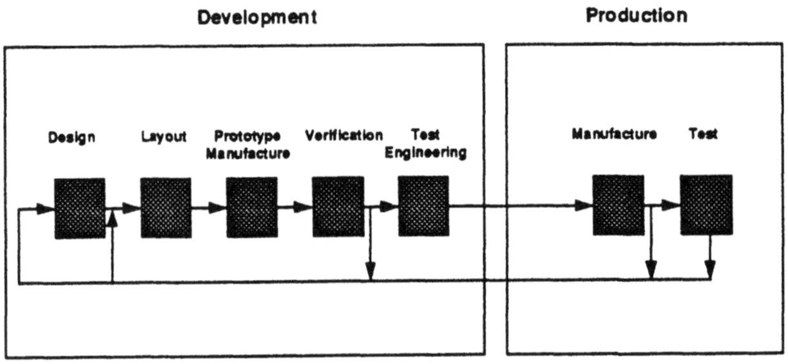

Figure 10. Flow diagram for board level phases.

- The prototype manufacture phase covers the construction of the prototype boards.
- In verification phase the evaluation of the board by verifying the specified functions with the prototypes. In many cases the addition of good testability and accessibility has significantly aided in the verification of the system.
- The test engineering phase covers the generation of test patterns, the generation of test programs, and the manufacturing of test tools, such as a bed-of-nails fixture for in in-circuit test.

Due to the test view the production phase is partitioned into two subphases:

- The manufacture phase includes the production preparation, fabrication, and assembly.
- The test phase comprises test application, which includes the costs for test, diagnosis, and repair.

In the test economics model structure the test engineering phase and test phase are combined into a single cost model because the cost structure of these costs is test method dependent. For each test method a separate cost model exists, which models the related costs for engineering and test application.

The life cycle of a board includes a chance that a redesign may become necessary. This can happen as a result of the verification or during the production phase. This fact is considered in the test economics model by the definition of the probability of a redesign and by iteration factors per development phase, which define the effort needed for the redesign as a percentage of the original effort.

The costs in the development phases include labor cost, equipment cost, and material cost. The labor costs are determined by the hourly labor rates and the predicted effort. The equipment costs are related to computer equipment, which are needed for the development of the design. They are also calculated by the hourly costing rates and the estimated time the equipment are needed. If the costs are included in the hourly labor rates of the designers, the equipment costs can be set to zero. Material costs occur only for the manufacturing of the prototypes.

The manufacture costs are composed of the production preparation costs, the material costs, and the assembly costs of the board. The material costs also include the total costs for the components, which are calculated by the ASIC test economics model. The calculation of the assembly cost is based upon the assembly cost per component and per assembly type, and the number of components per assembly type. In addition to the costs the number of defects per defect type are calculated. These data are needed for the test phase.

The test cost model includes the calculation of the test engineering costs and the test application costs. The test engineering costs are composed of test tool manufacture costs and test generation costs. The calculation of the test generation costs is based on the engineering effort, the engineering labor rate, the usage of equipment, and the equipment rate. The test application phase is built upon the test phase and the diagnosis phase as shown in Figure 11.

The test is applied to all boards coming from production. The diagnosis and repair are applied only to

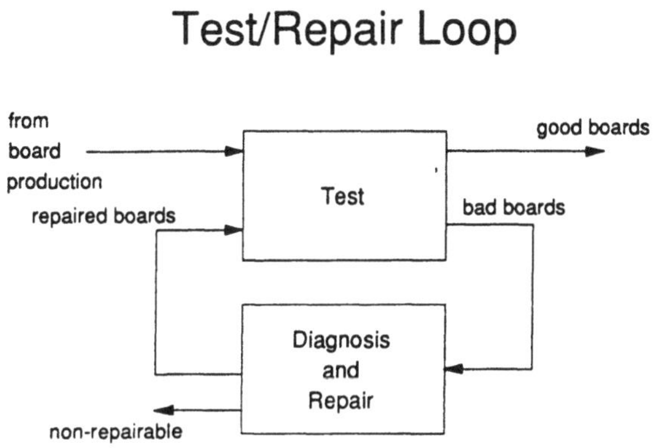

Figure 11. Test/repair loop for the test application phase.

those boards, for which a defect has been identified during the test. All repaired boards are retested in order to detect multiple defects and to detect defects, which are injected during the repair. It is not always possible to repair a board. This could be because of reliability reasons, for example if a board has been repaired more than the predefined number of times then it is considered to have a significantly lower reliability and thus not to be trusted. The scrap cost can be considerable for complex boards. Different companies will have different policies on this procedure. However, in some cases these boards are repaired and passed onto the field engineers to use as temporary replacements.

At this point it should be noted that not all boards declared bad by a test are in fact bad, and not all boards tested as good by a test are in fact good. These fault escapes can add considerably to the cost of testing. This is because the amount of extra test, diagnosis, and retesting that is required to detect a fault when none exists can be considerable. Likewise the cost of detecting a fault that has escaped to the next test stage or even the field is also not insignificant. In some cases these costs can outweigh all other test costs. The likelihood and risk of this occurring and its impact on the diagnosis time is considered in the modeling process.

Careful analysis of the test application phase is important. The application cost, the in-going quality and the out-going quality of boards to a test stage need to be considered. At board level test multiple test stages are used. Each stage is designed to detect a certain class of faults and therefore a detailed fault spectrum is requied [19]. An initial fault spectrum can be calculated from a board description (e.g., board components, technology, and production environment) in conjunction with historical data (e.g., Defect spectrum) that predict the likelihood of the different types of failure mode occurring. For example, the number of surface mount connections on a board and the probability that a connection will be poor can be used to determine the likely number of open faults on the board. Thus the in-going fault spectrum into the test phase can be calculated. As the board passes through the various test stages this fault spectrum is updated as faults are detected and repaired. It should be remembered that the likelihood of the repair introducing a fault onto the board also needs to be considered.

Each test stage will have different diagnosis times for different types of fault in the fault spectrum. For example, the time/cost of detecting escaped short/open failures on a functional tester will be greater than if an in-circuit tester was used. The likelihood of not diagnosing down to a single replaceable part is also considered as the resolution of different test stages for different failures will change.

In addition each type of failure will have a different repair cost. The material cost of each repair as well as the reapir time for each failure is modeled.

3.4. ECOvbs Economic Model Structure

The economics model is in fact a collection of many basic cost models that can be combined in different ways to create a single economics model for a specific test strategy. This has two main advantages:

- This structure allows the easy modifications and verification of the model. For example, if the production environment changes, only a few small models need to be changed. In addition it makes the data acquisition and model generation a more streamlined procedure, as individual experts can address each area of the model.
- Many of the basic models can be reused several times to build up a complex model. For example, models that define a test stage can be repeated as more test stages are added, with only minor changes made to the input data. This also has the advantage of minimizing the errors in the modeling process. If the model is intended to be used to compare the relative cost of test strategies and not calculate accurately the absolute design, production, and test cost, then it is more likely that any error in absolute cost will not effect the relative cost of test strategies.

To indicate how the collection of basic cost models interact a simplified structure is described as follows.

Board data	Physical design parameters and components on the board.

Design data	Design environment models and man effort data.
Labor rates	Basic labor models used to calculate labor costs.
Design costs	Addition of basic models and other parameters to calculate design cost.
Iteration factors	Prediction models to calculate the iteration factors for the design layout and verification loop.
Production data	Models to calculate the production environment and effort.
Production cost	Addition of basic models and other parameters to calculate production cost.
Assembly model	Contains models that detail the cost of assembling different types of components.
Fault spectrum	Models to calculate the number of faults per board of each fault model type.
Defect-to-fault	Models to calculate the likely faults from different manufacturing defects.
Test stage/ test equipment	Test generation and throughput models for a test stage.

The basic models can vary in complexity from a complex set of equations down to just pure data. This allows the modeling process to be flexible. If input data is identified to be highly important or needed earlier in the test strategy selection, process models can easily be generated to predict or define it more accurately.

Figure 12 shows the basic model when there are no test stages. Figure 13 shows the basic model interaction when a single test stage is added.

It can be observed that as a test stage is added the following interactions are possible: First, there is the addition of the test generation, preparation and throughput cost to be added to the overall cost. In addition, any modification to the board design could effect the design data. For example, the test strategy might require test sites or different components (e.g., boundary scan) to be added, and this will affect the design and layout of the board. This extra engineering time and effort will affect the design stage cost and the overall cost. In addition, the use of different or additional components might affect the assembly cost. The number of board connections is likely to also change. The addition of different components and increased number of connections will modify the fault spectrum as the likelihood of failures is modified. Thus the testing/repair throughput cost is modified. The repair

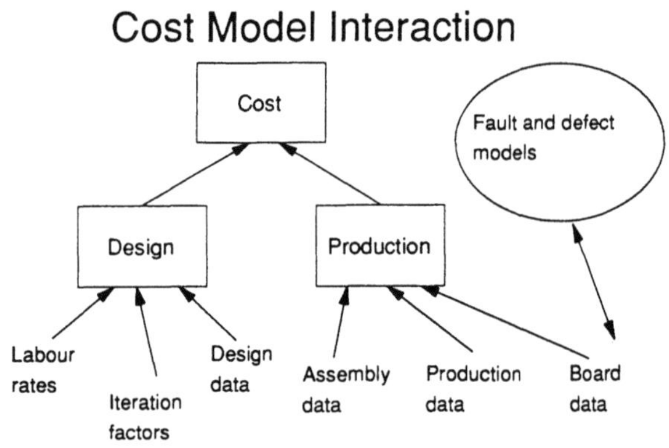

Figure 12. Cost model with no test stage.

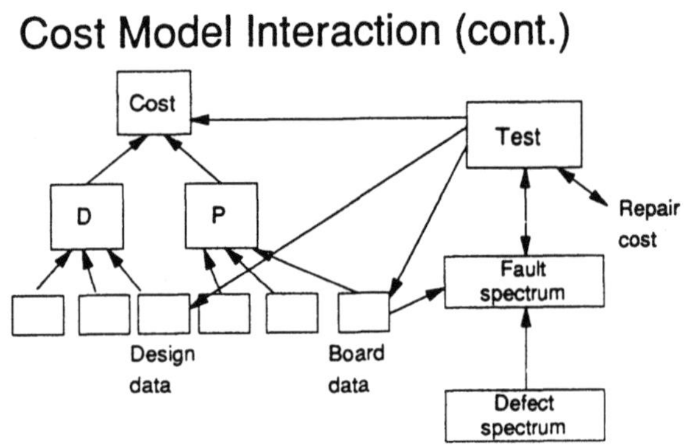

Figure 13. Cost model interaction when a test stage is added.

cost might need to be modified to take into account the time and effort for different failures and cost of replacement items. As more and more test strategies are added then the interaction and nesting of the cost model is increased.

3.5. Economic Analysis of Board Test Strategies

As an example of how the system operates four basic test strategies have been applied in different permutations so that the effects can be seen. The board used to compare different test strategies comprised the following:

— 40 ASIC devices that were available with and without boundary scan
— 8 256K RAM
— 64K RAM
— 4 64K RAM
— 16 1Mbit RAM
— 800 resistors
— 100 capacitors
— surface mount connections

— 15,000 solder joins
— 5000 production volume

The four test strategies applied are a mixture of a system test stage, a functional test stage, a boundary scan test stage, and a pre-screen test stage (MDA). The four strategies are defined below.

1. Functional test–system test
2. Boundary scan–functional test–system test
3. Pre-screen–functional test–system test
4. Prescreen–boundary scan–functional test–system test

The results of the cost and quality analysis are shown in Figure 14.

Test Strategy	Design Cost (ECU)	Prod. Cost (10^6ECU)	Test Cost (10^6ECU)	Total Cost (10^6ECU)	yield after test (%)
Functional test, system test	941391	85.3	11.1	97.4	99.65
B_scan, functional test, system test	957883	90.4	4.35	95.7	99.90
Pre-screen, functional test, system test	941397	85.3	9.38	95.6	99.73
Pre-screen, b_scan, functional test, system test	957883	90.4	4.93	96.3	99.90

Figure 14. Board test strategy planning results.

It can be seen that strategies two and three have very similar costs but strategy two has a slightly better overall quality. The boundary scan stage is the only stage that requires modifications to the board design. All the ASIC have boundary scan added with minimal cost due to the CAD environment used supporting automatic insertion of boundary scan. Thus the relative increase in design cost can be ignored in this case. The reason for strategy one having a high test cost is the poor detection and diagnosability of shorts and opens by a functional test stage. Thus larger amounts of time are spent in the interactive repair loop.

Figures 15 to 18 show how the fault spectrum is modified at each test stage.

In summary, the ECOvbs system enables economic comparisons of different DFT modifications for aiding board testing to be compared. It allows the mixture of different test stages and testers to be compared. It allows changes in the incoming component quality to be analyzed. The effect of different manufacturing quality on the most cost optimum test strategy can be analyzed. This enables a comparison of the relative effects of improving the process or improving the test to be made.

4. Field Maintenance Economics

To complete the life-cycle cost analysis, field test strategies need to be considered. In addition, the effects that chip, board, and system test strategies have on field maintenance costs need to be considered. In developing a field test and diagnosis strategy it is not always possible to access or use existing DFT that has been inserted to aid chip and board level testing. However, if consideration of field maintenance is given when selecting DFT methods many of these problems can be overcome. Indeed it is possible to add DFT in the design stages to merely improve the maintainability and serviceability of the system. The need for such a requirement can often impact on the cost effectiveness of DFT methods adopted at the board/system and sometimes even chip levels [20], [21].

Test economics at the field maintenance level is, again, company, market, and product specific. As with test stages the economics of scale can heavily influence the testing strategy. For example, when there is a large product market then the manufacture of dedicated field test equipment can be an economical approach or for a small market large amounts of BIST could be cost effective. However, the company infrastructure for field maintenance can affect the choice. A large company that manufactures many different systems that have low downtime costs, where there is a good level of technical support for a large pool of lowly skilled field engineer workforce, with a mature and efficient logistic support environment and well stocked repair depot sites, will have a very different viewpoint than a company manufacturing a few high tech systems that have high downtime costs (e.g., safety critical systems) and require highly trained field service engineers (or in some cases the original designers!).

The situation is made more complex as the guarantee and extended warranty costs for a system is considered. A company manufacturing, testing, and using its own equipment has a different perception of the cost and quality impact of field service than a company purchasing equipment and warranty from an Original Equipment Manufacturer (OEM) or a contract service company. In general it is only the OEM that has the ability to consider DFT modification to the design to reduce

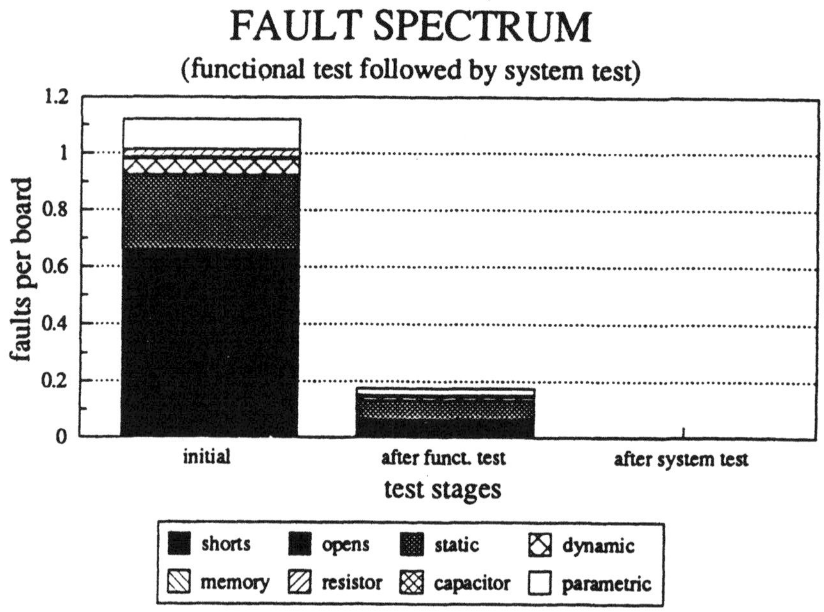

Figure 15. Fault spectrum for test strategy 1.

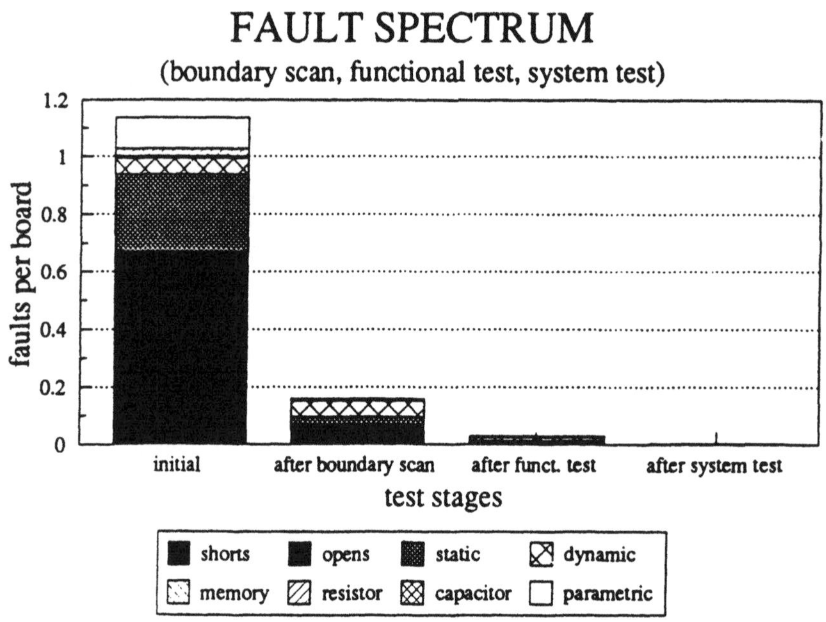

Figure 16. Fault spectrum for test strategy 2.

field service costs. The contract service companies need to rely on more efficient logistic support, technical ability, and test equipment to maximize profit by minimizing costs. The profit to be made in the service industry is high and thus some areas are becoming highly competitive.

In 1990 the estimated US service market stood at $46 billion. It was further estimated that 26% of the revenue and 50% of profits generated by service centers of equipment vendors. Obviously, if the service warranty is a fixed price contract then the more reliable or the cheaper to diagnose and repair a system, the greater the profits for the service centers. If not, then it would seem to indicate that the more unreliable the product, the greater the service center's profit. However, the reliability of the product can influence the

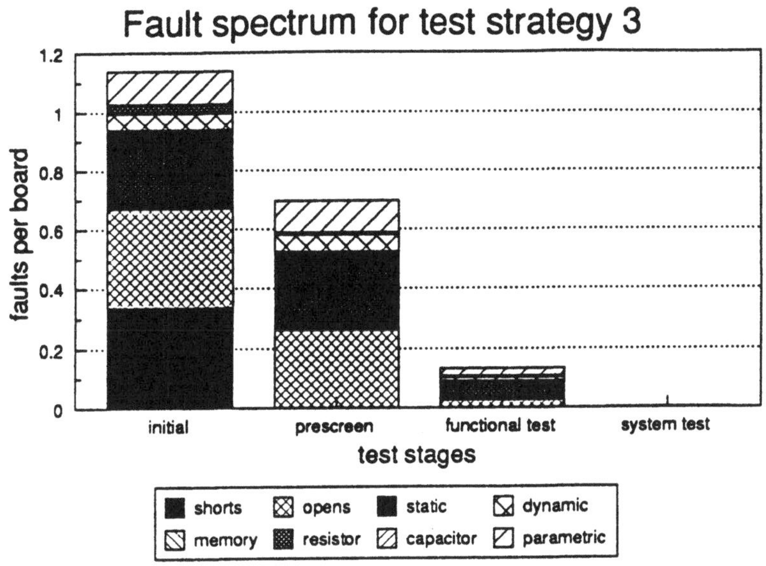

Figure 17. Fault spectrum for test strategy 3.

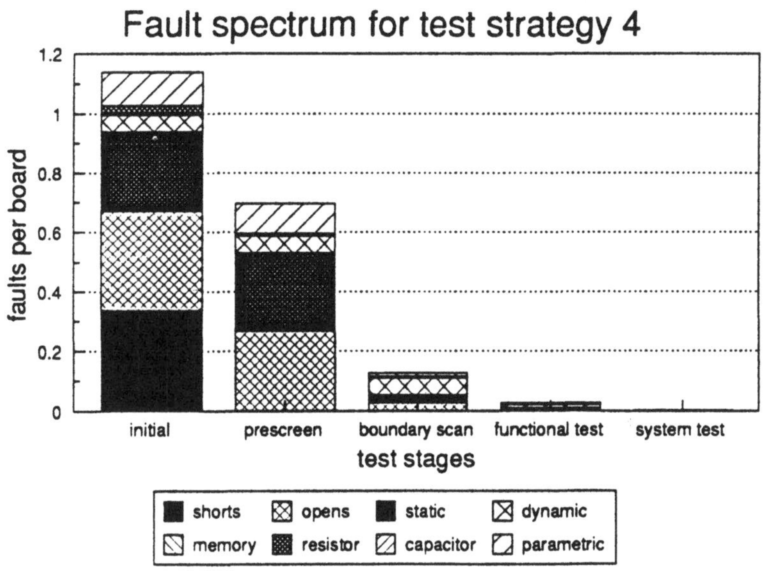

Figure 18. Fault spectrum for test strategy.

repeat sales of a company's product. In addition, it is unlikely that the downtime penalty associated with a field failure is a fixed price for the period of the warranty. Another factor to consider is the perception the user has on the quality of service that he obtains from a service center. A user might well prefer a less reliable product if strong guarantees can be obtained for the quality and speed of field service.

The above permutations can be considered part of a global field service test strategy but obviously to model accurately the many different permutations for different companies, contracts, and products is difficult. However, detailed analysis performed inside a specific and mature environment for a specific system can be achieved. This can be very valuable in assessing the cost of field service and the potential for profit.

Published studies for life-cycle costs are few and far between but some studies from a military environment have indicated that the field service costs can approach 60% of a product's total life cycle cost, production costs approach 30%, and design costs 10% [22]. This was for a product with a 30 year life in a military environment.

However, it does indicate the scale of the potential savings that can be achieved by reducing maintenance costs. Indeed, in some cases the cost of repairing old equipment can be more expensive than replacing it with new equipment. A US army study has shown that costs can be reduced by 25% if 70% to 80% of the items it currently repairs are discarded [23].

Analysis of other industries has also indicated the high cost of service contracts. For example, it has been quoted that 35% of a user's total computing costs are to be attributed to computer service costs [24], and 15–30% of a computer manufacturer's profits come from field service contracts. For a domestic video recorder, a 4 year warranty in the UK can cost around £70 and the purchase cost will be around £300, a cost of 23% of the purchase price.

Figure 19 shows a simple field service procedure. It is a two level strategy (i.e., there are two possible locations where failures can be corrected). Different industries and companies will adopt different levels of field service. Indeed, the US Army during the Vietnam war adopted a 6 level strategy for its equipment [23], [25].

Level 1 Soldier
Level 2 Organizational
Level 3 Direct support
Level 4 General support
Level 5 Depot
Level 6 Manufacturer

However, the current trend is towards reducing the number of levels of repair. As the complexity of systems increases the cost of repair centers increase and the diagnosability and repair of faults becomes an expert task. The currently adoped strategy is a three-level strategy, Unit (or field), Intermediate, and Depot (or OEM). Due to the complexity of its systems, the US Air Force adopts a two level strategy containing a Field Repair and Depot (OEM) repair. In the Digital-Avionics industry the removal of the intermediate level can produce savings estimated at 45% [26].

Figure 19 indicates the basic areas of a field repair. The economics modeling of the repair depot can be achieved by making use of the basic models within ECOvbs used to model the cost of a board level production test stage. In electronic systems, the field service requirement is to diagnose a fault down to replaceable units. The board is considered to be the lowest level replaceable unit in many designs. In many cases the depot is in fact the manufacturer. Indeed, the ECOvbs system can be used to consider the cost impact on the

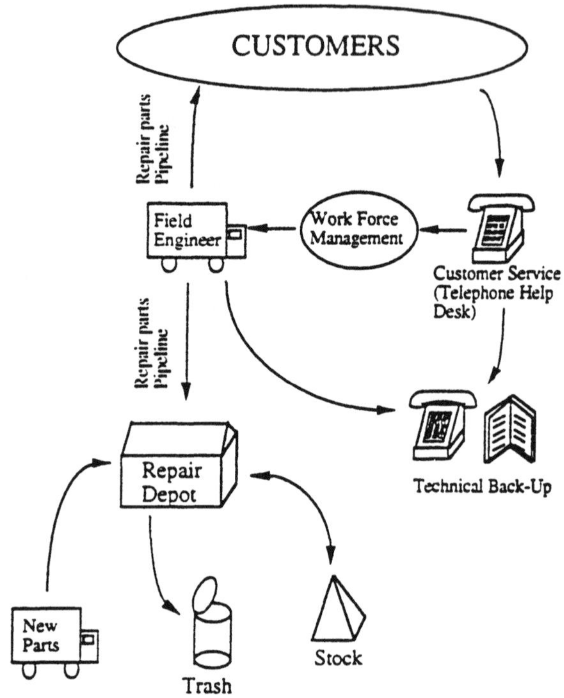

Fig. 19. Typical field service procedure.

repair depot costs. Thus the impact of the manufactured outgoing quality impact on the repair cost can be analyzed when selecting a board level test strategy. In addition the usefulness and cost of specific test strategies applied at the manufacturing stage to reduce diagnosis time in the field can be compared. Below we'll discuss some of the other issues and cost areas associated with field maintenance that are harder to model.

The harder areas to model are those that involve greater human interaction. Indeed the level of human interaction in the field service industry along with the delocalization of the field testing procedure are the main reasons for high field service costs. It is possible to generate logistic support models and compare the efficiency of different logistic structures. However, in most companies for the purpose of selecting a field level test strategy planning it is possible to consider the logistic support as a lump sum cost.

The cost per field can be calculated from knowledge of the field diagnosis and repair time for different failures and the cost of replacement parts. Each field engineer will have a slightly different profile, but due to the sensitivity of the modeling and data collection procedure it is possible to group engineers into groups of different skill levels. Thus the diagnosis time and repair time can be associated to a particular skill of engineer or the field test equipment carried. The identification of the number of field engineers required in each skill group and type of field test equipment carried could be considered part of the global field service strategy plan. However, this level of analysis will not be discussed in any great detail in this article.

In addition to the cost per repair there is the overhead cost per field engineer to be considered. For example, each engineer can have associated a significant investment in spare parts, test equipment, transportation, training costs, etc.

The knowledge of the likely number and distribution of field visits is important. This will affect the number of field engineers required. However, the number of field visits is not equal to the number of field failures. Our analysis has shown that in the order of 20% of field visits are repeat visits to service the same fault. Repeat visits can be attributed to several factors:

- Poor diagnosis by badly equipped or unskilled staff
- Poor repair by badly equipped or unskilled staff
- Incorrect or inadequate documentation of the reported fault.
- Inadequate/out-of-date field service service documentation

Poor logging or poor diagnosis of the fault by the initial fault logging procedure, and poor logistic support can result in the wrongly skilled engineer being sent to repair a fault with the incorrect documentation, test equipment, and replaceable parts. The use of remote diagnosis can have a significant effect on the number of repeat visits along with a reduction in the number of initial visits required and the size of the pipeline. AT&T has developed a system to perform remote diagnosis and maintenance monitoring that is estimated to produce a reduction in service calls of 50%. An estimated reduction of 10–40% in service calls for hardware faults could be achieved by remote diagnosis. Indeed 90% of software faults could be corrected remotely [27].

On-line documentation could be a cost effective solution to overcome the problem of a service engineer not having the correct documentaiton. This could reduce the number of repeat visits, training time, the need for on-site specialists, the number of good boards swapped, the pipeline inventory, and repair time.

The lack of automation and thus high dependency on human interactions causes many inconsistencies in service. Indeed services are prone to error. Any incorrect diagnosis not only results in repeat visits and downtime but an increase in the number of parts being sent back to the depot for repair. The cost of diagnosing and ensuring a board reported "bad" by a field engineer is in fact "a good board" is not insignificant. There is also the consideration of the degradation in reliability of the board through test and repair stress to be considered.

The downtime cost is another factor that can drastically affect the cost of field service, particularly as safety safety critical systems are considered. Recent disruptions to finance companies have resulted in signficant loss of revenue. It was estimated in 1984 that downtime on an airline reservation system can cost $24,000 a minute [28]. Due to the increased amount of air traffic and inflation, costs could be much higher today. Even if the downtime cost is small, long downtimes can indicate a poor service quality, thus affecting system resales or future maintenance contracts.

The pipeline cost (i.e., the cost of spares and faulty units held outside the depot) is not insignificant. Approximately 80% of the value of the service pipeline is returned through the depot repair operation annually [27]. The spares pipeline, in some cases, can tie up 5% of a company's assets [29]. As the complexity of replaceable units increases the cost of the piple will go up and not down, unless new practices for repair are considered.

The above has considered a few reasons for high field maintenance costs. Models can be generated to predict trends but the models need to be highly specific to obtain any meaningful trends for the total field maintenance cost. However, the effects that board level test strategy planning has on depot repair cost can be usefully modeled.

The continued rapid explosion of field service costs should be expected for several reasons:

- The complexity of systems and boards is increasing.
- Cost of replacement units is increasing.
- The repair of boards becomes more complex.
- The level of system customization is increasing.
- Working life of systems is decreasing.

It will thus become more important to consider design for maintainability and other methods some of which have been mentioned above to constrain maintenance costs.

5. Conclusion

This article has presented a set of tools that is being used successfully in an industrial environment. Indirectly, it has also shown that economic analysis per se can be a powerful aid in the selection of an optimal set of design for test strategies, and in the organization of production test processes.

Use of the tools by DFT experts has shown that the results are as good as or better than what would otherwise have been obtained. It has been extremely useful as an educational tool to both management and designers, and also pointed to further areas where these techniques could be useful.

There are those that say that the cost of test should not be a consideration, that test should be an integral part of the design specification in the same way as basic functionality, performance, etc. We would totally agree, but in a commercial environment if there are two or more ways of doing the same thing, it makes sense to use the cheapest.

All this must still be kept in its correct context. Use of the results from an economic analysis must be handled skillfully. Our opinion is that to assume that the analysis is 100% correct is wrong—there will inevitably be errors, hopefully small, in the models themselves, but also in the data that are collected for input to the model. We have therefore used the results from the calculations in a comparative way, e.g., DFT method "A" is cheaper than DFT method "B," to mitigate these effects.

Creation of the economic model is relatively simple although experience does make it simpler to create new ones. Data collection is often the most time consuming task. Data might exist, but access to it might be restricted. Some data may simply not be available anywhere. In these cases, it is still beneficial to have created the model as a sensitivity analysis will show to management how crucial it is to have access to the data, or that data must be collected where it isn't already—if you don't know it, then you can't control it!

Time-to-market has already been mentioned as something that has not been discussed in detail in our analysis but which is of critical importance. So too is product liability. To summarize Prof. Steve Szygenda [30], "Don't think about test costs—if you don't test and the product fails, then you can be sued!" Obviously there is a financial impact here that you may wish to consider.

References

1. B. Kitchenham and B. De Neumann, "Cost Modeling Estimation." Chapter 11, *Software Reliability Handbook*, Paul Rock (Ed)., Elsevier Science Publishers, 1990, pp. 333-376.
2. J. Turino, *Design to Test — A Definitive Guide for Electronic Design Manufacture, and Service*, Van Nostrand Rienhold, New York, 1990.
3. CIAS, Technology Strategies, pp. 1-15, May 1990.
4. B. Davis, "Economic Modeling of Board Test Strategies," Second International Workshop on the Economics of Design, Test and Manufacturing, Austin, Texas, 1993.
5. P. Williams, "Deisng for Testability-A Survey," *IEEE Proceedings*, vol. 71, pp. 98–112, Jan. 1983.
6. X. Zhu, M. Breuer, "Analysis of Testable PLA Designs," *IEEE Design & Test of Computers*, vol. 5, pp. 14–28, Aug. 1988.
7. C. Dislis, I.D. Dear, J.R. Miles, S.C. Lau, A.P. Ambler, "Cost Analysis of Test Method Environments," *Proc. IEEE International Test Conference*, pp. 875–883, August 1989.
8. I.D. Dear, C. Dislis, J.R. Miles, S.C. Lau, A.P. Ambler, "Hierarchical Testability Measurement and Design for Test Selection by Cost Predictions," European Test Conference, pp. 50–57, Paris, 1989.
9. I.D. Dear, C. Dislis, A.P. Ambler, J. Dick, "Economic Effects in Design and Test," *IEEE Design and Test of Computers*, vol. 8, no. 4, pp. 64–77, December 1991.
10. C. Dislis, J. Dick, A.P. Ambler, "An Economics Based Test Strategy Planner for VLSI Design," *Proceddings of the European Test Conference*, pp. 437–446, Munich, April 10–12, 1991.
11. C. Dislis, J. Dick, A.P. Ambler, "Algorithms for Cost Optimized Test Strategy Selection," *Proc. IEEE International Test Conference*, p. 234, March 1993.
12. J.H. Dick, E. Trishler, C. Dislis, A.P. Ambler, "Sensitivity of test costs to individual cost modeling parameters," Second International Workshop on the Economics of Design, Test and Manufacturing, Austin, Texas, 1993.
13. K.K. Saluja, "A Built-In-Self-Testing PLA Design with Extremely High Fault Coverage," Technical Report EE8517, Dept. of Electrical and Computing Engineering, University of Newcastel, NSW, 2308, Australia, March 1985.
14. R.J. Illman, "Design of a Self Testing RAM" *Silicon Design Conference*, 1986, London, pp. 439–446.
15. Treuer, Fujiwara, Agarwal, "Implementing a Built-in Self-Test PLA Design," *IEEE Design & Test of Computers*, vol. 2, pp. 37–48, April 1985.
16. C. Dislis, J. Dick, I.D. Dear, A.P. Ambler, "Economics Modeling for the Determination of Test Strategies for Complex VLSI Boards," *Proc. IEEE International Test Conference*, pp. 210–217, Oct. 1993.
17. B. Davis, *The Economics of Automatic Testing*, McGraw-Hill, 1982.
18. C. Pynn, *Strategies for Electronic Test*, McGraw-Hill, 1986.
19. J.H. Dick, E. Trishler, A.P. Ambler, "DOM: A Defect Occurrence Model for Evaluating the Lifecycle Costs of Test Strategies," chapter in *Economics of Design and Test*, Ellis Horwood, 1992.
20. J. Perone, "Reducing Test Costs Through Strategic Changes in Maintenance and Service," *International Test Conference, IEEE* 1986, pp. 205–212.
21. M. Fuller, "How Design for Service Speeds Production Development Time and Reduces Service Costs," *ATE & Instrumentation Conference East 1988*, pp. 549–554.
22. A. Deway, T. Gadient, "VHDL Motivation," *IEEE Design and Test of Computers*, pp. 12–16, April 1986.
23. W. Murray, "Is Two-Level Repair Feasible," *ATE & Instrumentation Conference East*, pp. 571–681, 1988.

24. G. Fink, "On-board diagnostics lower service costs for OEMs and system integration," *Mini-Micro Systems*, May 1984.
25. W. Hughes, "Two Levels vs. Three Levels of Maintenance: The Cost," *Autotestcon89, IEEE* 1989, pp. 19–25.
26. W. Bryant, "Logistics Impacts and Influences of ATE Design to Meet the TLM Concept," *Proceedings of Autotescon89*, IEEE, p. 32, 1989.
27. R. Comerford, "Automation Promises to Lighten the Field Service Load," *Electronics*, April 1982.
28. W. Warwick, "Keynote address to International Test Conference 1984," *IEEE Design & Test of Computers*, Nov. 1984.

Tony Ambler holds the Racal-Redac Chair in Test Technology at Brunel University. He currently serves as a member of the editorial board of JETTA, is Programme Chair for the 1994 IEEE International Conference on Computer Design (ICCD) and will be General Chair for the 1995 European Design and Test Conference. He is the co-founder of the International Workshops on Economics of Design, Test and Manufacturing held in Austin, Texas. His research interests include design for testability, test economics, CAD for electronic systems and concurrent engineering. He is a member of the IEEE Computer Soeicty, ACM, and IEE. He is a Chartered Engineer.

Chryssa Dislis is a Research Fellow in the Department of Electrical Engineering and Electronics at Brunel University and currently works on economics modeling in the design of MCMs. Her research interests include design for testability, CAD for VLSI devices, and the economics of test at chip and board levels. Dislis holds a PhD from Brunel University. She is a member of IEEE and IEE. She is a Chartered Engineer.

Ian Dear is a Lecturer in Digital Systems in the Department of Electrical Engineering and Electronics at Brunel University. His current research interests include design and test economics, test strategy planning and CAD for VLSI devices. Dear holds a PhD from Brunel University in test strategy planning and economic modeling, a BSc in Electronic Engineering from the University of Sussex, and an MSC in Digital systems from Brunel University.

Jochen Dick is a staff member of Siemens-Nixdorf-Informationssysteme in Munich. His work has included research in automatic test pattern generation and the development of a DFT rule checker. He gained his PhD from Brunel University on test economics and test strategy planning and holds an MS in Electrical Engineering from the Technical University of Munich.

JOURNAL OF ELECTRONIC TESTING: Theory and Applications, 5, 157-170 (1994)

Economic Modeling of Board Test Strategies

BRENDAN DAVIS
BDConsulting, 22 Cliveden Mead, Maidenhead, Berkshire, SL6 8HE, England

Abstract. A long and deep recession, coupled with continuous competitive pressure to reduce costs, is forcing many companies to review their test strategies. Testing costs have become a more significant proportion of the overall manufacturing cost even though manufacturing yields have increased dramatically over the past ten or twelve years. This causes attention to be focused on testing costs as a key source of cost reduction. The increased use of DFT and the integration of design and test are very positive moves towards controlling testing costs but other methods employed can often backfire. The increased use of low priced testers is one such method. The pressure to reduce costs, higher process yields and exhortations that "testing adds no value" can lead the test engineering manager to take the "cheap" route. In reality this can often turn out to be an expensive decision. The only way to avoid expensive mistakes is to perform an economic analysis of the alternative courses of action. In most cases this is done, but not always in the right manner or with the necessary amount of detail to make the comparisons meaningful. This article discusses the need for effective cost analysis of test strategies and highlights some of the pitfalls.

Keywords: Economic modeling, printed circuit board test, test economics, test strategies, value added testing.

1. The Need for Economic Analysis

Engineering decisions are almost always economic decisions. Arthur Wellington, who used an economic analysis to plan the layout of the US railway system, defined engineering as "...the art of doing that well with one dollar which any bungler can do with two, after a fashion." So an economic analysis is the logical thing to do and it needs to be performed correctly for the specific task in hand. It becomes particularly important when you consider that the major driving forces in the electronics industry are cost, quality, time to market, and technology. The common denominator for all of these is cost or economy. The cost benefits of quality and time to market are well documented, and the impact of decisions taken about design, manufacturing, and test technologies will have a predominantly economic impact.

Secondary to these fundamental needs for economic analysis, but important all the same, is the need to develop the figures required to calculate an ROI on any investment needed to implement the test strategy.

2. The Need for Life-Cycle Cost Analysis

The comparison of alternative board test strategies is frequently performed using a cost-of-ownership approach that only considers the direct costs relating to the test process. For a commercial board tester this would usually involve four cost areas: the purchase price, the maintenance cost over the economic life of the tester, the cost of test program development and fixtures, and the cost of operating the tester. These operational costs will be the testing, diagnosis, and repair costs. Repair or rework costs are frequently left out of the analysis because they are assumed to be the same for all testers or because they are deemed to be a manufacturing cost as opposed to being a test cost. They should in fact be included because the repair costs will vary from tester to tester for a variety of reasons.

This kind of analysis does not really go far enough because the choice of test strategy and the choice of the testers to implement it will have an effect on other cost areas such as the following test stages and the field. Failure to include these cost areas can lead to the analysis favoring a tester with a low price and a poor fault coverage performance. Indeed if you take a silly example of a tester with no fault coverage then there would be no diagnosis and repair costs associated with it! All of these costs will have been pushed downstream to the more expensive test stages including the most expensive test stage of all, the customer. Life-cycle cost analysis is essential if the true differences between alternatives are to be seen. Not including all of the relevant cost areas in the analysis is one of the two most common mistakes made in the comparison of test strategies. The second of these common mistakes is to over-simplify the analysis and this problem is addressed in the next section.

3. The Rule of Tens

The concept of the rule of tens, as illustrated in Figure 1, has been with us in electronics testing since the early 70s, and it has made a valuable contribution to helping non-technical managers understand that it costs more to detect and rectify a defect as you progress through the manufacturing process. However, the ten to one relationship has also been inaccurate since the early seventies and yet people still rely on it to help them to make their testing decisions. The originators of the concept, Kemon Tashioglu and Jim Prestridge, would be the first to argue that you should not make important testing decisions based upon such a global concept. Much more detail is required. Around 1971, when they were both with the Teradyne company, they recognized that there was an approximate ten to one cost difference for the detection and the rectifying of a defect between the four major test stages of component test, board test, system test, and the field. At the component (incoming) test stage and the board test stage this typically translated into costs of about $1 to $3 per defect at incoming inspection and $10 to $30 per defect at board test. By 1973 this relationship had changed significantly. The high cost at board test had come about because of the lengthy diagnosis times involved. Board testers had no automated diagnostics until guided probes were introduced in early 1973. This reduced the diagnosis times from 15 to 20 minutes per defect down to about 3 minutes per defect with the result that the cost per defect at the board test stage fell considerably. Under some conditions it could be less than the cost at incoming inspection. Figure 2 shows this changed situation.

The low cost at the incoming inspection test stage came about as a result of the high defect rates for com-

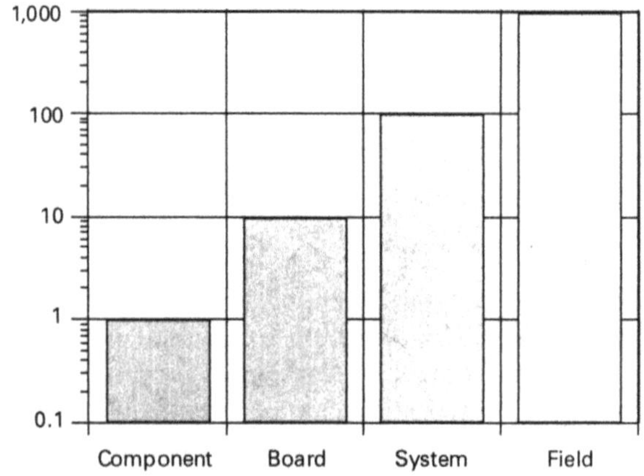

Fig. 1. The Rule of Tens as it appeared in 1972.

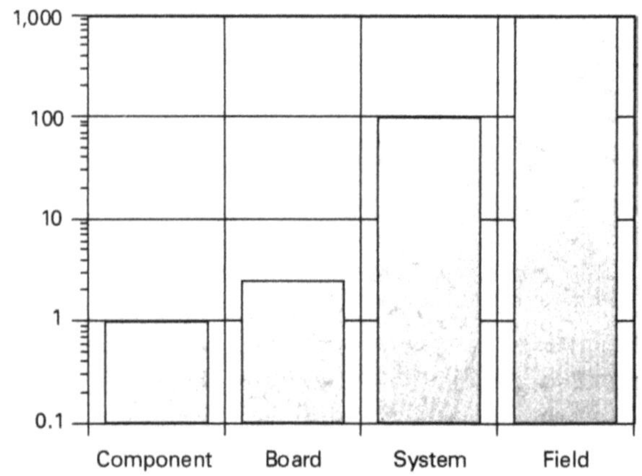

Fig. 2. The change in the cost per defect relationship between component test and board test in 1973.

ponents in the early 70s. Digital ICs had failure rates of around 1% and linear devices were much worse. In most cases you only needed to test one hundred devices to find a defective one. Today this sort of defect rate would be almost unheard of and you would have to test 10,000 or more devices before finding a defective one. This puts the cost per defect up and makes incoming inspection uneconomic for the majority of devices. This change in component quality has led to the sort of cost relationships seen today and illustrated in Figure 3.

The actual cost factors at each test stage vary widely depending on the test strategy, the yield, the fault spectrum, and other variables. You cannot use the same global concept for all situations, and you cannot use the same methodology to calculate the average cost of a defect at each test stage. Because of this, any analysis that is based upon the average cost of detecting and

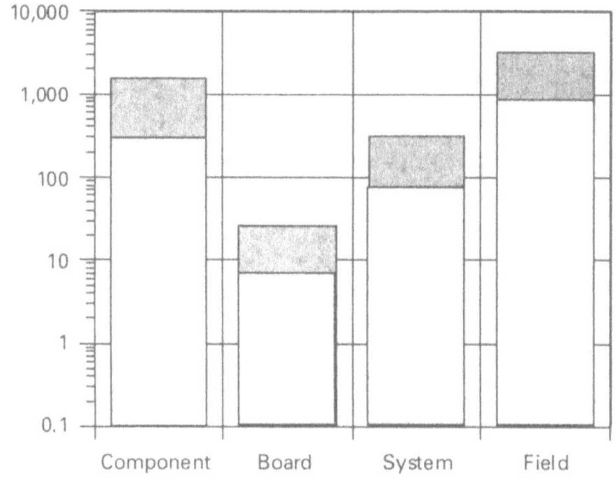

Fig. 3. Typical ranges of cost relationships today (1993).

rectifying a defect can be very misleading. For example, the calculation of the average cost of detecting and rectifying a defect at the component test stage (incoming inspection) should include the cost of testing all of the defect free devices as well as the bad ones. However, you cannot apply this same logic to the final system test stage. In an ideal world very few defects should escape to the final test stage. This stage should be reserved for proving that the product meets its operational performance specifications. Now if we assume that this test stage costs one million dollars per year to operate, and ten defects per year are detected, then the average cost per defect is $100,000. Now if we improve the effectiveness of the preceding tests such that only five defects are detected at the final test, then the cost per defect will rise to $200,000. However, the justification for improving the previous tests may have been based upon saving $500,000 by preventing five defects from reaching the final test stage, where it costs $100,000 per defect! This is quite obviously a load of nonsense. In a simple and exaggerated example such as this it is easy to spot that the logic of the argument is invalid, but in the real world it's not usually quite so obvious. I have seen numerous examples of this incorrect approach used to justify expensive test systems that will reduce the number of defects that reach the final test stages. The real savings that will be made will relate purely to the cost of diagnosis and repair. Including the cost of testing the defect free units, will exaggerate the savings because these tests would not be eliminated even if you could reach zero defects at these stages. In contrast the incoming inspection tests can, and should, be eliminated when the defect rate reaches a very low level.

4. Loaded Labor Rates

Another common cause of major errors comes from using fully loaded labor rates in determining costs and savings. This is a similar problem to the one described above in that some of the costs involved will not actually be saved. Fully loaded labor rates should only be used when trying to determine manufacturing costs for the purpose of pricing the product or determining its profitability if, as is usually the case, the pricing is set by what the market will bear. To use fully loaded labor rates to determine the savings made by a given test strategy relative to an alternative, or the current strategy, will result in grossly overstated savings because most of the overhead elements that result in the loading factor will not be saved. The cost of the buildings the utilities, the management, and other overheads will not generally be saved. In many cases the only savings will come from direct labor costs and the basic overheads associated with that such as bonuses, insurance, pension contributions, etc. I have lost count of the number of times that I have seen worked examples using loading factors of four to six times the direct labor cost, that have resulted in calculated return on investments (using the IRR method) of several hundred percent!

5. Simple Models

The rule of tens approach is the simplest type of analysis we can perform. The next step up from this is the use of simple strategy models but these can also result in an oversimplification of the analysis. Over the years there have been many models produced by in-house groups and by equipment vendors that have made too many simplifying assumptions. Such models can be very useful for performing quick comparisons with a minimum number of variables but they should not be used as a substitute for the detailed analysis needed for final decision making.

6. The Wrong Formula

A problem that has afflicted many of these simple models has been the use of an incorrect formula to model the testing process. This formula falls into the same situation as the rule of tens in that it may have been right at the time but it rapidly became inaccurate. Before board testers got automatic diagnostics, testing was a two-stage process. A low-skilled operator would sort good boards from bad boards (the go/no-go test) and then a skilled technician would work on the defective boards to manually diagnose the defects. This would be done on the board tester with the aid of the logic diagram, a logic probe, and possibly an oscilloscope. Once the testers got diagnostics, the low-skilled operator did the entire job in one operation. If the board passed the test it would be placed in one box, if it failed the test then the diagnostic routine would be invoked, the diagnosis would be printed out, and the board would be placed in another box. One operation and one operator. Unfortunately this old formula, which attempted to model the "two operator" approach, will not die. I still see it, written in a variety of forms, in quite recent books and articles. The formula determines the average time that each board will spend on the test system. This is a vital piece of information that is central to the model because the cost, throughput, and capacity information will all be derived from it. The most common form of this formula is shown below.

$$TTT = Tt(2 - Y) + Td(1 - Y)$$

TTT = Total time on the tester
Tt = Test time (for a good board)
Td = Diagnosis time per defect
Y = Yield (expressed as a probability of the boards being defect free, e.g., 0.9 is equivalent to a 90% yield and $1 - Y$ is the probability that the boards are defective)
NB: Tt and Td include the board handling time

The formula was originally used when functional board testers were the only type available and as such it is quite wrong because functional testers detect one fault at a time. This means that the number of diagnostic actions will be related to the average number of faults per board rather than $1 - Y$. The $2 - Y$ term implies that some boards will get more than one passing test, which cannot happen. Once the board passes the test, it will migrate to the next stage of the process. This term was correct for the two operator situation because the defective boards did receive two tests. The first test would have taken place when the low-skilled operator did the go/no-go testing (since the testers had no diagnostics, a failing test took the same time as a passing test since the handling time was usually the dominant factor), and the second test would have occurred when the board was returned for retest after the repair had been performed. So each good board gets one passing test and each failing board gets a failing test at the go/no-go test stage, a diagnosis from the skilled operator, and a retest after repair. Boards containing two defects would simply pass around the diagnosis/repair loop twice. There is no likely scenario today where a board receives two passing tests at the same test stage so we need to use formulae that more accurately model what really happens.

7. Getting the Formulae Right First Time

With functional diagnostic testers each board will get one passing test, either because it was good to begin with or because the defect was repaired, and one diagnosis for each defect. So the more accurate basic formula for functional test is simply

$$TTT = Tt + Td(FPB)$$

where

$$FPB = \text{average number of faults per board}$$

For in-circuit testers and most MDAs (manufacturing defects analyzers) the situation is rather different because they can detect multiple faults in one operation. However, if a short circuit is detected the testing will usually be terminated. This is because the presence of a short may cause damage when power is applied and it will also affect the validity of other tests made on components connected to the shorted nodes. The basic formula for an in-circuit test stage is therefore

$$TTT = Tt + Tds(1 - Ys) + Tdo(1 - Yo)$$

where

Tds = Diagnosis time for a short
Ys = Yield at shorts test
Tdo = Diagnosis time for other defects
Yo = Yield for other defects

In all cases it is necessary to account for the fact that all test stages have four possible outcomes rather than simply two (pass or fail).

1. The test may pass because the UUT is defect free.
2. The test may pass even though there is a defect present.
3. The test may fail because a detect is detected.
4. The test may fail even though there is no defect present.

Outcomes one and three are catered for by the basic formulae, and outcome two will usually be handled by defining the fault coverage of the tester and making the calculation based upon the calculated *detected faults per board* figure. Outcome four is frequently overlooked in modeling but it can be quite important, especially if diagnosis times and repair times are long. There are many reasons why a test may fail when there is no defect present but they usually fall into three groups. Problems with the tester, the fixture, or the test program. Measurement range, accuracy, and guarding limitations in the tester can all cause problems, as can jammed test pins and other causes of poor contact in the fixture. The test program may not reflect the latest revision of the board or it may have been created with inaccurate or marginal library elements.

The end result of incorrect diagnosis action such as this is a general lowering of efficiency. Boards pass round the diagnosis/repair loop more often than necessary so that costs rise and the capacity of the test stage is reduced. This can be an important performance difference between testers in different price ranges. The better testers tend to have better accuracy, wider ranges, and more accurate and robust library elements both in the functional and in the in-circuit arenas. The test program preparation software will also impact this problem as will the availability of any diagnostic enhancements that can eliminate ambiguity or misdiagnosis. Outcome four can be handled in several ways but the simplest is to add a correction factor to the diagnosis term of the formula and to the calculation of repair time.

Determining a formula that accurately models what happens in the process is key to the success of the model. The usual approach to defining the formula is to develop it from a flow diagram of the process and then to test it with manual calculations. Markov processes and matrix calculations can be used to determine process flows but this level of sophistication is not really necessary. The yield information at each test stage will determine what proportion of the production will migrate to the next stage and what proportion will go to rework etc.

Regardless of whether the formulae are correct or not, the main problem with simple models is their simplicity. You cannot reduce a complex problem to a simple model without losing something. What is usually lost is the ability to see the impact of fault spectrum changes, the impact of tester performance variation, and the ability to include all of the relevant cost areas. Such models are quite adequate for global comparison of very different strategies but they lack the detail to see the effects of alternative implementations of the same strategy. In other words they can compare strategies but they cannot compare tactics.

8. Test Strategies and Test Tactics

We borrow the term *strategy* from the military. In both a military context and a testing context a strategy by itself is quite meaningless. The process begins by defining an objective, selecting the strategy that gives the best chance of reaching the objective, and then defining the tactics that will give the strategy the best chance of success. Although every organization's objectives will vary there has to be a common element that relates to the four main market forces of cost, quality, time, and technology. Whatever products you decide to produce, and whatever markets you choose to sell them in, you have to plan to meet these common goals in order to be competitive. Figure 4 shows the relationship between strategies and tactics and illustrates how a good strategy can get screwed up by using the wrong tactics. It also shows that it is possible to get a more effective result by good tactical implementation of a poor strategy. What we really need of course is both. We should be aiming to maximize effectiveness with the best strategy and the best tactics. Therefore the analysis model needs to be able to see differences in tactics as well as strategies. Probably the key tactic is the choice of board tester to implement the strategy. For example your chosen strategy may be to use an in-circuit tester with some boundary scan capabilities and to follow this by a sub-assembly test stage and then a final system test. There is a very wide range of testers commercially available to implement this strategy. They vary a lot in terms of price and performance and some will be more suited to your strategy, with your set of problems, than others. If you can model the price and performance differences you stand a much better chance of selecting the best product for the job.

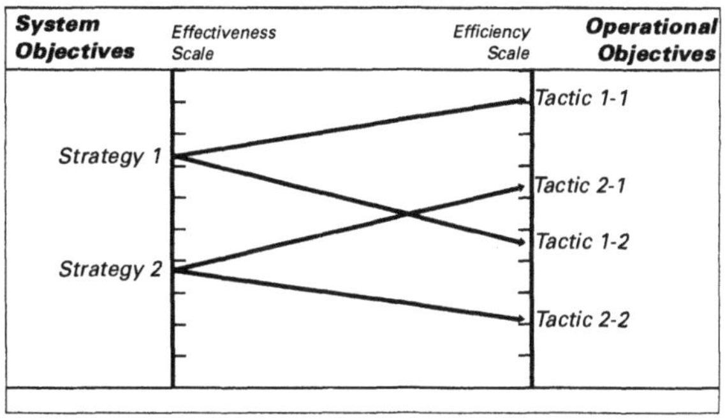

Fig. 4. For maximum efficiency we need to select the best strategy and implement it with the best tactics.

9. The Search for Cheaper Board Testing

The test engineering manager is under pressure to reduce the cost of test. At the same time he is being told that testing adds no value, so he is really having a hard time. The only ray of sunshine he has been seeing is the increased yields that the quality improvement process has generated. Unfortunately the search for cheaper testing and the high manufacturing yields, has occasionally resulted in some less than optimal decisions being made. Cheaper testing does not necessarily come from cheaper testers unless those testers still have good fault coverage, and higher yields do not mean that you can get by with a lower level of fault coverage. In fact the reverse is true. The fault coverage capabilities of a tester are partially dependent on the fault spectrum that it sees. If all of the faults are simple ones then the coverage can be very high. But as the fault spectrum moves towards defects that are more difficult to detect then the fault coverage will fall. Most quality improvement processes are based upon the Pareto method of eliminating the biggest causes of defects first, so there is a tendency for the fault spectrum to get biased towards the more problematic defects as the yield improves. The laws of mathematics cannot be bent. If you attack a 90% yield with a 70% fault cover the escape rate will be the same as attacking a 70% yield with a 90% fault cover. The net result will be no improvement in the shipped quality of the product. All the time, money, and effort taken to improve the yield will have been wasted.

The cheapest testing will be achieved with a test strategy that is optimized to meet your *cost, quality, time to market, and technology implementation* objectives with the lowest possible overall life-cycle cost. This and the fact that the board test stage is currently the cheapest place to detect defects may easily justify additional costs being incurred at this stage in order to minimize the total costs. Looking at the board test stage in isolation may give you the wrong answers.

10. Does Testing Add Value?

The concept that testing adds no value stems from the quality revolution that started in the early 1980s. Most of the quality experts talk about the need to eliminate test and inspection because these activities add no value. "Do everything right the first time, every time, and there will be no need for test." They're quite right of course. It is the correct concept and it is what we should all be striving for. However, I believe that their comments have been misinterpreted. What the quality experts want to eliminate is unnecessary test and inspection and the testing of quality into a product rather than eliminating the sources of defects. Also, we should not forget that these quality experts are talking about quality in general for all kinds of manufacturing and service industries. The electronics industry is a little different to the average of everything else. That may sound like a bit of an excuse but the electronics industry really is different. We have to live with a situation where complexity is constantly increasing at the same time that physical dimensions are constantly decreasing. This presents problems for design, manufacturing, test, and quality, that are unmatched by any other industry. "Doing it right first time—every time" implies that we need defect free designs and that we build the product with no manufacturing defects, using defect free materials and components. Is this realistically achievable at present? The problem is that the complexity is so great that we cannot fully verify our designs and we cannot exhaustively test the components and the boards. The complexity problems are enormous to a degree that people outside the industry find hard to comprehend. I occasionally lecture on quality issues to audiences made up mainly from mechanical and automotive companies. To give them a feeling for the complexity of electronics testing I use the following example:

"You have a VLSI tester that can perform one hundred million tests per second." This simple fact, though not exactly state of the art, usually causes mutterings of disbelief to begin with. "Now you need to test a digital device that has 64 input pins by applying every possible combination of *ones* and *zeros* and monitoring the results of the tests on the output pins. At the one hundred million tests per second rate this will take 5,850 years!!" It takes a bit of believing but it's true. Two raised to the power of sixty-four is about eighteen million million million. Of course we wouldn't attempt to test it that way, we would be much smarter than that. We would make use of a fault simulator and possibly some automated test pattern generation to define and verify a shorter set of test vectors. Unfortunately the complexity and the available computing resources prevent us from doing a complete job, so we have to restrict our fault modeling to single stuck-at-one stuck-at-zero faults. We then have to hope that the effectiveness of our test vectors will not be reduced for multiple fault conditions, or for defects we did not simulate. Kenneth Parker [1] gives an excellent example by describing the problem of simulating all combinations of stuck-at-one and stuck-at-zero faults on a simple device or board with only 70 electrical nodes. Since there are three possible conditions at each node this would result in 2.5×10^{33} fault possibilities (3^{70}). To name each fault condition uniquely would require about 14 bytes of storage for each condition. If the mass of the memory is one gram for every 60 kilobytes stored then the total memory required to store the list of names would be the size of the planet Saturn! Storing the simulation data as well would require a little more memory!

So we cannot yet achieve perfection in design, we can't guarantee perfection in the components, and we still have the opportunity for human error even in highly automated manufacturing processes. So how close can we get. A target that many companies have set themselves is the six-sigma quality level of 3.4 defects per million opportunities for defects. This is a level of quality that would have been considered out of reach ten years ago but is already being achieved in some areas. The problem however is the complexity issue. There may well be thousands or even tens of thousands of opportunities for defects in a complex product. A mid-sized PBX system designed and built with all processes and components at the six-sigma level could well have a system turn on yield below 20% if testing of the boards and subassemblies were eliminated.

Testing adds value so long as it is cheaper to test than to not test. Incoming inspection of the better quality devices may have already moved beyond this point as indicated in Figure 3 but even here there is some room for doubt. Figure 3 still shows a lower cost for incoming inspection than for field service. Many of the defects detectable at incoming inspection are not detectable at board test and may have a low detection rate at final system test, so these have a high probability of reaching the field. Under these conditions incoming inspection might still be cost justified! Again, the analysis of such a situation requires a fairly detailed model with fault coverage definable on a defect by defect basis rather than the application of some global fault coverage figure for a test stage.

11. The Golden Rule of Testing

If the golden rule of quality is to

"Do it right first time—every time."

then the golden rule of testing is to

"Do the right testing right—first time—every time."

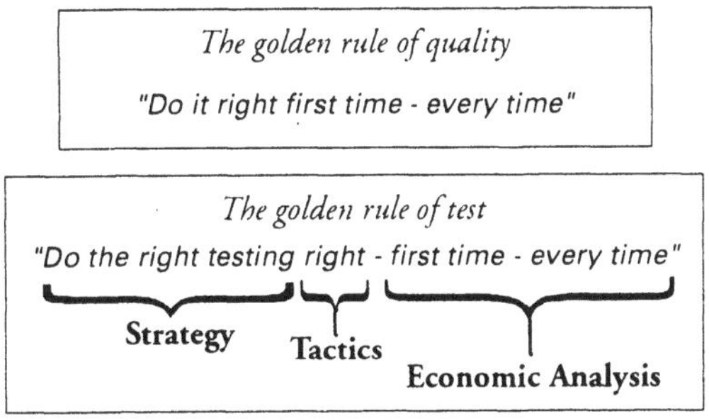

Doing the right testing requires the selection of the most appropriate test strategy, and doing it right requires the selection of the best tactics to implement the strategy. The only way to determine what is right, and to do so the first time and every time, is to perform a detailed economic analysis of the various alternatives.

12. Developing the Model

The easiest way to show the structure of a model is to describe an existing one. Figure 5 shows an overview diagram of the EVALUATE model where the input data is divided into two categories. There is production information that defines the unit under test and other production variables, and there is information that defines the performance of the various test stages that make up the test strategy. From these two sets of input parameters the model can calculate times, costs, throughputs, loading, spare capacity, etc. A measure of the quality of the strategy and tactics under consideration can be produced from the calculated escape rate from the overall process. If required, a cost can be applied to this so that a full life-cycle cost can be derived for each analysis. This model runs on a PC under Microsoft Windows and each set of data is input to an on-screen form. The board data form defines the complexity of the UUT and also computes the number of component defects per board from the quantities and the defect rates for each component type. The number of board level component defects calculated in this manner often causes some discussion since many people report lower levels of component defects. There are a number of reasons for possible discrepancies here. If you buy components to a 100 ppm specification from a reputable supplier then the chances are that the defect rate is better than 100 ppm. This is the positive view. In many cases however, the lower apparent levels of component defects result from a poor fault coverage at the board tester. The product design may also be tolerant of certain defects so the subsystem and final

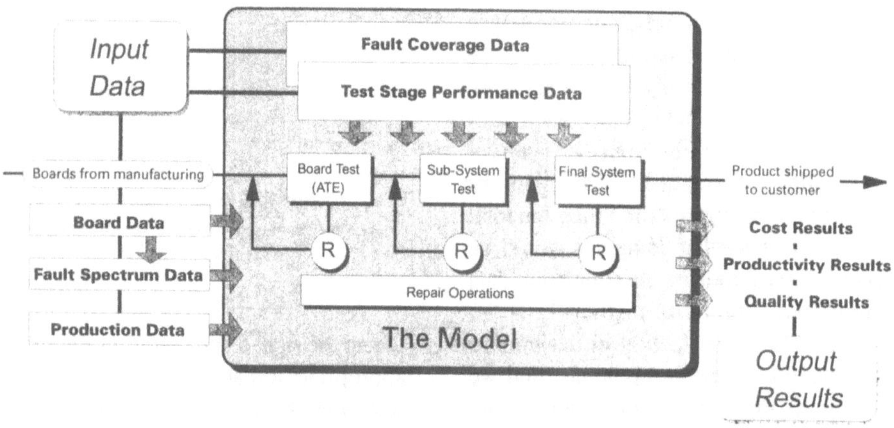

Fig. 5.

system test stages may have an even poorer fault coverage and the defects escape to the field. When these defects degrade enough to cause failures they will usually be classified as *early life failures* with no attempt to determine if they could have been detected at an earlier stage. The more tolerant the design is to component parameter variations, the more likely it is that this will happen. The enormous improvement in the quality of components that has taken place over the past decade, and the publicity that has accompanied it, have led us to expect all components to have very low defect rates. The fact remains, however, that there are still classes of devices that suffer with quite poor defect rates. This board-data form is shown in Figure 6. The individual component types are then reduced into a smaller number of subgroups and the defect data is fed forward to the fault spectrum form which is shown in Figure 7. Here the remainder of the defects making up

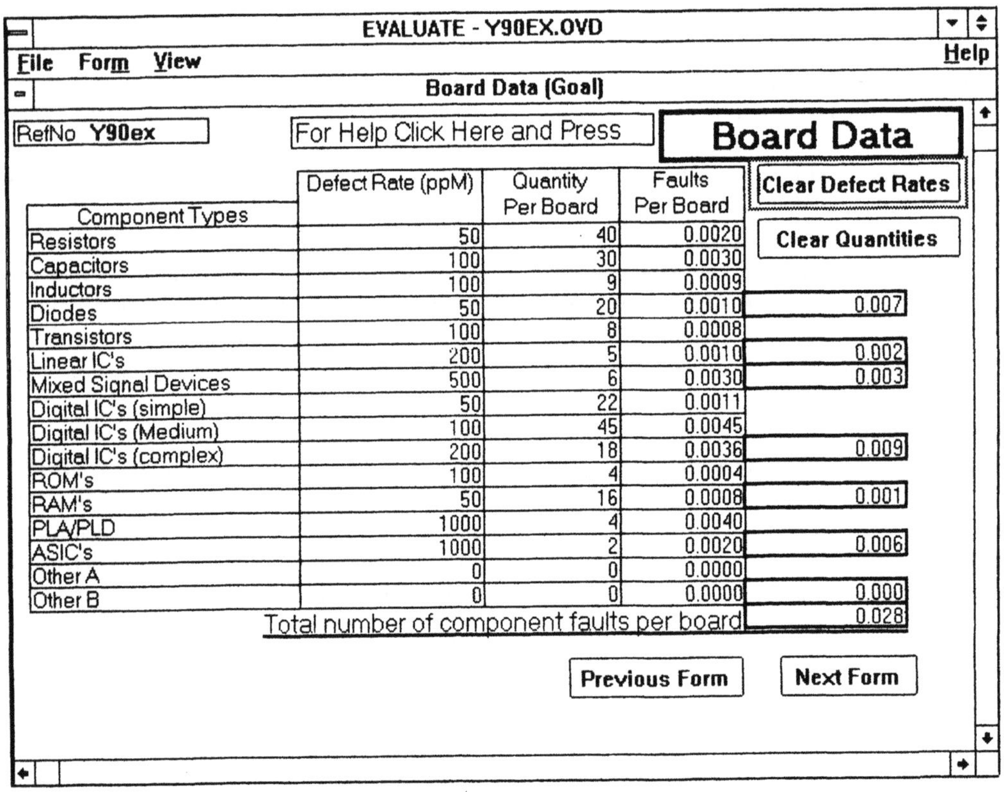

EVALUATE - Y90EX.OVD

File Form View Help

Board Data (Goal)

RefNo Y90ex | For Help Click Here and Press | **Board Data**

Clear Defect Rates | Clear Quantities

Component Types	Defect Rate (ppM)	Quantity Per Board	Faults Per Board	
Resistors	50	40	0.0020	
Capacitors	100	30	0.0030	
Inductors	100	9	0.0009	
Diodes	50	20	0.0010	0.007
Transistors	100	8	0.0008	
Linear IC's	200	5	0.0010	0.002
Mixed Signal Devices	500	6	0.0030	0.003
Digital IC's (simple)	50	22	0.0011	
Digital IC's (Medium)	100	45	0.0045	
Digital IC's (complex)	200	18	0.0036	0.009
ROM's	100	4	0.0004	
RAM's	50	16	0.0008	0.001
PLA/PLD	1000	4	0.0040	
ASIC's	1000	2	0.0020	0.006
Other A	0	0	0.0000	
Other B	0	0	0.0000	0.000
Total number of component faults per board				0.028

Previous Form | Next Form

Fig. 6.

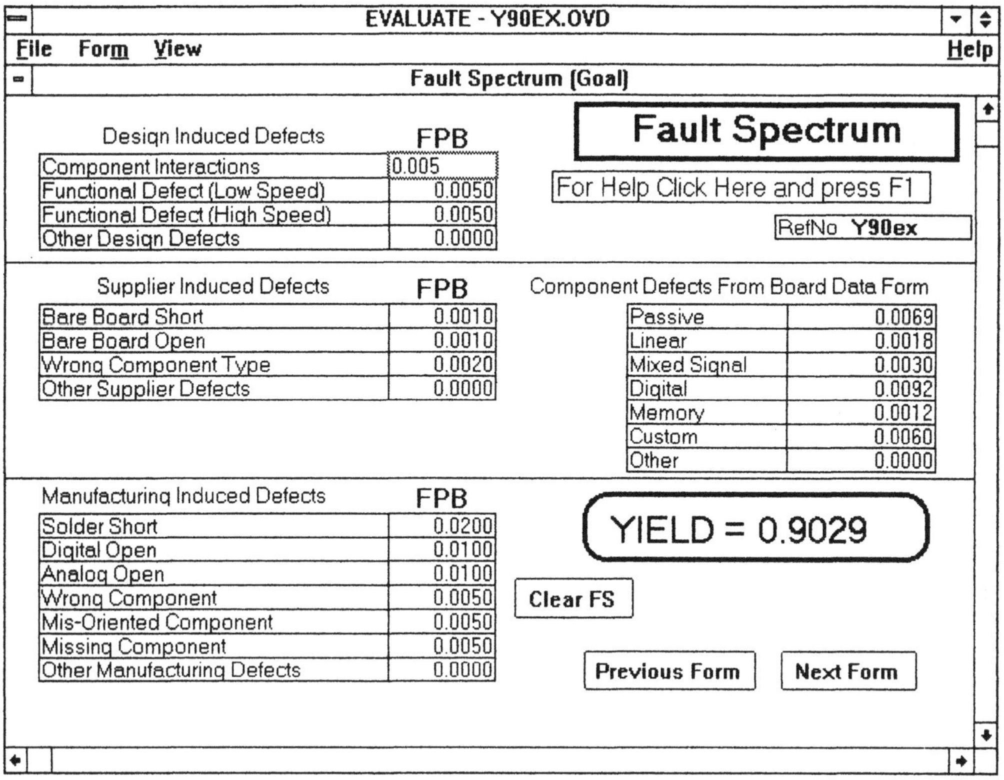

Fig. 7.

the overall fault spectrum are entered and the yield is calculated. This acts as a quick cross-check that the defect data have been entered correctly.

Production variables such as volumes, the number of hours available, the labor rates of the various operators, etc. are then entered into another form (not illustrated) and then the definition of the performance of the various test stages begins. The Test Stage Data 1 form (not shown) collects data about two in-circuit and two functional testers in a similar manner to the Test Stage Data 2 form, shown in Figure 8, which collects data about two alternative combinational testers as well as the subsystem and the final system test stages. These forms define the test and diagnosis time, the repair times, the program preparation workload, and the cost of fixtures and the testers. At the bottom of the Test Stage Data 2 form there is a data field to collect the average cost of a field service call. This enables a cost to be assigned to the defects that escape to the field. There is no attempt in this model to include early life failures in this field cost. The escapes calculated are based upon the defects existing during the manufacturing process.

Figure 9 shows the Fault Coverage Matrix 2 form that is used to define the fault coverage performance of the two different combinational testers. Similar forms collect the same data for other types of test stages. Once all of the relevant input data has been collected the required test strategy is selected from a list and the results of the calculations are presented on a series of screens. The Defect Detection Matrix forms, Figures 10 and 11, show a summary of the faults entering the first stage of the test process, the number detected, and the number that escape. These *escapes* become the input faults to the next test stage where again some get detected and some will escape. This process continues to the final system test stage where any escaping defects are deemed to result in a field failure. At the bottom of the form is a summary of the Yield Progression, which shows how the yield improves as the boards progress through the process. The escapes from each stage are also totaled and the global fault coverage for the stage is also shown. A more detailed version of this data showing the complete fault spectrum is also available from the model. In the example shown, the objective was to compare the performance of two alternative

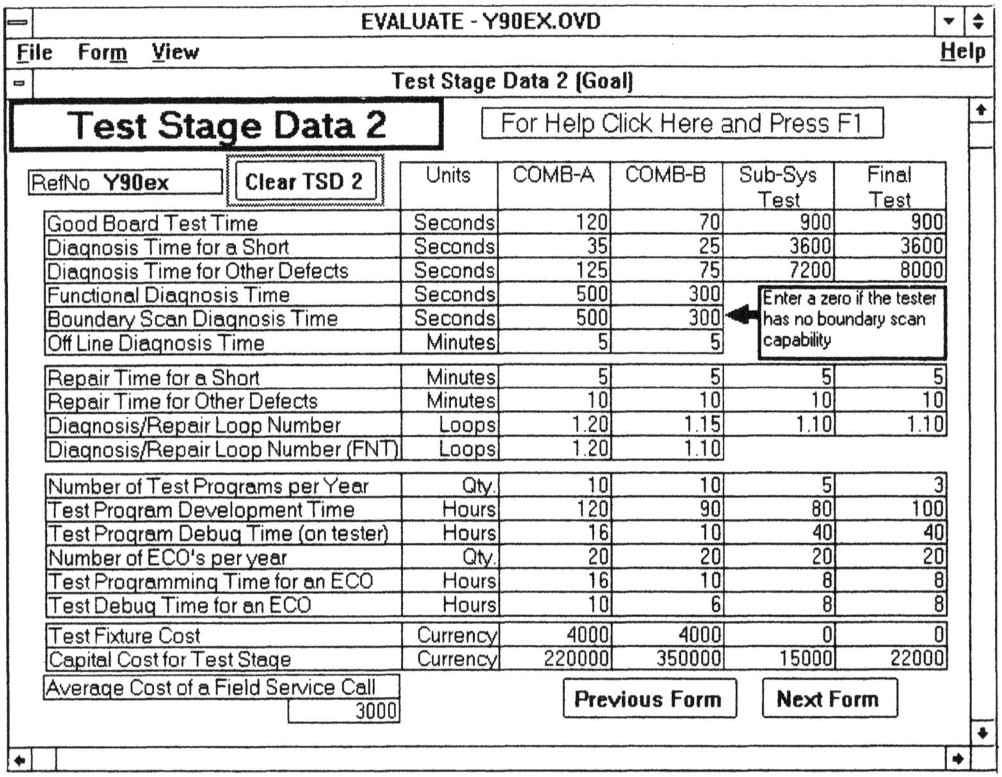

EVALUATE - Y90EX.OVD

File Form View Help

Test Stage Data 2 (Goal)

Test Stage Data 2

For Help Click Here and Press F1

RefNo Y90ex — Clear TSD 2

	Units	COMB-A	COMB-B	Sub-Sys Test	Final Test
Good Board Test Time	Seconds	120	70	900	900
Diagnosis Time for a Short	Seconds	35	25	3600	3600
Diagnosis Time for Other Defects	Seconds	125	75	7200	8000
Functional Diagnosis Time	Seconds	500	300		
Boundary Scan Diagnosis Time	Seconds	500	300		
Off Line Diagnosis Time	Minutes	5	5		
Repair Time for a Short	Minutes	5	5	5	5
Repair Time for Other Defects	Minutes	10	10	10	10
Diagnosis/Repair Loop Number	Loops	1.20	1.15	1.10	1.10
Diagnosis/Repair Loop Number (FNT)	Loops	1.20	1.10		
Number of Test Programs per Year	Qty.	10	10	5	3
Test Program Development Time	Hours	120	90	80	100
Test Program Debug Time (on tester)	Hours	16	10	40	40
Number of ECO's per year	Qty.	20	20	20	20
Test Programming Time for an ECO	Hours	16	10	8	8
Test Debug Time for an ECO	Hours	10	6	8	8
Test Fixture Cost	Currency	4000	4000	0	0
Capital Cost for Test Stage	Currency	220000	350000	15000	22000

Enter a zero if the tester has no boundary scan capability

Average Cost of a Field Service Call: 3000

Previous Form — Next Form

Fig. 8.

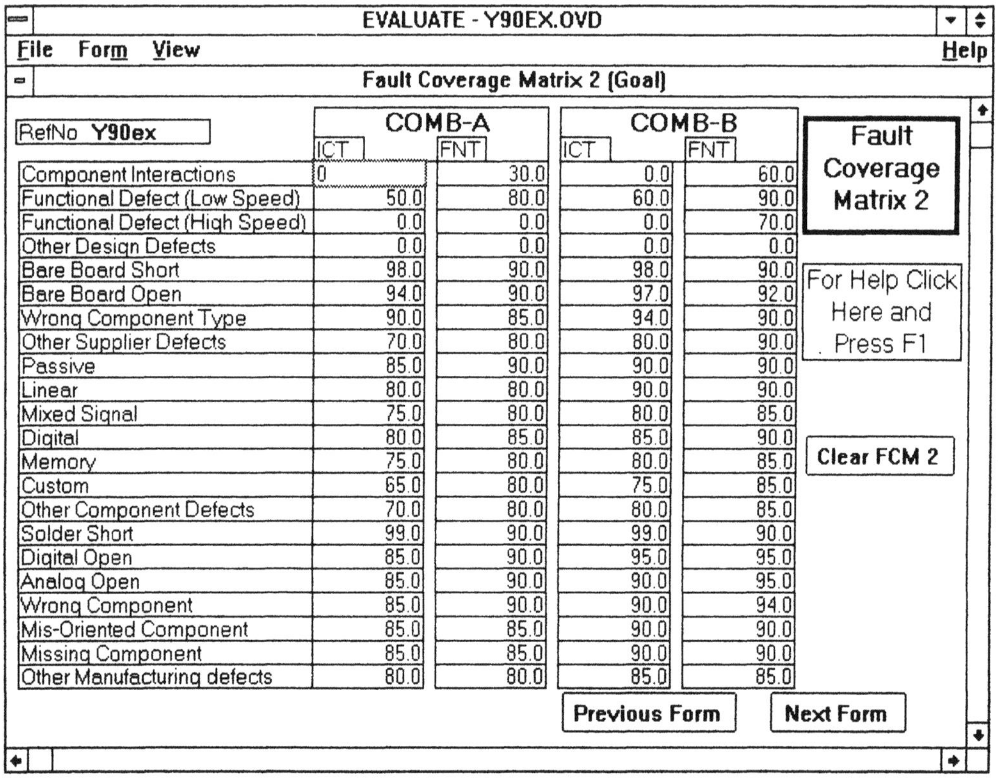

EVALUATE - Y90EX.OVD

File Form View Help

Fault Coverage Matrix 2 (Goal)

RefNo Y90ex

	COMB-A ICT	COMB-A FNT	COMB-B ICT	COMB-B FNT
Component Interactions	0	30.0	0.0	60.0
Functional Defect (Low Speed)	50.0	80.0	60.0	90.0
Functional Defect (High Speed)	0.0	0.0	0.0	70.0
Other Design Defects	0.0	0.0	0.0	0.0
Bare Board Short	98.0	90.0	98.0	90.0
Bare Board Open	94.0	90.0	97.0	92.0
Wrong Component Type	90.0	85.0	94.0	90.0
Other Supplier Defects	70.0	80.0	80.0	90.0
Passive	85.0	90.0	90.0	90.0
Linear	80.0	80.0	90.0	90.0
Mixed Signal	75.0	80.0	80.0	85.0
Digital	80.0	85.0	85.0	90.0
Memory	75.0	80.0	80.0	85.0
Custom	65.0	80.0	75.0	85.0
Other Component Defects	70.0	80.0	80.0	85.0
Solder Short	99.0	90.0	99.0	90.0
Digital Open	85.0	90.0	95.0	95.0
Analog Open	85.0	90.0	90.0	95.0
Wrong Component	85.0	90.0	90.0	94.0
Mis-Oriented Component	85.0	85.0	90.0	90.0
Missing Component	85.0	85.0	90.0	90.0
Other Manufacturing defects	80.0	80.0	85.0	85.0

Fault Coverage Matrix 2

For Help Click Here and Press F1

Clear FCM 2

Previous Form — Next Form

Fig. 9.

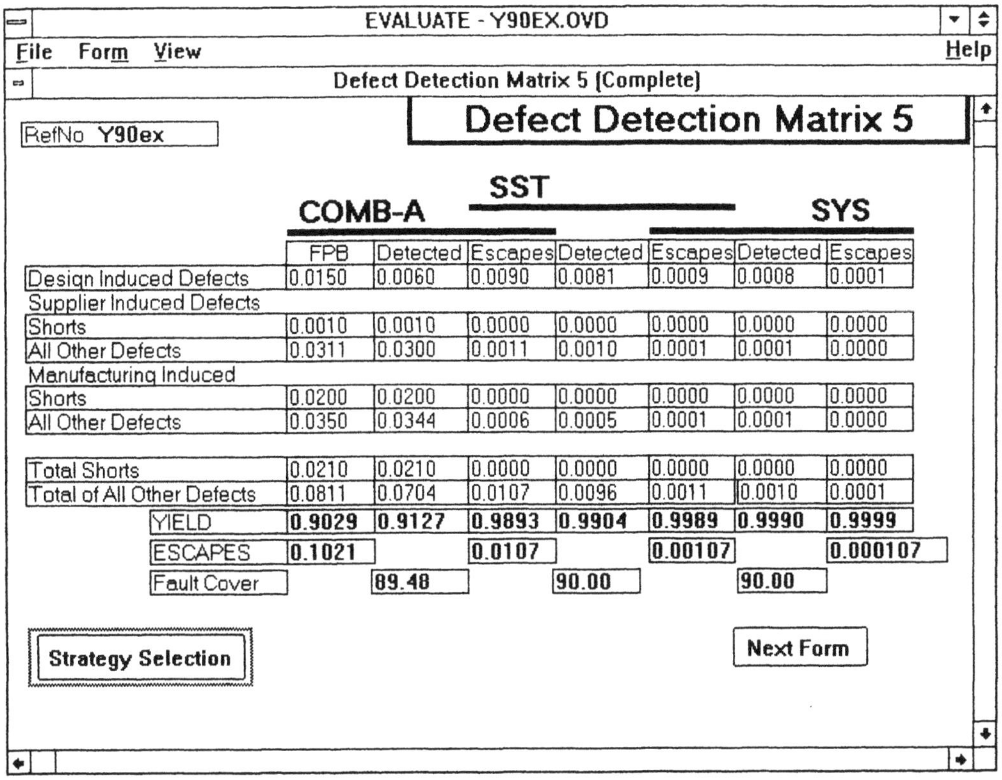
EVALUATE - Y90EX.OVD

File Form View Help

Defect Detection Matrix 5 (Complete)

RefNo Y90ex

Defect Detection Matrix 5

	COMB-A		SST		SYS		
	FPB	Detected	Escapes	Detected	Escapes	Detected	Escapes
Design Induced Defects	0.0150	0.0060	0.0090	0.0081	0.0009	0.0008	0.0001
Supplier Induced Defects							
Shorts	0.0010	0.0010	0.0000	0.0000	0.0000	0.0000	0.0000
All Other Defects	0.0311	0.0300	0.0011	0.0010	0.0001	0.0001	0.0000
Manufacturing Induced							
Shorts	0.0200	0.0200	0.0000	0.0000	0.0000	0.0000	0.0000
All Other Defects	0.0350	0.0344	0.0006	0.0005	0.0001	0.0001	0.0000
Total Shorts	0.0210	0.0210	0.0000	0.0000	0.0000	0.0000	0.0000
Total of All Other Defects	0.0811	0.0704	0.0107	0.0096	0.0011	0.0010	0.0001
YIELD	0.9029	0.9127	0.9893	0.9904	0.9989	0.9990	0.9999
ESCAPES	0.1021		0.0107		0.00107		0.000107
Fault Cover		89.48		90.00		90.00	

Strategy Selection Next Form

Fig. 10.

combinational testers when used in a test strategy that included a subsystem test stage and a final system test stage. For this reason the fault coverage for each defect type at each of these stages following the board tester was arbitrarily set to 90% on the relevant Fault Coverage Matrix forms. As a result the global fault coverage calculated on the Defect Detection Matrix forms is also 90%.

The final form, Figure 12, shows the cost, productivity, and capacity results for the selected strategy. The cost data are broken down into the major cost elements for each of the test stages and the field cost is added in to the totals as a separate item. The capital cost of the equipment needed to implement the strategy is calculated based upon the whole number of test stages required to handle the workload. If the board, or boards, that are the subject of the analysis do not represent the total production volume of the factory then the capital cost of the ATE is applied in proportion to its loading as determined on this results form.

The productivity information consists of throughput calculations for each test stage in boards per hour derived form the calculation of TTT. This is possibly the most important reason for getting TTT as accurate as possible and why the formula should reflect what really happens in practice and include some of the correction factors mentioned earlier. I have seen examples developed using simple models that could be out by two or more testers for a high volume case!! The number of testers required is calculated from the throughput figure, the volume, the test program debug time, the ECO debug time, and the total number of available hours per year. The available capacity is calculated from the free time available and the throuhput figure. This assumes that no further program debug time will be required. Finally the number of repair stations needed is also calculated.

13. ROI and Incremental Investment

In the example shown, tester B (COMB-B) is a clear winner because it has annual running costs that are $60,000 less than tester-A, and it also requires less of a capital investment because the workload can be handled with one system due to its superior throughput. In many cases, however, the choice is less obvious. A mistake that is often made is to determine the ROI for

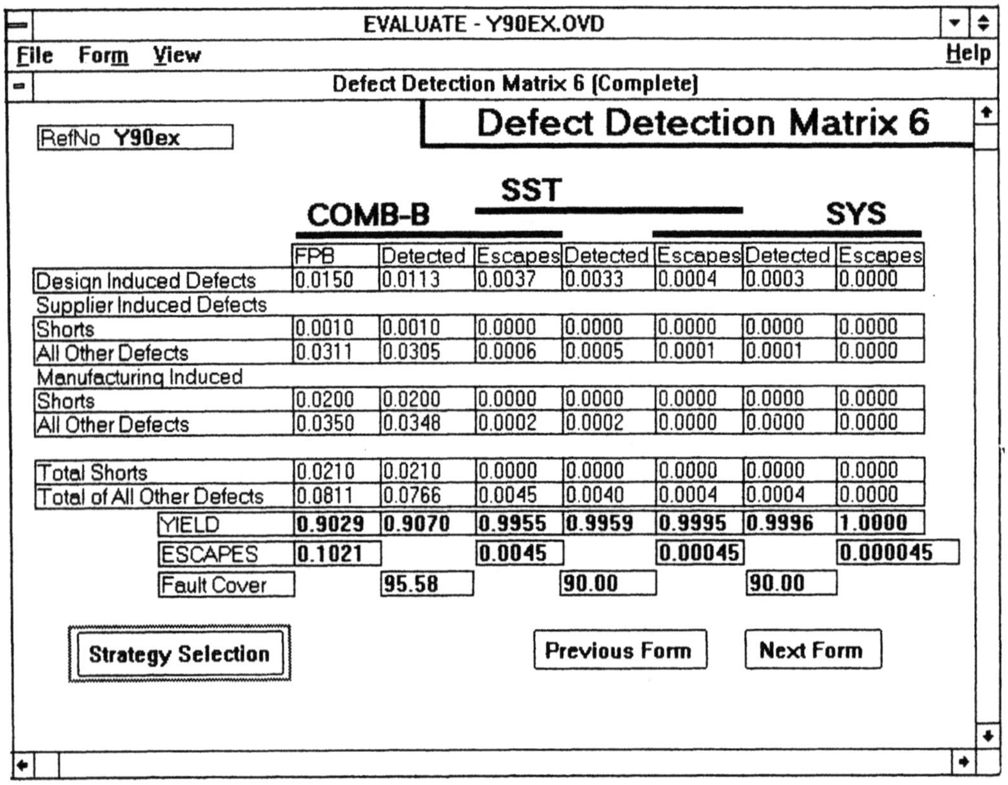

EVALUATE - Y90EX.OVD

File Form View Help

Defect Detection Matrix 6 (Complete)

RefNo Y90ex

Defect Detection Matrix 6

	COMB-B			SST		SYS	
	FPB	Detected	Escapes	Detected	Escapes	Detected	Escapes
Design Induced Defects	0.0150	0.0113	0.0037	0.0033	0.0004	0.0003	0.0000
Supplier Induced Defects							
Shorts	0.0010	0.0010	0.0000	0.0000	0.0000	0.0000	0.0000
All Other Defects	0.0311	0.0305	0.0006	0.0005	0.0001	0.0001	0.0000
Manufacturing Induced							
Shorts	0.0200	0.0200	0.0000	0.0000	0.0000	0.0000	0.0000
All Other Defects	0.0350	0.0348	0.0002	0.0002	0.0000	0.0000	0.0000
Total Shorts	0.0210	0.0210	0.0000	0.0000	0.0000	0.0000	0.0000
Total of All Other Defects	0.0811	0.0766	0.0045	0.0040	0.0004	0.0004	0.0000
YIELD	**0.9029**	**0.9070**	**0.9955**	**0.9959**	**0.9995**	**0.9996**	**1.0000**
ESCAPES	**0.1021**		**0.0045**		**0.00045**		**0.000045**
Fault Cover		**95.58**		**90.00**		**90.00**	

Strategy Selection | Previous Form | Next Form

Fig. 11.

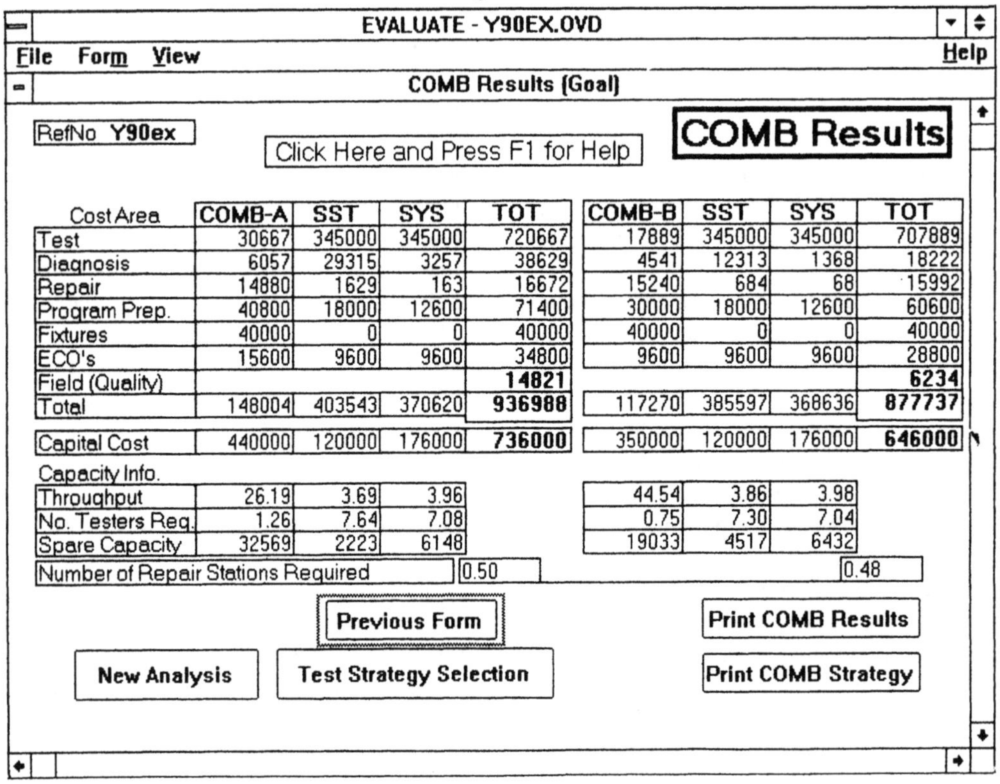

EVALUATE - Y90EX.OVD

File Form View Help

COMB Results (Goal)

RefNo Y90ex

Click Here and Press F1 for Help

COMB Results

Cost Area	COMB-A	SST	SYS	TOT	COMB-B	SST	SYS	TOT
Test	30667	345000	345000	720667	17889	345000	345000	707889
Diagnosis	6057	29315	3257	38629	4541	12313	1368	18222
Repair	14880	1629	163	16672	15240	684	68	15992
Program Prep.	40800	18000	12600	71400	30000	18000	12600	60600
Fixtures	40000	0	0	40000	40000	0	0	40000
ECO's	15600	9600	9600	34800	9600	9600	9600	28800
Field (Quality)				**14821**				**6234**
Total	148004	403543	370620	**936988**	117270	385597	368636	**877737**
Capital Cost	440000	120000	176000	**736000**	350000	120000	176000	**646000**
Capacity Info.								
Throughput	26.19	3.69	3.96		44.54	3.86	3.98	
No. Testers Req.	1.26	7.64	7.08		0.75	7.30	7.04	
Spare Capacity	32569	2223	6148		19033	4517	6432	
Number of Repair Stations Required				0.50				0.48

Previous Form | Print COMB Results | New Analysis | Test Strategy Selection | Print COMB Strategy

Fig. 12.

the better tester using the calculated annual savings and the purchase price of the tester. This sounds logical enough but the savings will be generated by spending the additional money on the better tester so the ROI should be calculated using this incremental investment and the calculated savings. The overall ROI will be determined relative to not using any ATE at all, but since this may not be a viable alternative the incremental approach has to be used. In essence, if some form of ATE has to be purchased because there is no viable alternative, then we have to assume that the basic ROI requirements will be met. Thereafter the tester that gives the best ROI relative to the costliest alternative will be the correct financial choice.

14. Summary

The example shown illustrates the degree of detail needed in a model in order to be able to compare test tactics as well as test strategies. Production variables such as yield and fault spectrum are key to the choice of strategy and tactics and by nature they tend to be rather variable. They vary by product, by production line, by supplier, by week, by shift, by season, and so on. If the model cannot see the effect of such variations then it will be of little help as a decision support tool. The test systems needed to implement a test strategy are also variable. They vary in price and in all aspects of performance so it is also necessary to be able to model these parameters.

The logic used to arrive at these conclusions is as follows:

The key competitive driving forces in the electronics market are *cost, quality, time to market,* and *technology.* The overall strategic plan of any company, and all the sub elements of it, should be geared to meeting these objectives. Money is the common denominator of these four forces so an economic analysis of all alternatives is a logical approach to take. Testing has been said to be a *non-value adding activity* but I believe that this is a misinterpretation of the philosophies of the quality experts. "Eliminate unnecessary test and inspection" is what they are really saying. Electronics has a complexity problem that is far greater than almost all other industries and this works against our being able to do everything "right first time, every time." As a result we cannot yet eliminate the need to test. What we must strive to do is "the right testing right, first time, every time." The right testing is the selection of the optimum test strategy. Doing it right means implementing the strategy with the best possible tactics. Knowing what is right and doing it first time and every time requires a detailed economic analysis. Selecting the right strategy and tactics within the bounds of reasonable financial investment also requires a detailed economic analysis.

References

1. K.P. Parker, *Integrating Design and Test*, IEEE Computer Society Press. 1987.
2. B.P. Davis, *The Economics of Automatic Testing*, Second Edition, McGraw-Hill Book Company, 1994.
3. C. Pynn, *Strategies for Electronics Test*, McGraw-Hill Book Company, 1986.

Brendan Davis has been involved in electronics testing since the early 60s and in automatic testing (ATE) since 1969. He ran a test instrument maintenance and calibration operation for the Plessey Company in the UK prior to joining the UK subsidiary of the General Radio Company, now GenRad, where he held a number of sales and marketing positions. He now runs his own consultancy in the UK providing various marketing services and also specializing in test strategy selection using economic analysis techniques. In this specialized field he provides consulting services, training courses, and analysis tools. He is the author of several books on the subject of test economics and has contributed material to others as well as authoring conference papers and magazine articles. His latest book *The Economics of Automatic Testing* (second edition) was published by McGraw-Hill in January 1994.

JOURNAL OF ELECTRONIC TESTING: Theory and Applications, 5, 171-177 (1994)

Economics of "Design For Test" to Remain Competitive in the 90s

FARZAD ZARRINFAR
LSI Logic Corporation, 1525 McCarthy Blvd., Milpitas, CA 95035

Abstract. This article emphasizes the criticality of maximizing "value adders" and minimizing the costs of "design for test" (DFT) in order to remain competitive in ASIC manufacturing in the 90s.

Keywords: Bed-of-nails method, DFT strategy, value adders.

1. Introduction

In order to remain competitive in the 90s, ASIC manufacturers must focus on improving "value adders" and describing the cost of "design for test."

"Value adders" include offering cells with maximized performance, an easy method for monitoring process performance, supporting testing standards, providing test alternatives, and shortening time to market. Reducing the cost for "design for test" involves the issues of escape rate, board level rework, silicon overhead, test vector length, test development, failure analysis, and hidden costs.

2. Cells with Maximized Performance

Choosing the proper DFT strategy can have an impact on die size and performance of complex ASICs. Two phase scan flip-flops and special boundary scan cells can optimize the performance of a design.

The performance of basic scan designs erodes as a result of metal loading required for scan path interconnect. Such performance degradation can be minimized or eliminated by using advanced two-phased scan, which involves no setup time penalty caused by multiplexer in scan design. The implementation difference between basic scan cell (FD1S) and optimized scan cells (FD1S2) is illustrated in Figure 1.

Figure 2 shows a typical data path that includes scan path enhancement produced by the use of scan cells with optimized performance. In this example, the metal RC loading for the scan interconnect would not degrade the functional performance of the design.

Figure 3 shows the logic diagram for I/O buffers wtih built-in multiplexers for boundary scan technology. This type of implementation minimizes the imposed delay of boundary scan cells. The loading effect of boundary scan circuit has also been isolated from functional logic. Such a design supports operating modes such as CAPTURE, SHIFT, UPDATE, and SET.

The most optimum boundary scan buffer for high speed signals such as "system clock" has also been illustrated in Figure 3. This configuration will support just CAPTURE and SHIFT modes of operation. In addition to scan cell implementation, the interconnect delay resulting from poor placement is the major cause of performance erosion. With submicron technology, optimum placement and use of laout information to reorder the scan chain plays a principal role in ensuring that a design's performance can be maximized. The order of scan chains will also affect the chip's routeability.

LSI Logic provides optimized internal and boundary scan methodology, which allow designers to maximize system performance while achieving high fault-coverage testing. These optimization features are used in high performance applications such as supper scalar processing, SONET, or ATM designs.

3. Capability to Monitor Process Performance

Special structures also can give IC users the ability to monitor process performance factors. ASIC designers can benefit from using a process monitoring cell, such as LSI Logic's PROCMON, to monitor the speed characteristic of their ASICs in low-cost environment.

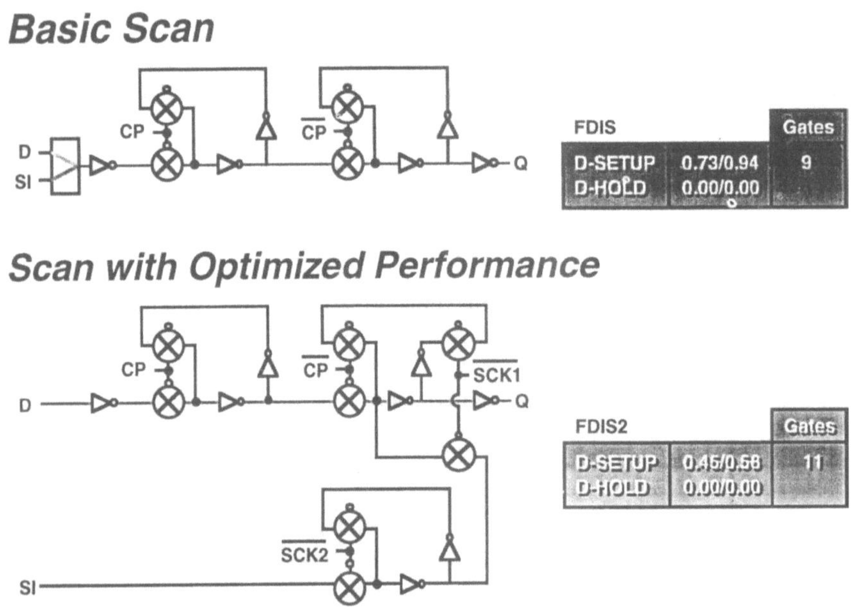

Fig. 1. FD1S2 with no speed degradation.

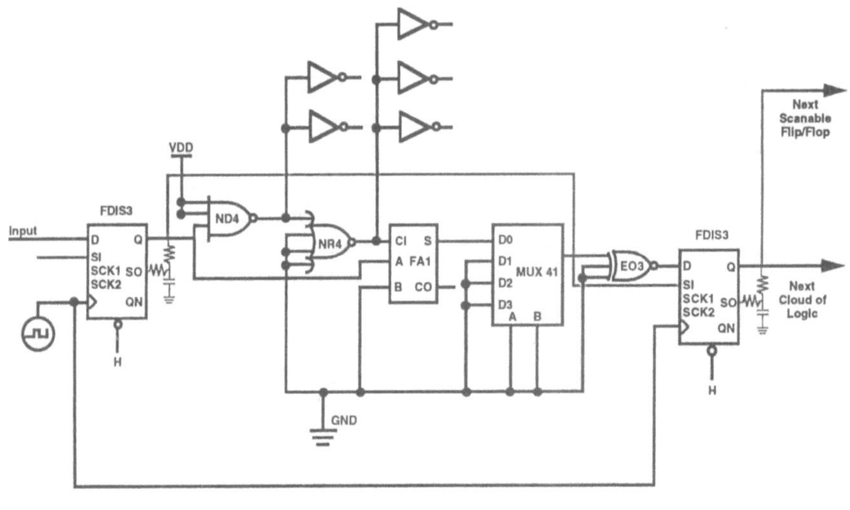

Fig. 2. Optimized scan design.

The PROCMON cell is a chain of gates (a fixed layout of 360 gates). Best-Case (BC) and Worst-Case (WC) process can be monitored easily using PROCMON data. Figure 4 illustrates the distribution of PROCMON data.

Semiconductor manufacturers can also provide customers with best-case/worst-case process multipliers for allowable voltage and temperature variations. These can be used in simulation environment. Thus customers have a cost-effective method for monitoring the product process and thus insuring that process characteristics are within their required simulation windows.

4. Supporting Testing Standards (JTAG & BSDL)

The traditional bed-of-nails method for testing PCBs is no longer effective. Smaller traces, higher speed ICs, and surface-mount components in today's PCBs require other testing method such as boundary scan testing. In addition to using boundary scan for checking connectivity and functionality, other applications including prototype debugging and system diagnostics have been cited. For example, JTAG users benefit from:

- Shorter prototype test time

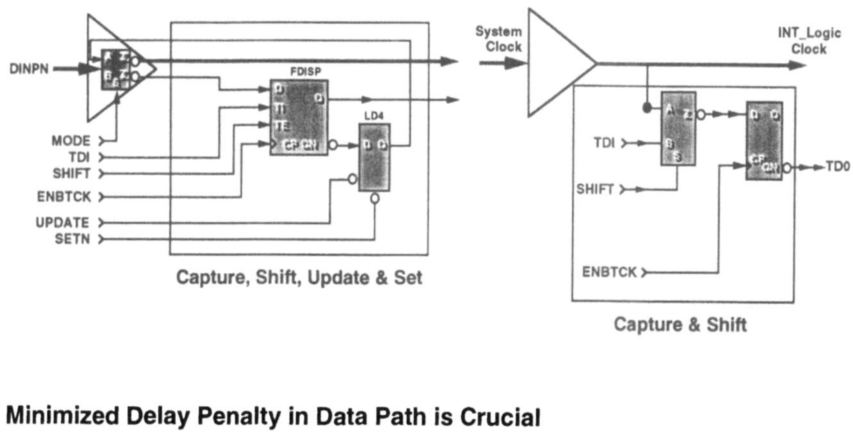

Fig. 3. Logic diagram for input and clock cells.

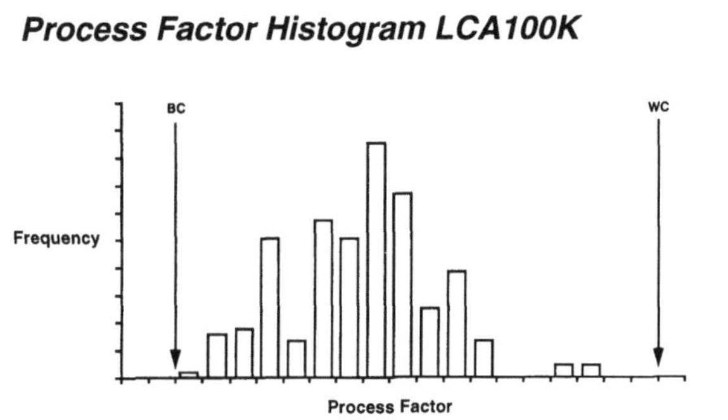

Fig. 4. PROCMON is an enabling solution for LSI Logic as well as system designers to monitor IC's process performance.

- Lower test equipment costs
- Lower cost for board-test generation
- Improved test quality
- The ability to detect and locate opens and shorts between boards and ICs
- Standardized test access to chips, boards, and system during emulation or stress screening)

Also, board-level test tools such as Teradyne's Victory get knowledge of a device's boundary scan circuitry from a model written in BSDL, a subset of VHDL and provide test vectors for board-level testing. BSDL is a standard for talking about devices (for board testers, ASIC verifiers, and device vendors). BSDL is tailored to the 1149.1 standard. Standard syntax for BSDL describes op codes, bit fields, the topology of the test access port (TAP), the order of boundary scan registers, and correspondences to physical device pins. Physical mapping of signal pads to package pins requires access to the packaging data base. Therefore the logical place to generate a BSDL file is the ASIC foundry.

BSDL and boundary scan cells can support various IC or package types, but according to recent survey by *IEEE Design & Test of Computers* [1], the greatest demand for use of boundary scan testing is for ASICs and complex processors. From a packaging point-of-view, system designers most often want to use boundary scan for high or medium pin-out packages, such as PGA (pin grid array), PQFP (plastic quad flat pack), QFP (quad flat pack), and PLCC (plastic leaded chip carrier). Boundary scan is also most useful for BGA (Ball Grid Array) packages, because most solder joints in BGA application are hidden beneath the package and therefore cannot be visually inspected. Electrical inspection by means of boundary scan is a cost-effective way for checking solder ball contacts.

According to Teradyne, the test development time for 80386 microprocessor was about seven weeks, at a cost of $14,000. A complex ATE costing about $500,000 with $5,750 bed-of-nails fixture was used. But for the next generation of microprocessor (80486), which included boundary scan, only ten hours were required for test pattern generation, at a cost of merely $500. The required ATE was a less expensive version costing $150,000 and with $2,250 bed-of-nails fixture. This comparison shows how boundary scan can significantly reduce the cost of board testing.

In the near future, two newly proposed instructions for IEEE1149.1, CLAMP and HIGHZ will further reduce the costs of board testing.

5. Alternative Test Strategies

There are various product market segments. The example of these target segments including high-end work stations, high-end telecommunication, consumer electronics, etc. The design methodology in these segments are different, and required DFT strategy for these various market segments are not the same. Therefore designers require proper DFT solutions such as optimized scan, full scan, boundary scan, partial scan, CrossCheck, BIST, and fault simulation on hardware accelerators.

LSI Logic's emphasis has been to provide the alternative DFT solutions for designers in a transparent fashion. In this case, designers have been able to focus on the functionality of their designs, and proper DFT methodology automatically integrate into the final design. Therefore designers can implement the most optimized DFT solution to minimize the development cost of their ASICs. With new innovations in submicron CMOS technology, other test enhancement such as scan reordering from layout database, IDDQ testing, delay fault testing, and using optimized boundary scan cells would also be paramount.

6. Predictability

Engineering managers need to meet their project schedule to prevent delay in arrival to market. Assuming designers follow the proper scan methodology (such as the one that LSI logic has been providing for ASIC designers), test synthesis process can be done predictably in a few days [2]. On the other hand, the sequential test generators for partial scan ATPG, will break down for high gate count ASICs [3].

The severeness for required computation dilemma, which fits into exponential model, will be the function of final design topology. Therefore, for high gate count ASICs, the risk of selecting partial scan can be extremely costly.

7. Time to Market

As the semiconductor industry continues to introduce innovative devices with increased design complexity, existing products are accelerated toward their rapid maturity and decline. Thus making time to market is a crucial aspect of the entire designing and manufacturing loop.

Figure 5 shows a revenue model for market entry. If we assume that a product offers a conservative total revenue of 50 million dollars and product life of 18 months, then each day late to market means a revenue loss of 200 thousand dollars.

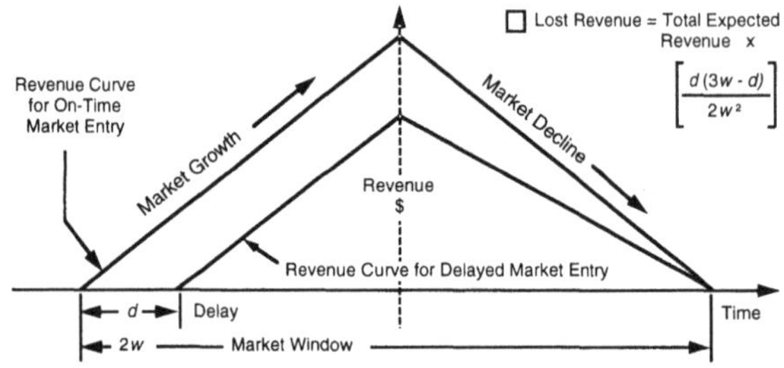

Typical Examples

Total Product Revenue	$25M	$50M
Product Life	18 Months	18 Months
Revenue Lost for Each Day Late to Market	$100K	$200K

Source: Electronics Magazine April 1989 (Ateq Model)
MC1287Q. 12
3/93

Fig. 5. Economics model for time to market.

Achieving system-level integration in the face of competitive time to market pressure requires ATPG tools, which not only reduce design cycles, but also improve the quality of test vectors and the resulting shipped products. These benefits offer significant economic incentive for system designers.

8. Impact of Escape Rate

High-reliability and high-volume ICs demand minimal escape rate. Williams and Brown proposed theoretical prediction for the defect level and escape rate of an IC [4], which was also used by McCluskey and Buelow [5]. This model relates the defect level to fault coverage according to the equation

$$DL = 1 - Y^{(1-t)}$$

where Y is the functional yield and t is an expression of the test fault coverage. This model, which predicts the quality level for shipped products, agrees well with experimental data.

Figure 6 illustrates a graph based on actual data, which represents six different high-volume products that were shipped to LSI Logic's customers. These designs contain similar design methodology, as well as the same process, with fault coverage ranging from 74% to 98%. The variation in DPM levels experienced in the customer's boards are also illustrated. Increases in fault coverage show approximately a ten-fold improvement in the quality of shipped products.

9. Cost of Board Level Rework

For some market segments, such as consumer or PCMCIA, board rework is often impractical. It has

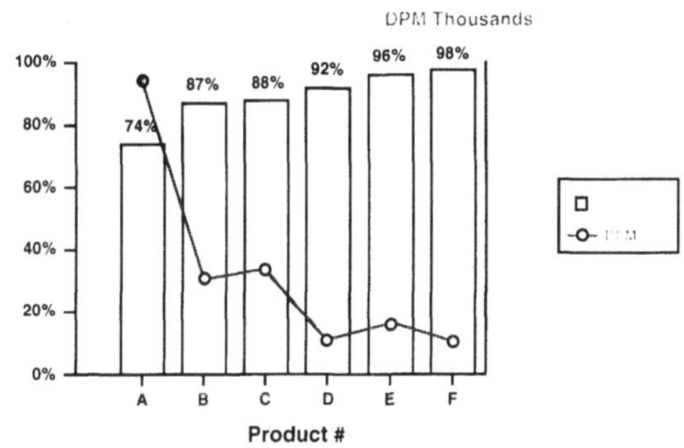

Fig. 6. Fault coverge vs. DPM.

been shown [6] that production parts with high fault coverage, can significantly reduce the cost of board rework.

If we assume an average number of four ASICs per board, with an average repair cost of $500, with fault coverages ranging from 65% to 82%, then the range of ASIC-related repair costs per board would be between $60 and $110. If 1,000 boards are delivered per year, board repair could run between $60,000 and $110,000. LSI Logic's Test Builder tool achieves over 99% fault coverage for ASICs. In the given example, had Test Builder tool been used, rework costs could have been reduced to less than $14,000. Therefore using ASICs, which have high fault coverage tests, significantly reduces the cost of field service and rework.

10. Silicon Overhead

From 1992 to 1994, proportional silicon areas allocated to Design For Test are expected to remain about the same. According to Prime Data [7], area allocation for DFT will vary from 11 to 14 percent. (These are average values that depend on the design type, such as custom ICs, ASICs, and MCMs.)

Product category also is a major parameter for determining the budget for DFT. Standards RISC or CISC processors, for example, employ bus multiplexing to minimize pin count and are heavily core-limited; therefore, DFT circuitry can increase the required die size. One of the most advanced CISC processors of 1993 that contains over 3 million transistors, packs the two 64-bit-wide CPU buses into one 64-bit-wide bus. In contrast, high-performance ASICs typically cannot afford to limit bus bandwidth by multiplexing. ASICs could include greater than 128-bit bus for just a graphic-accelerator card, as well as having independent buses for floppy disk, SCSI, ethernet, I/O serial/parallel ports, real-time clock, mass storage, a nonvolatile RAM, boot PROM, and a keyboard/mouse. Such bus-intensive ASICs can become pad limited; therefore, DFT circuitry would not increase the device die size or die cost.

11. Test Vector Length

Figure 7 shows trend in ASICs, from 1986 to 1993, for LSI logic's technologies with the extent of the testability challenge indicated by the ratio of a design's gate count to the number of device pins. Note that the challenge of test generation is increasing almost exponentially. This increase also reflects the local memory required for testing complex ASICs.

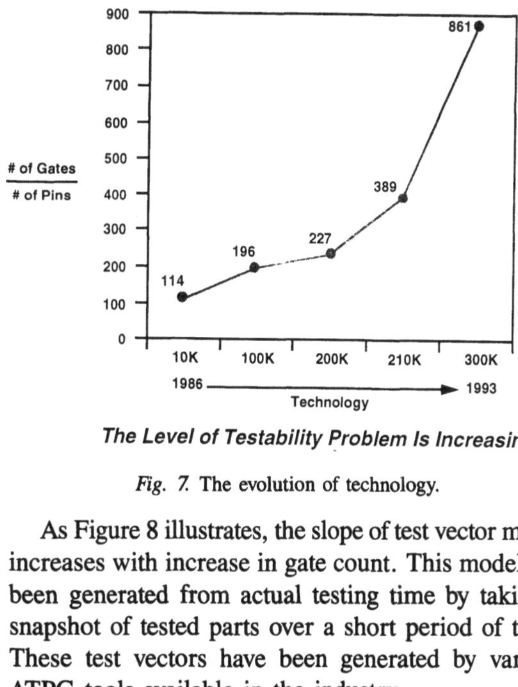

The Level of Testability Problem Is Increasing

Fig. 7. The evolution of technology.

As Figure 8 illustrates, the slope of test vector model increases with increase in gate count. This model has been generated from actual testing time by taking a snapshot of tested parts over a short period of time. These test vectors have been generated by various ATPG tools available in the industry.

Managing the rate of increase shown by this model, will require using multiple scan chins, which will accelerate the testing throughput on the manufacturing floor.

The cost of testing also can be reduced by effective design methodology for test vector truncation. One method is to provide a mode of operation to eliminate the serialization of boundary scan from internal scan paths. For example, in a design with 305 boundary scan

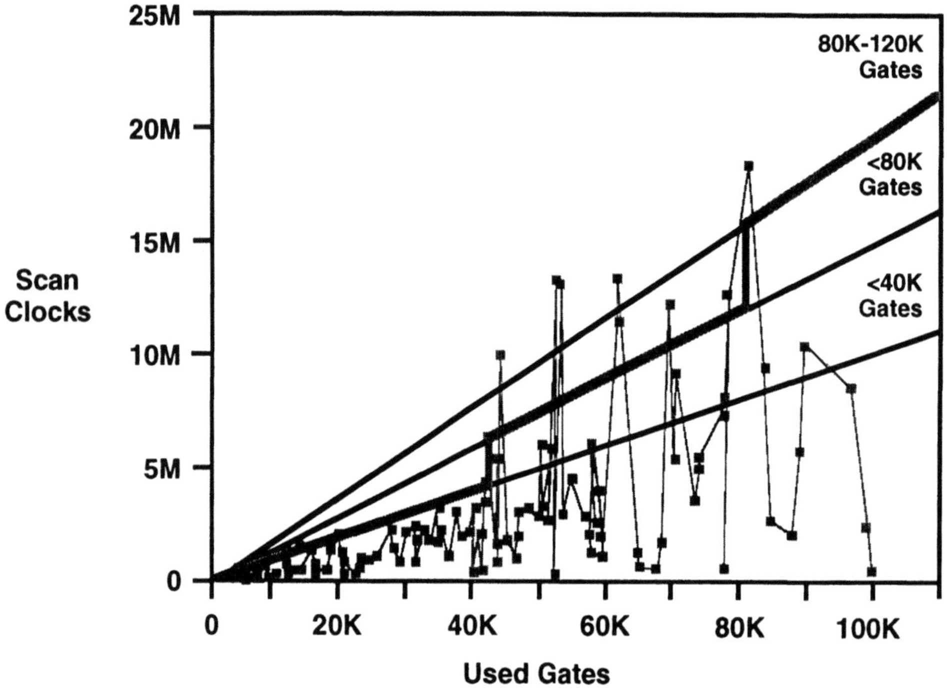

Fig. 8. Model for test time.

cells and 3835 internal scan elements in serial, if the ATPG generates 1902 scan vectors, the overall test cycle is (305 + 3835) * 1902 = 7,874,280. Instead, test clock cycle of 3835 * 1902 = 7,294,170 results from not shifting through boundary scan for IC-level testing. Thus the saving is 8% in test time and usable scan memory on the tester. Such test vector reduction also minimizes the cost of test vector loading as well as test vector compiling.

Similarly with scan design including RAM blocks, the RAM scan chain can be anywhere within the chip's scan chain, but must also be accessible separately from the primary output pins of the chip. It is recommended that the memory scan chain should be at the beginning or at the end of the chip scan chain [6]. This efficient design methodology will minimize the size of IC test vectors.

12. Test Development

Trends show that ASIC starts for designs greater than 40,000 gates are growing and that ASIC starts for designs less than 40,000 gates are declining. This trend toward ever larger ASIC designs demands the development of powerful test implementation and motivates decreasing the cost of test development.

LSI Logic's scan Test Builder is a product that automatically generates test vectors for ASICs in the range 100,000 gates and more, within hours, in contrast to the months of work that might be required by manual test generation. This saving, by using ATPG tool, would reflect on the labor cost for test development.

13. The Cost of Failure Analysis

Failure analysis is more difficult where multilevel interconnect is involved. The die may be widely covered by a ground plane that blocks both the visual search for underlayers and electrical measurement through the shielding plane. Even in the absence of a shielding plane, measurement through the thick dielectric layer is difficult. However, proper DFT implementation makes it possible to access a device's internal nodes so that physical defects can be detected for failure analysis. For today's sub-micron technologies, links to the CAD database are also crucial for helping determine the location of defect.

DFT tools such as Scan, BIST, CrossCheck, and IDDQ testing can be used for silicon debugging and failure analysis in a nondestructive way. These DFT techniques allow for both sensing and driving of internal nodes in the circuit, which is the key for minimizing costs in failure analysis. By reducing the cone size, defect isolation, and defect analysis can be performed down to a few gates.

Without using structured DFT techniques, the time required for failure analysis of three-layer device (and those of the future's four or higher level technologies) will be extremely expensive.

14. Hidden Costs of Analysis and Failure Detection

Typically costs for increasing the reliability of design, component engineering, and re-engineering are hidden. Also, the capital expenditure costs such as $100,000 for SMT rework fixture of $75,000 for e-beam prober can easily be overlooked.

15. Conclusion

In order to remain competitive in the ASIC market, one must minimize the costs of system development, continually concentrating on maximizing the added values discussed in this article. In addition, one must consider all aspects of DFT to reduce the costs of test and improve the quality of the large designs that are seeking system-level integration. This article emphasizes that a proper DFT strategy, can significantly reduce the development effort and time to market, as well as the time to volume of complex ICs while providing high coverage and superior diagnostics.

Acknowledgment

The author is grateful to John Daane, Barney Tanner, John Chin, Bob Miller, Terry Barrette, Joe Dibish, Mark O'Brien, Nick Sporck, Marla Kertzman, and Susan Zarrinfar for their many valuable comments.

References

1. C. Maunder, "Boundary Scan: An End-of-Term Report," *IEEE Design & Test of Computers*, June 1992, pp. 82–85.
2. LSI Logic, "Chip-Level Full-Scan Design Methodology," Jan. 1992.
3. Solution Strategies, Testable ASICs, pp. 3–8.
4. T.W. Williams and N.C. Brown, "Defect Level as a Function of Fault Coverage," *IEEE Trans. Computers*, vol. C-30, pp. 987–988, Dec. 1981.
5. E.J. McCluskey and F. Buelow, "IC Quality and Test Transparency," *Proceedings of the IEEE CS Press*, 1988, pp. 295–301.
6. F. Zarrinfar, "The Four Major Forces That Drive ASIC Testability," *Proceedings of the 1993 DESIGN & TEST EXPO.*
7. Prime Data, "Addressing the Integration of Design & Test," Feb. 1993.

Farzad Zarrinfar received his MSEE degree from SMU in Dallas, Texas in 1984. He is the Product Marketing Manager for LSI Logic Corporation, currently responsible for design methodology, focusing on design verification and CoreWare solutions. Methodologies supports ASICs, DSP, and Standard products.

His past experience includes over 14 years of design, DFT, and marketing for system, microprocessor, and ASICs. This experience has been in several companies such as LSI Logic, Martin Marietta, and Texas Instruments.

JOURNAL OF ELECTRONIC TESTING: Theory and Applications, 5, 179–193 (1994)

The Economics of Scan-Path Design for Testability

PRAB VARMA AND TUSHAR GHEEWALA
Crosscheck Technology Inc., 2833 Junction Ave., Suite 100, San Jose, CA 95134

Abstract. In this article, scan design for testability (DFT) methods are categorized based on the percentage of storage elements made scannable. The non-scan element state retention problem that occurs in partial scan design methods, in which not all of the storage elements are implemented as scan elements, is discussed. Solutions to this problem are described and the overheads associated with them are discussed. An economic model that allows the costs of a range of scan methods that differ in the percentage of storage elements made scannable to be compared with each other is presented. It is shown that, for systems produced in low volumes, the adoption of full scan DFT can be more cost-effective than partial scan DFT when life-cycle costs are considered if it results in significant reductions in the time taken to get the product to market.

Keywords: Design for testability, life-cycle costs, partial scan, scan, state retention problem, test economics.

1. Introduction

The use of sequential Automatic Test Pattern Generators (ATPG) for even small circuits of a few thousand gates is highly unpredictable and can result in excessive test generation times. In addition, despite long computation times, often only an unacceptable fault coverage level is achieved within the time available for test generation. For this reason, scan-path based Design For Testability (DFT) techniques, which reduce the sequential circuit test generation problem to the simpler combinational circuit problem, and on-chip test approaches [1] have grown in popularity.

The use of structured DFT techniques, such as the scan-path or Level-Sensitive Scan Design (LSSD) methods, considerably simplifies the test generation problem by dividing the circuit into combinational blocks and sequential storage elements. The standard full-scan technique involves designing all system storage elements (except RAMS/ROMS) so that they can be reconfigured in the test mode to form one or more serial shift-registers. Test patterns can be shifted (or scanned) into these registers and applied to the combinational portion of the circuit. The test responses of the circuit can then be captured in the scan-registers and shifted out to be observed. A problem with scan-path DFT techniques is that they tend to result in an increase in area and a degradation in performance of the circuit. Thus, partial-scan design techniques, in which not all of the storage elements are made scannable, are gaining in interest.

In this article, a taxonomy of scan techniques is presented, and the costs associated with the various techniques are discussed. One of the major barriers to the acceptance of partial scan techniques is that, to attain adequate fault coverage, ATPG tools may require that non-scan elements retain their state during scan operations. This is not usually the case in the standard single clock scan design methodology. Solutions to the non-scan element state retention problem [2] are described, and the area/performance costs associated with these solutions compared. Finally, an economic model that allows the range of scan solutions to be compared with each other is presented. It is shown that full scan DFT techniques can be more economical than partial scan techniques, despite their potentially higher area overhead, if their use results in a significant reduction in the time taken to get the product to market.

2. Comparison of Scan Techniques

The most widely used scan based DFT approach utilizes single clock multiplexed D-type flip-flops. Another well known approach is Level Sensitive Scan Design (LSSD). However, this is less popular as the use of a standard LSSD element [3] can result in a significantly larger area overhead than if a standard multiplexed D-type flip-flop scan element is used. It typically requires 48 transistors to implement as opposed to 32 for a regular multiplexed D-type flip-flop scan implementation. In addition, LSSD requires the routing of three clock lines throughout the design and this increases the area overhead of LSSD over a single clock, edge-triggered, scan approach. Lower overhead implementations of both types of scan elements are possible. For example, the area overhead of the multiplexer in the standard scan-path approach can be reduced if it is implemented using transmission gates, so resulting in a multiplexed D-type scan element with only 24 transistors, as compared to 18 transistors for a typical D-type flip-flop. In a typical ASIC family, the inputs and outputs of storage elements are buffered and a QBAR output is provided. This has the effect of increasing the transistor count of a D-type flip-flop by 8 to 26 and so reducing the area overhead of adding scan.

A typical overhead for making a D-type flip-flop scannable is approximately 20–30% in terms of gates. However, the actual area overhead incurred will be dependent on the gate to latch ratio of the design and the extra routing overhead, which is influenced by the order of the storage elements in the scan-chain and the number of extra test related global signals that have to be routed through the design. While area overheads will be very design and implementation dependent, increases in area of about 10–20% are typically incurred [4]. The significant area overhead associated with making all storage elements scannable has resulted in an interest in techniques that rely on scanning only a portion of the design.

Scan techniques are categorized as falling into three types based on the percentage of elements that are made scannable:

i. Minimal Scan
ii. Medial Scan
iii. Maximal Scan

In minimal scan, typically up to a third of the storage elements are made scannable. Minimal scan techniques usually involve progressively adding scan capability to storage elements in the circuit, which is initially not scannable, either based on the circuit topology or based on iterative ATPG runs. Minimal scan techniques are generally used when area is at a premium and minimizing area overhead is considered to be more important than maximizing fault coverage. They involve the selection of a minimum number of scan elements, usually without considering the effect on circuit timing [2], [5]–[8]. Thus, the performance of the circuit may still be degraded.

In medial scan, typically between a third to two thirds of the storage elements are scanned. Medial scan techniques are usually used when quality is an important issue and a certain amount of overhead is acceptable to reach a higher fault coverage. Most medial scan techniques are similar to minimal scan techniques in that they try to select a minimum number of scan elements without regard to the impact on circuit performance; they differ from minimal scan techniques only in the amount of scan allowable. Thus, medial scan techniques may result in a degradation in circuit performance. To reduce the impact on circuit performance, minimal and medial scan techniques should take path timing slack information into account [9].

In maximal scan techniques, over two thirds of the storage elements are typically made scannable. Maximal scan methods often involve progressively removing the scan capability from performance critical storage elements in an initially fully scannable design. Maximal scan techniques tend to be used when area overhead is much less of a concern than the achievement of high quality with minimal impact on performance. In maximal scan, to achieve desired quality levels, as many storage elements as possible are usually made scannable. To minimize the impact of scan on performance, storage elements are typically excluded from the scan-chain if they are in the critical paths of the circuit. Full scan is at the extreme end of the maximal scan range and is mainly used when very high quality and/or system debug is a requirement. Partial scan techniques involve scanning more than 0% but less than 100% of the storage elements. In practice, many so-called full-scan designs are actually partial-scan designs as a few storage elements are often excluded from the scan-chain.

Recently, the use of minimal scan, in which only a small percentage of the storage elements are made scannable, has seemed more attractive than the use of full scan to many circuit designers because of the lower performance and area overhead it promises to involve. However, the effectiveness of partial-scan techniques in terms of reaching required fault coverage levels is heavily dependent on both the sequential complexity/sequential depth of the circuit and the design/test methodology used.

It has been shown that the sequential complexity of the circuit (in terms of the number of cycles in the circuit) is the main determinant of test generation difficulty [2]. By making a subset of the storage elements in the circuit scannable such that all cycles are broken, test generation difficulty is significantly reduced. However, there are many circuits in which every storage element has to be made scannable to break every cycle. For some circuits, to reduce the sequential complexity of the circuit to a level where 100% coverage of detectable faults is obtained, nearly all of the storage elements must be made scannable. The difficulty of predicting the fault coverage achievable for a given design with a certain percentage of storage elements made scannable is a problem, and typically leads to several test generation iterations to achieve the desired fault coverage.

In this article the costs associated with the various scan techniques will be presented, and the fault coverages typically achieved will be discussed.

3. The Non-Scan Element State Retention Problem

One of the barriers to the adoption of partial-scan techniques is the non-scan element state retention problem that occurs when the same clock is used during scan and test modes. The most popular scan design style involves the use of a single clock multiplexed scan D-type flip-flop. If such a design style, in which the same clock is used for both scan and system purposes, is used then even though high coverage test vectors may purportedly be generated, these vectors may result in simulation mismatches and lower than expected fault coverages when actually applied on the tester unless precautions are taken during the design. The values in the non-scan storage elements can be corrupted during the scan operations [2] if care is not taken in the design stage to ensure that non-scan elements can hold their state. If the state of the non-scan storage elements is not retained, the test generator should assume that non-scan elements contain unknown values at the end of scan operations. This will have the effect of significantly degrading fault coverage.

In this section solutions to the state retention problem are discussed and the area/performance costs associated with these solutions are compared.

3.1. Solutions

The solutions to the state retention problem discussed are the use of:

1. Separate scan and system clocks
2. Data hold capability
3. System clock gating
4. System clock bursting

3.1.1. Separate Scan and System Clocks. The use of a scan clock that is separate from the normal system clock avoids the problem of the scan operation changing the value of the non-scan elements. However, this approach requires either a multiplexer or gating in the clock tree and the routing of an extra clock signal throughout the design. In addition to the increase in area overhead because of routing, the extra logic in the clock tree results in an increase in the clock to Q delay of the scan elements and potential clock skew problems. The area of the scan cell can be reduced by sharing the clock gating/multiplexing logic between multiple scan elements. However, the requirement to route an extra high priority clock signal throughout the design is an onerous one. Sharing clock multiplexing or gating circuitry while reducing the area overhead would exacerbate clock skew problems and so would tend to increase design time.

Assuming that, in an attempt to minimize the additional routing and clock skew problems that can be caused by the use of a separate scan clock, the clock multiplexing logic is not shared then an additional 26 transistors per scan element over a regular D-type flip-flop will be required if NAND-NAND multiplexers are used to multiplex the clocks and multiplex the scan and data signals. If clock gating rather than multiplexing is used, an additional 20 transistors will be required over a standard non-scannable D-type flip-flop. The use of transmission gates to implement the multiplexers would reduce the transistor overhead to 12 per scan element in each case. Routing the complement of the system and scan clock signals to the scan elements allows NAND clock gating to be used so reducing the transistor overhead to 10 per scan element for the scan clock gating approach. No extra logic is required for the non-scan elements in either the scan clock multiplexing or gating cases.

If transmission gate multiplexers are used, care must be taken to ensure that storage elements directly feeding the multiplexer inputs are buffered to prevent corruption of data in those storage elements [10]. For this reason most ASIC families buffer the inputs (and sometimes the output) of transmission gate multiplexer cells. This has the effect of increasing the transistor count by 6–8. However, if the transmission gate multiplexer is built into the D-type flip-flop cell, the overhead of buffering can be reduced.

The test enable signal (TE) is used to select the system or scan clock in the clock multiplexing approach, which is illustrated in Figure 1, while if scan clock gating is performed using an OR gate, as shown

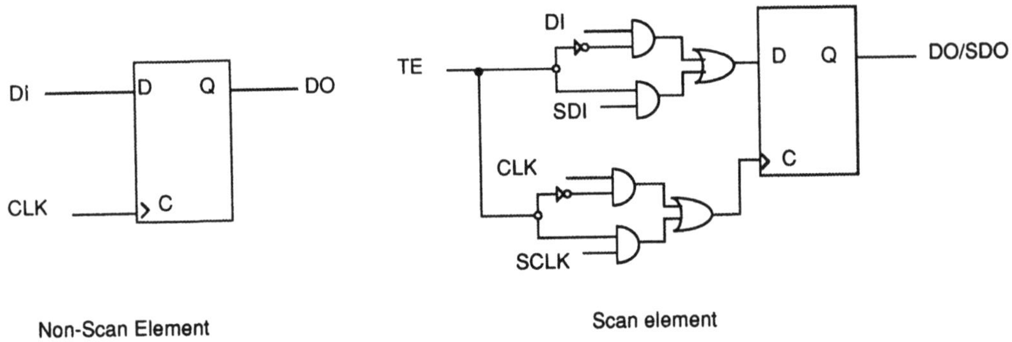

Fig. 1. Scan clock multiplexing approach to state retention.

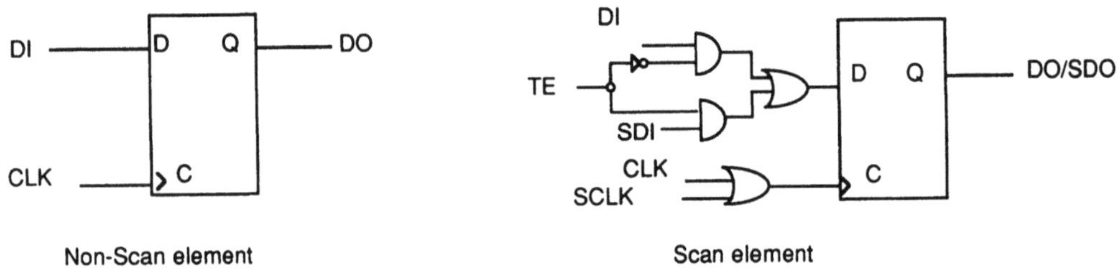

Fig. 2. Scan clock gating approach to state retention.

in Figure 2, the system clock must remain low during the scan operations and the scan clock must remain low during system operation.

Another clock gating approach involves the use of a test enable signal that doubles as both a scan clock and a scan mode selection signal during scan operation [2]. However, while this approach removes the need to route an extra scan clock, it has the problem that races can occur between the scan data and the scan "clock."

Separate scan and system clocks are used in the LSSD approach. Thus, the use of LSSD allows the state of non-scanned storage elements to be retained. A potential disadvantage of LSSD is that the standard LSSD element [3] results in a significantly larger area overhead than the standard multiplexed D-type flip-flop scan element. However, lower overhead implementations are possible. For exampled, a low overhead LSSD element that requires only 26 transistors is shown in Figure 3. Although such an implementation may be possible in a full custom design, in an ASIC the input/output logic of the storage element will typically be buffered and this increases the transistor count by 6 to 32.

3.1.2. Data Hold Capability. A possible solution to the state retention problem is to implement all non-scan elements such that they have a state holding capability.

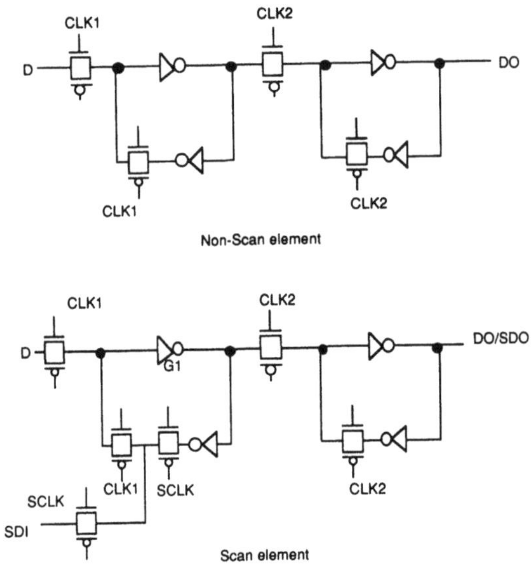

Fig. 3. LSSD approach to state retention.

This can be achieved by feeding the data output of the non-scan storage elements back to their data inputs via a multiplexer and enabling this path during the scan mode of operation of the circuit. However, this approach has the same area and performance impact of full scan without any of the benefits. The use of a multiplexer to hold state is shown in Figure 4.

If transmission gates are used to implement the multiplexers, both the non-scan elements and scan

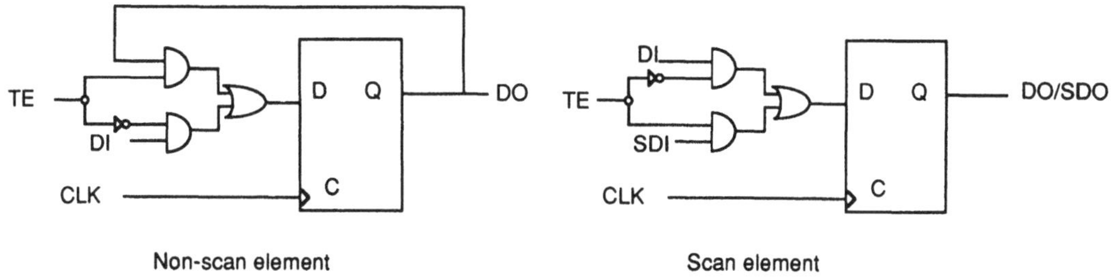

Fig. 4. Data holding approach to state retention.

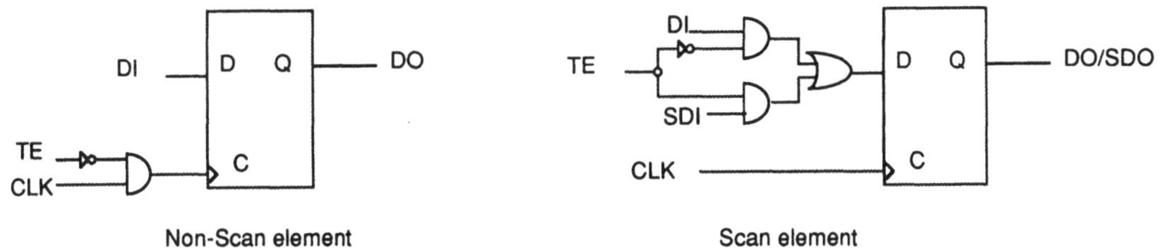

Fig. 5. System clock gating approach to state retention.

elements each require an additional 6 transistors over regular storage elements.

3.1.3. System Clock Gating. The addition of gating in the clock path of non-scan elements allows the system clock to be disabled during the scan in and out operations. However this approach, which is illustrated in Figure 5, results in an increase in the clock to Q delays of the non-scan elements and requires the test enable signal to be routed to each storage element.

If it is assumed that, to simplify routing and clock skew problems, clock gating is not shared between storage elements, an additional 6 transistors will be required for each non-scan element, assuming AND clock gating and that scan operation is enabled when the test enable (TE) signal is low, and an additional 6 transistors will be required per scan element. This results in the same area overhead for partial scan as for full scan. In addition, the clock to Q delay of the non-scan elements will increase by approximately one gate delay.

3.1.4. System Clock Bursting. In a single clock design, the possible corruption of the states of the non-scan elements during scan-in and scan-out can be taken into account by assuming that the non-scan elements are at unknown states after the scan operations. A burst of *n* system clocks can then be applied to initialize them to the required values for the test and propagate the fault effects to an observable point. However, this approach constrains the test generator to scan in and out of the scannable storage elements only at the beginning and end of the test sequence. After the initial scan operation, the test generator has to derive a sequence of primary input vectors that puts all the storage elements into the sequence of states required to excite the fault and to propagate the effects of the fault to an observable output. This method tends to result in a significant reduction in fault coverage for most designs. However, it does not result in any increase in area or performance overhead for non-scan elements. The transistor overhead for the scan elements is the same as in regular scan implementations; an additional 6 transistors will be required per scan element.

3.2. Overhead Comparison

The overheads associated with the various solutions to the non-scan element state retention problem that have been discussed are described here for a hypothetical 50K gate non-pipelined design containing both datapaths and control circuitry with an operating frequency of 50 MHz. We assume that the design contains 3,000 storage elements, which occupy one third of the area of the chip. We also assume that a one micron fabrication process with two layers of metal is used and that routing an extra global signal results in a 1% area overhead.

The extra delay caused by the use of a scannable D-type flip-flop in a 1.0 micron technology is assumed to be 0.8 ns. This takes into account the delay caused by a transmission gate based scan multiplexer, the increase in loading and the increase in setup time. Thus, the performance degradation for full scan will be approximately 4% while for partial scan approaches it will be between 0–4%, depending on whether scan selection is timing driven or not.

Figure 6 compares the overheads of the scan clock multiplexing, scan clock gating, LSSD and system clock bursting state retention methods to the area overhead of full scan assuming that to minimize clock skew problems state retention circuitry is not shared. The data hold and system clock gating partial scan solutions result in area overheads similar to those for full scan.

It can be seen from Figure 6 that the lowest area overhead partial scan implementation is achieved by the use of system clock bursting. However, it should be noted that this approach usually results in a significant degradation in fault coverage achieved. The area overhead for the other partial scan approaches is actually greater than for full scan in this example if greater than 60% of the storage elements in the design are made scannable.

For ASICs, the actual cost of an increase in area will depend on the granularity of arrays offered in the ASIC family. Consider an ASIC family (such as the LSI Logic LCA100K product) that includes arrays with 45,500, 51,000 and 58,000 usable gates. For a design that contains 50,000 gates without DFT, the use of a DFT technique with a gate overhead of more than 2% will result in the need to move from the 51,000 gate array to the 58,000 gate array. A DFT technique with a gate overhead of 16% or less will also fit into the 58,000 gate array. Thus, in this case, the silicon cost of both the 2% and 16% area overhead techniques will be the same.

4. An Economic Model for Scan DFT Techniques

Recently, there has been increasing interest in the use of economic models to analyze the effects of test on life-cycle costs [11]–[14]. In this section, an economic model that allows the range of scan based DFT solutions to be compared with each other is presented. The model proposed takes into account the major factors that effect the life cycle costs of scan based DFT strategies:

area overhead,
performance degradation,
packaging costs,

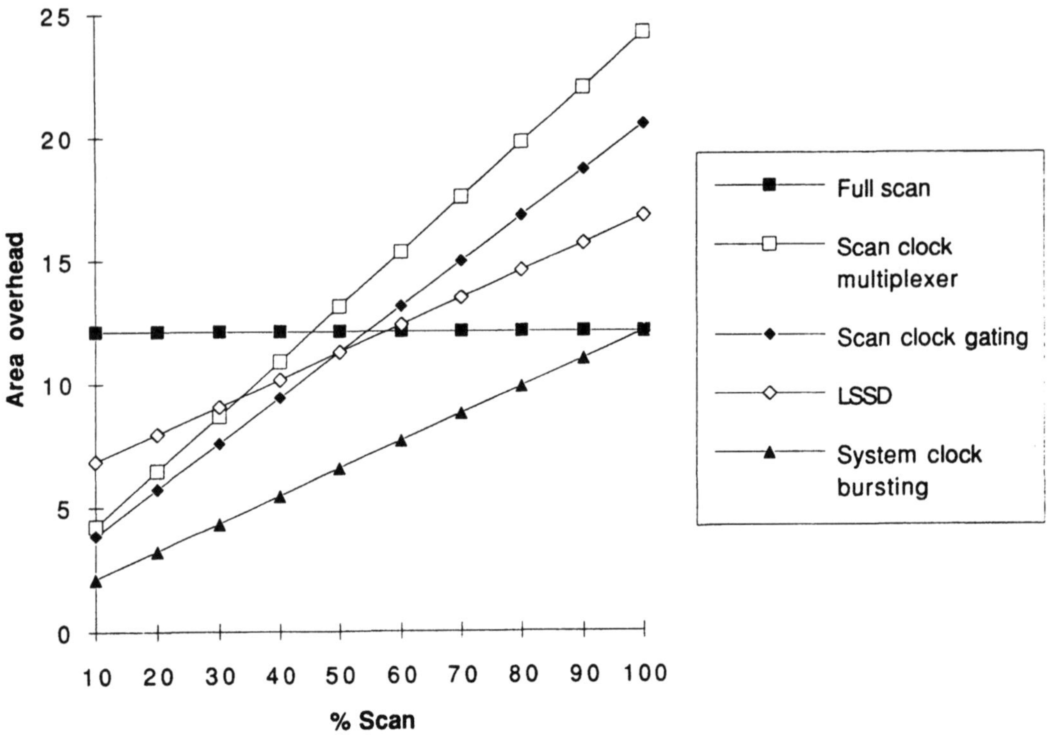

Fig. 6. Comparison of area overheads of state retention solutions.

design costs,
time to market costs,
test development time,
test application time,
board test and rework costs,
system debug time, and
field test costs.

The effect of DFT on reliability is not considered as it is assumed that burn-in is used and that the increase in early life failures because of the additional area overhead of DFT is small. An analysis of the effects of early life failures on life-cycle costs is given in [11]. The use of supplemental test techniques, such as Iddq testing, may have effects on both test escapes and reliability. However, the use of such techniques is not considered here.

It is not intended that the model proposed in this article be used to predict actual costs of a particular DFT strategy but that it be used only to examine relative trends and to compare relative costs of different scan DFT strategies.

4.1. Area Overhead

The use of DFT results in an increase in area overhead. For vertically integrated companies, the costs, C_a, caused by an increase in area from A_0 to A_1 can be modeled using Seeds equation [15]:

$$C_a = \frac{C_0 A_l Y^{1 - \sqrt{\frac{A_1}{A_0}}}}{A_0} - C_0$$

where C_0 is the cost for the original die of area A_0 and Y is the yield.

For companies without internal manufacturing capability, the costs associated with a choice of test technology and its inherent area overhead do not have to be modeled using yield-area relationships and can be simply modeled using the vendor's price differential for the different test approaches. The use of scan may result in additional recurring (RE) and non-recurring (NRE) engineering costs. However, many ASIC designs are pin limited and so an increase in gate count may not result in the need to use a larger array. For 0.8 micron ASIC families, NRE costs, though usually negotiable, are typically \$1 per gate (in 1993) for a low volume design of under 150k gates, falling to a flat rate of \$20,000 for high volume designs. While recurring costs for ASIC manufacture are typically in the region of 0.03–0.05 cents per gate for low volume designs.

4.2. Performance Degradation

A degradation in the performance of the design can result if DFT techniques are used. For example, scan DFT techniques typically result in a 5% performance degradation, although scan elements with negligible performance penalty are available from some ASIC vendors. A lower performance ASIC tends to result in a product that is less competitive and this can translate into a reduction in the price of the product. However, the performance of a system is just one of the factors that affects price, and we assume that the relationship between performance and price is such that a small decrease in performance has little impact on price, but that as performance decreases beyond a certain competitive point the price of the system decreases at a faster rate. Inordinate performance degradation could result in a redesign of the product, which would have an effect on the time taken to get the product to market.

The cost of a performance reduction is assumed to be given by C_{pr}:

$$C_{pr} = N_u P_u e^{-\frac{K_s S_r}{1 - S_r}}$$

where N_u represents the number of units, P_u represents the expected unit price, S_r represents the speed reduction factor caused by the use of DFT (i.e., performance with DFT / performance without DFT) and K_s is the performance factor. K_s is determined by factors such as the nature of the product and the speed achieved by competing products. It is assumed to typically lie in the region of 0.3–0.4. However, a lower value of K_s can be used for performance sensitive products whose price would be affected by even a small decrease in performance.

4.3. Packaging Costs

DFT techniques typically require the use of extra package pins. This results in an increase in packaging costs C_p:

$$C_p = K_p N_t N_c$$

where K_p is a packaging cost factor, N_t is the number of additional pins required for test and N_c is the number of components.

As with area related costs, for companies without internal ASIC manufacturing capability, the increase in packaging costs can be simply modeled using the vendor's pricing model. For core limited designs, there may not be an increase in packaging costs if DFT is used. However, for pad limited designs, a larger array may be required.

4.4. Design Costs

Ensuring that the design complies with DFT rules results in an increase in engineering design time. The time required is highly dependent on the DFT knowledge and expertise of the designer. The extra design cost of designing for testability is given by C_{de}.

$$C_{de} = D_{de} L_{de}$$

where D_{de} is the number of engineering days required to insert DFT structures into the design and to ensure that the design does not violate the DFT rules, and L_{de} is the loaded daily labor rate of a design engineer.

4.5. Time to Market Costs

The lost revenue (R_l) caused by a delay (d) in getting a product with expected revenues of R_e and a market window of $2w$ to market is widely estimated using the ATEQ model described in [12] and in [16]:

$$R_l = \frac{R_e d(3w - d)}{2w^2}$$

Although the ATEQ model is widely used, it may be too pessimistic, and it breaks down when d approaches or exceeds w. A simpler model is proposed here. In this model, the market window is fixed and the revenue that would be earned in a period is lost if the product is delayed by that period:

$$R_l = \sum_{i=1}^{i=n} \frac{d_i R_e^i}{P_i}$$

where d_i is the delay in period P_i, R_e^i is the expected revenue in period P_i, and the market window is $\Sigma_{i=1}^{i=n} P_i$, where n is the number of periods.

The loss in future revenue has to be compared to costs that must be incurred in the present. Discounted cash flow (DCF) techniques can be used to analyze the present value (PV) of sums of money to be received in the future at a given interest rate R_i:

$$PV = \sum_{i=1}^{i=n} \frac{C_i}{(1 + R_i/m)^{im}}$$

where C_i is the cash to be received in period i, m is the frequency of compounding and n is the number of periods [17].

4.6. Test Development Costs

The cost of developing tests is given by C_{td} where

$$C_{td} = L_{td}(D_{tg} + D_{tpd}) + C_{tt}$$

and L_{td} represents the loaded daily labor rate of the test development engineer, D_{tg} represents the number of days required to generate test vectors, and D_{tpd} represents the number of days required for test program development. The cost of any additional test tools that must be purchased for automatic test pattern generation and fault simulation are given by C_{tt}.

$$C_{tt} = \frac{C_{atg} + C_{fs}}{N_a}$$

where C_{atg} is the cost of automatic test generation software and C_{fs} is the cost of a software or hardware fault simulator. N_a is the number of different ASIC designs for which these tools will be used.

Test vector development time (including the time taken to fault grade the design) is assumed to be related to the test generation complexity TG_c:

$$D_{tg} = K_{tg} N_g^{TG_c}$$

where K_{tg} is the test generation factor. The test generation complexity is dependent on a number of factors: the number of gates (N_g) in the design, the number of loops, the sequential depth, and the clock control complexity. For synchronous circuits, clock complexity is determined by the amount of clock gating, while in asynchronous circuits the percentage of the storage elements that are asynchronously clocked must be taken into account.

4.7. Test Application Time Costs

If the ATE is depreciated over N_y years and the fraction of the purchase price of the tester (C_{te}) incurred in maintenance costs each year is given by C_m then the cost per hour of using the ATE is given by

$$C_{ta} = \frac{C_{te}\,(1 + C_m N_y)}{N_h N_d N_m N_y}$$

when it is operated for N_h hours per day, N_d days per month during N_m months per year. The labor costs of the ATE operator are given by

$$C_{tal} = \frac{L_{te}N_m}{N_t}$$

where L_{te} is labor cost/hour and N_m is the number of chips manufactured. N_t is the number of chips tested per hour and is given by

$$N_t = \frac{T_{su} + T_t}{3600}$$

where T_{su} represents the setup and handling time in seconds per device, and T_t represents the test time.

$$T_t = T_{tct}(N_{pv}N_{cpv} + N_{sv}N_{se} + N_{sv}N_{csv}) + T_p$$

where T_{tct} is the test-cycle time, N_{pv} is the number of parallel vectors, N_{cpv} is the number of system clock cycles per parallel vector, N_{sv} is the number of scan vectors, N_{se} is the number of scan elements N_{csv} is the number of system clock cycles per scan vector, and T_p is the parametric test time.

ASIC manufacturers typically allow a certain number of test vectors to be used and charge extra if their limit is exceeded. Thus, for companies without captive manufacturing facilities, as with yield and packaging costs, the costs of test application can be modeled using the ASIC manufacturers pricing model. Typically, ASIC manufacturers will allow test times up to t_t seconds but will charge C_{ct} extra if this limit is exceeded according to a formula such as:

$$C_{ct} = \frac{(C_{ta} + L_{te})(T_t - t_t)}{3600}$$

For example, some ASIC manufacturers allow test times of up to 3–4 seconds per device but may charge about 5 cents per second, assuming a fully loaded test cost of \$180 per hour, in excess of this.

4.8. Board Test Costs

ASIC manufacturing processes are imperfect and so manufacturing yields, even for mature processes, are less than 100%. A defect coverage of less than 100% at the manufacturing test stage will result in some faulty ASICs escaping to the board or system manufacturing stages. Thus, it will be necessary to attempt to identify and replace defective devices on the board or system.

Board test costs can be modeled in a similar way to that used for component test costs. However, board diagnosis and repair costs also need to be taken into account.

Boards may contain a mixture of assembly and component faults. Assembly faults may be detected using a combination of visual, in-circuit, functional and boundary-scan test techniques. The number of apparent faults per board is given by N_{fb}:

$$N_{fb} = N_{afb}FC_{ba} + N_{cfb}FC_{bc}$$

where the number of assembly faults per board is given by N_{afb}, the fault coverage of assembly faults is given by FC_{ba}, the fault coverge of the component faults at the board test stage is given by FC_{bc} and the number of component faults per board is given by N_{cfb}.

$$N_{cfb} = -N_{cb}\log_e Y_{cb}$$

where N_{cb} is the number of components per board and Y_{cb} represents the actual yield of components passing to the board test stage. Y_{cb} can be estimated using a defect level equation, which relates the defect level DL to the process yield Y and the fraction of faults detected f, such as that proposed by Williams and Brown [18]:

$$Y^{1-f} = 1 - DL$$

The total number of board faults is represented by

$$N_{tbf} = N_b N_{fb}$$

where the number of boards is given by N_b.

The costs associated with diagnosing and repairing board test failures (C_{drb}) are given by

$$C_{drb} = K_l T_{drb} L_{drb} N_{tbf}$$

where the time taken to diagnose and repair a board test failure is given by T_{drb} and L_{drb} represents the loaded labor rate of a board debug engineer. Incorrect diagnosis may result in some boards passing around the testing loop a few times. In addition, a repair may result in the introduction of new faults. Thus, a loop factor K_l is introduced to account for this.

4.9. System Debug Costs

One of the major advantages of structured DFT techniques is the ability to use the controllability and observability of internal storage elements during the system debug stage. Leveraging DFT during system debug can result in significant reductions in debug time, so reducing the time taken to get to the market with the product.

The costs of system debug are given by C_{sd}:

$$C_{sd} = D_{sd}L_{de}$$

The number of days required for system debug is given by D_{sd}:

$$D_{sd} = S_{dc}N_g$$

where S_{dc} represents the system debug complexity expressed in terms of the number of system debug days required per thousand gates.

The use of scan reduces the system debug complexity, and it is assumed that there is an inverse relationship between the amount of scan and the debug complexity of the system:

$$S_{dc} = \frac{S_c}{1 + K_{sdc}PS}$$

where K_{sdc} represents the scan debug complexity factor, S_c represents the system complexity without scan, and PS represents the fraction of storage elements that are scanned.

4.10. Field Test Costs

Field test costs are given by C_f:

$$C_f = N_f(C_{drf} + C_{sf})$$

where N_f represents the number of number of field failures, C_{drf} represents the cost of field diagnosis and repair, and C_{sf} represents the cost of carrying spares.

The costs associated with diagnosing and repairing field test failures (C_{drf}) are given by

$$C_{drf} = T_{drf}L_{fe}$$

where L_{fe} represents the loaded labor rate of a field engineer and the time taken to diagnose and repair a field failure is given by T_{drf}.

4.11. Cost of Lack of Quality

The lack of quality in a product can result in an intangible loss of revenue because of the loss of customer satisfaction and the poor company and product image that may result. This lack of quality is difficult to model but can have a significant effect on total revenues.

5. Case Study

In this section, the costs associated with minimal (30% scan), medial (50% scan), and maximal (80% scan) DFT strategies are compared with the costs of full scan for a system consisting of a single board with two ASICs. The ASICs contain both control machines and data-paths and comprise 20k gates and 80k gates. It is assumed that the ASICs contain internally gated clocks, internally generated sets/resets and bidirectional buses and ports. It is also assumed that 33% of the chip area of the 20k ASIC and that 40% of the chip area of the 80k ASIC is occupied by storage elements.

The price of the system (without structured DFT) is assumed to be \$10,000 (with a 15% annual price erosion) over a two year product life cycle, and a sales volume of 2,000 units is expected. The present value of future revenue is calculated using a rate of 10% assuming annual compounding.

It is assumed that the system manufacturer does not have captive ASIC manufacturing facilities and that a commercially available 0.8 micron array family that includes bases with usable gate capacities of 20K, 25K, 78K, 88K and 100K has been selected from an ASIC manufacturer.

It is assumed that for the partial scan approaches, time is spent during the design stage to experiment with partial scan selection. In addition, it is assumed that the fault coverage goal is 95% and that a maximum of two weeks is allowed for the final generation of patterns. Thus, several ATPG iterations may be possible with different scan selections in an attempt to reach this goal. ATPG for both ASICs is assumed to be carried out in parallel, and so more iterations are possible for the smaller design. It is assumed that a software solution based on system clock bursting (rather than a hardware solution) is used to solve the non-scan element state retention problem for the partial scan approaches.

Another assumption is that all the scan approaches require the same number of test pins and that the ASICs are area limited so that there is no increase in costs related to the test pins. The use of any of the DFT strategies results in the requirement to move from an array with 20K usable gates to 25K gates for the 20K gate design. For the 80K gate design, the use of an 80% or 100% scan strategy results in the requirement to move from an array with 88K usable gates to one with 100K usable gates while the use of a 30 or 50% scan strategy still allows an 88K gate array to be used. The increase in recurring ASIC manufacturing costs for the 20K design are considered to be zero independently of which DFT strategy is used. For the 80K design, they are considered to be the same for the non-scan, 30% scan and 50% scan strategies, but to increase from \$40 per unit to \$42 per unit for the 80% and 100% scan strategies.

Tables 1 and 2 summarize the basic parameter assumptions that are made for the two ASICs.

PS represents the percentage of elements made scannable,
A represents the area overhead,
NG represents the total number of gates required with DFT,
FC represents the fault coverage obtained,
NI represents the number of test generation iterations required, and
TTG represents the time taken to generate test patterns.

Although it is assumed that the system clock bursting partial scan technique is being employed, the fault coverages used for the partial-scan approaches in this case study are more in line with those achieved by the use of state retention design techniques. The partial-scan fault coverage assumptions may seem to be pessimistic. However, results less favorable than these are typically achieved on most real complex designs with current ATPG technology. For example, it was reported that scanning approximately 50% of the storage elements in a small block of a microprocessor, assuming non-scan element state retention, resulted in close to 5% degradation in fault coverage, a 200–300% increase in test set size and approximately a 400% increase in ATPG run time over a full scan design, and that even a 1600% increase in run time failed to achieve fault coverages approaching that of full scan [19]. The difference between full-scan and partial-scan run times and fault coverages tends to become even more significant as design size and complexity increase.

Table 1. Cost model parameters for 20K design.

Parameters for 20K Design					
PS	A	NG	FC	NI	TTG
30%	4%	20.8k	90%	3	3 days
50%	7%	21.4k	95%	3	2 days
80%	10%	22.0k	95%	3	1 day
100%	12%	22.4k	100%	1	0.5 hours

Table 2. Cost model parameters for 80K design.

Parameters for 80K Design					
PS	A	NG	FC	NI	TTG
30%	5%	84.0k	85%	2	1 week
50%	8%	86.4k	88%	2	1 week
80%	12%	89.6k	90%	2	1 week
100%	14%	91.2k	99%	1	4 hours

The time taken to design for testability, perform ATPG and debug the system all affect the time to market costs. The increase in design time to check for DFT rule compliance and to perform scan selection experiments and final scan insertion is assumed to be two weeks for full scan and four weeks for partial scan. In addition, the time taken to perform the final ATPG runs is assumed to be one day for full scan and two weeks for the partial scan approaches. It is assumed that the time taken to perform system debug is halved if full scan is used but that this reduction decreases as the amount of scan in the circuit decreases. The increase in time taken to get the product to market is illustrated in Figure 7.

We assume that the same ATPG tool is used to generate tests for the various scan approaches, and we conservatively assume that the tester program development time is the same for full and partial scan. However, the costs of developing the tester program could be higher for the partial scan approaches as the complexity of the test program is potentially higher. The test times required for all the scan approaches are assumed to be similar as although the scan chain lengths may be smaller in the partial scan designs more scan vectors will be required and the number of parallel vectors per scan vector will also increase. However, in practice, partial scan test lengths have been found to be significantly higher than those for full scan [19].

The lower fault coverages achieved by the use of partial scan result in an increase in the number of bad devices passing to the board test stage and so results in an increase in board diagnosis and repair costs. We consider only ASIC related board faults and assume that the actual yield of packaged components passed to component test is 70%.

We assume that the board test loop factor, K_l, is 1.1, the board diagnosis and repair loaded labor rate, L_{drb}, is \$100 per hour, the average time taken to diagnose and repair a board failure, T_{drb}, is 15 minutes and that the fault coverage of the ASICs attained by the board test program is 75%. It is assumed that the actual board test costs of the various DFT strategies do not differ but that the diagnosis and repair costs related to the ASIC faults will vary.

Field test costs will also increase for scan methodologies that result in lower fault coverages than that for full scan as the fault coverage at board/system test is less than 100% of the ASIC faults. We assume the loaded labor rate of a field test engineer, L_{fe}, is \$100 per hour and that the average time taken to diagnose and repair a field failure, T_{drf}, is 1 hour. The

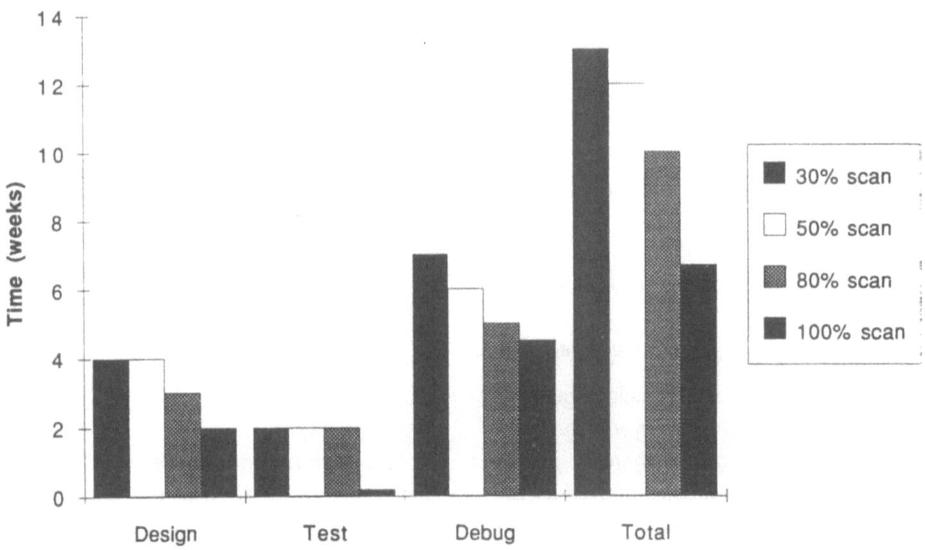

Fig. 7. Time to market increases of scan DFT strategies.

cost of spares, C_{sf}, is taken to be an average of $100 per failure.

It has been argued that the use of logic optimization on circuits will tend to result in an increase in the number of the paths in the circuits becoming critical [20]. In circuits where there are many critical paths, even scanning only a small percentage of the storage elements could have an impact on performance. However, it is assumed in this case study that the full scan approach results in a 5% performance degradation, while there is no performance degradation for the partial scan approaches.

In performance sensitive markets, a reduction in performance of the system caused by the use of full scan may be considered to be equivalent to a reduction in price or a decrease in number of units sold. Similarly the increase in time to market caused by the use of partial scan approaches translates into fewer units sold during the fixed market windown. Thus, both a performance degradation and product delays can result in lost revenue.

The differential in engineering and manufacturing costs of the partial scan designed systems over the full scan system is illustrated in Figure 8, while the effect of lost revenue caused by product delays and performance related price reductions on the total cost differential of the partial over the full scan approaches is shown in Figure 9.

Time to market costs are calculated using the model proposed in this article rather than the more widely used ATEQ model. This model assumes that the penalty for product delays is less harsh than would be estimated using the ATEQ model. LR1 and TC1 represent the lost revenues and total costs respectively if the degradation in performance of all the DFT approaches is considered small enough to have no impact on system price. If a 5% degradation in performance results in a $200 unit price reduction for the full scan approach then the lost revenues are represented by LR2 and the total costs by TC2. In this case full scan still results in lower total costs than the partial scan approaches. However, if the price reduction is $500 per unit for the full scan approach then a comparison of the lost revenues and total costs (represented by LR3 and TC3 respectively) shows that the 80% scan strategy now results in lower lost revenues and total costs than full scan. Thus, for performance sensitive markets, a performance driven maximal scan strategy may be more economical than a full scan strategy.

RE represents the difference in recurring ASIC manufacturing costs.

NRE represents the difference in ASIC manufacturing non-recurring costs.

DT represents the difference in design engineering costs.

TDT represents the difference in test development time costs.

BD represents the difference in board diagnosis and repair costs.

SDT represents the difference in system debug time costs.

F represents the difference in field test costs.

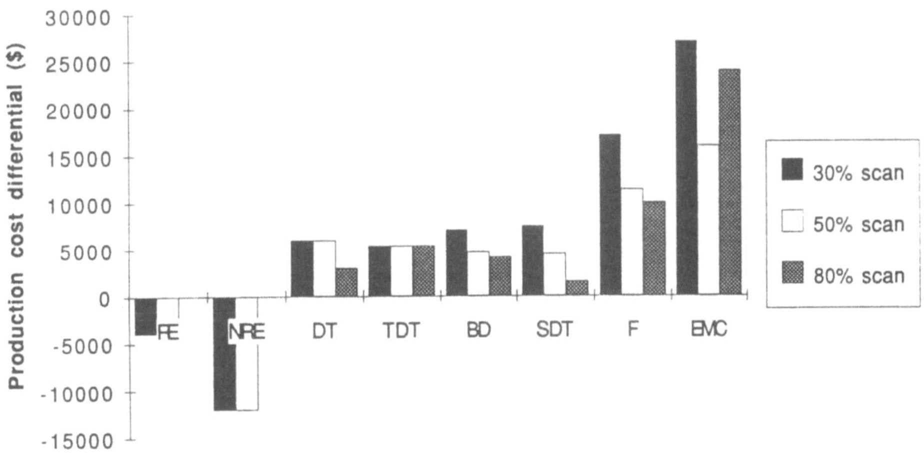

Fig. 8. Difference in engineering and manufacturing costs between partial and full scan.

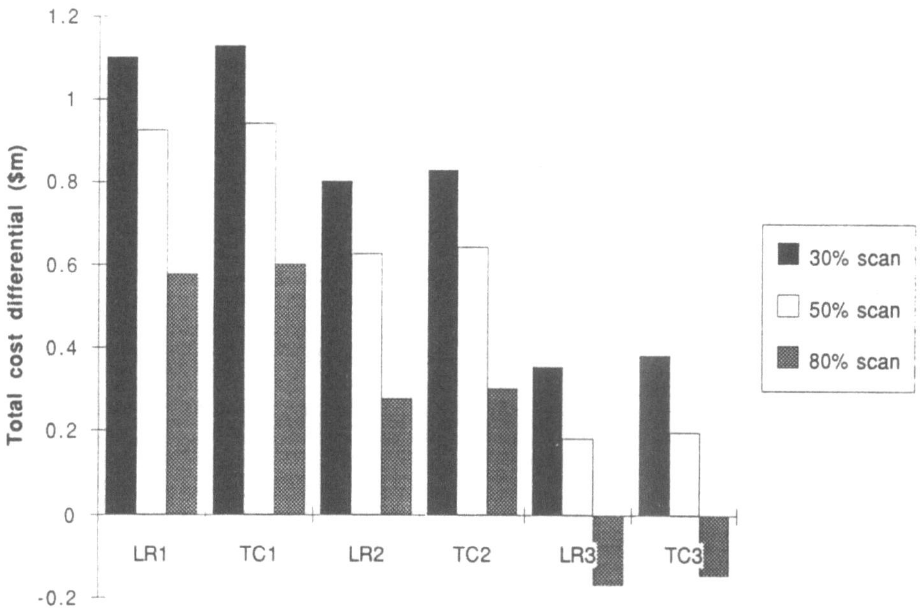

Fig. 9. Difference in lost revenue and total costs between partial and full scan.

EMC represents the difference in engineering and manufacturing costs.

*LR*1 represents the difference in time to market costs assuming P_u = 10k.

*TC*1 represents the difference in total test costs assuming P_u = 10k.

*LR*2 represents the difference in lost revenue assuming P_u = 9.8k for full scan.

*TC*2 represents the difference in total test costs assuming P_u = 9.8k for full scan.

*LR*3 represents the difference in lost revenue assuming P_u = 9.5k for full scan.

*TC*3 represents the difference in total test costs assuming P_u = 9.5k for full scan.

It can be seen that, for this example, the decrease in silicon costs of adopting partial scan can be dominated by the increase in the time to market costs. For systems produced in larger volumes, the effect of an increase in silicon costs will be more pronounced. However, except for the largest volume systems, the increase in time to market costs of a less automated test and debug strategy could still negate any savings in the cost of silicon.

In performance sensitive markets, a faster system can often command a price premium. Thus, while the use of partial scan approaches often may not offer significant savings in silicon costs, designers may sometimes still elect to trade-off fault coverage and run-time for performance by backing-off from full scan in the critical paths of the circuit. For high performance ASICs, the use of performance driven maximal scan based approaches may result in the most economic trade-off between performance, quality, and time to market costs. In many other cases, for low volume ASICs, full scan may be more cost-effective than partial scan when life-cycle costs are considered.

6. Summary

In this article, scan methods are categorized based on the percentage of storage elements made scannable. Solutions to the problem of retaining the state of non-scan storage elements during the scan operation in partial-scan designs are described and the overheads associated with these solutions are discussed. Finally, an economic model that allows the costs of a range of scan solutions to be compared with each other is presented. It is shown that, for systems produced in low volumes, the adoption of full scan DFT can be more cost-effective than partial scan DFT if it results in a reduction in the time required to get the product to market.

References

1. T. Gheewala, "CrossCheck: A Cell Based VLSI Testability Solution," 26th *ACM/IEEE Design Automation Conference*, pp. 706–709, 1989.
2. K.T. Cheng and V. Agrawal, "An Economical Scan Design for Sequential Logic Test Generation," *Proceedings of International Fault Tolerant Computing Symposium*, pp. 28–35, 1989.
3. T. Williams and K. Parker, "Design for Testability — A Survey," *Proc. IEEE*, vol. 71, pp. 98–112, Jan. 1983.
4. S.C. Seth and V.D. Agrawal, "Cutting Chip Testing Costs," *IEEE Spectrum*, pp. 38–45, April 1985.
5. V. Chickermane and J. Patel, "An Optimization Based Approach to the Partial Scan Design Problem," *IEEE International Test Conference*, pp. 377–386, 1990.
6. V. Chickermane and J. Patel, "A Fault Oriented Partial Scan Design Approach," *IEEE International Conference on Computer Aided Design*, pp. 400–403, 1991.
7. R. Gupta et al., "The Ballast Methodology for Structured Partial Scan Design," *IEEE Trans. Computers*, vol. C39-4, pp. 538–544, April 1990.
8. D.H. Lee and S.M. Reddy, "On Determining Scan Flip-Flops in Partial-Scan Designs," *IEEE International Conference on Computer Aided Design*, pp. 322–325, 1990.
9. J.Y. Jou and K.T. Cheng, "Timing Driven Partial Scan, *IEEE International Conference on Computer Aided Design*, pp. 404–407, 1991.
10. *Logic Design Manual for ASICs*, LSI Logic, September 1989.
11. P. Varma et al., "An Analysis of the Economics of Self-Test, *IEEE International Test Conference*, pp. 20–36, 1984.
12. T. Gheewala and K. Pierce, "Economics of ASIC Test Development," *Economics of Design and Test for Electronic Circuits and Systems*, A.P. Ambler, M. Abadir, and S. Sastry (eds.), Ellis Horwood Ltd., 1992, pp. 39–51.
13. M.E. Levitt, "Economic and Productivity Considerations in ASIC Test and Design for Test," *IEEE Compcon*, pp. 440–445, 1992.
14. B. Davis, *The Economics of Automatic Testing*, McGraw Hill, U.K., 1982.
15. D. Ong, *Modern MOS Technology Processes, Devices & Design*, McGraw-Hill, 1984, p. 342.
16. W. McClean (ed.), *ASIC Outlook 1994*, ICE, Arizona, 1993.
17. T.M. Carroll, *Microeconomic Theory, Concepts and Applications*, St. Martins Press, N.Y. 1983.
18. T.W. Williams and N.C. Brown, "Defect level as a Function of Fault Coverage," *IEEE Trans. Computers*, vol. C-30, pp. 987–988, Dec. 1981.
19. G. Ganapathy and J. Abraham, "Selective Pseudo Scan — Combinational ATPG with Reduced Scan in a Full Custom RISC Microprocessor," *30th ACM/IEEE Design Automation Conference*, pp. 550–555, 1993.
20. T.W. Williams, Bill Underwood, and M.R. Mercer, "The Interdependence Between Delay-Optimization of Synthesized Networks and Testing," 28th *ACM/IEEE Design Automation Conference*, pp. 87–92, 1991.

Prab Varma received an M.Sc. in VLSI Design in 1982 and a Ph.D. in Microelectronics in 1984 from the University of Manchester, U.K.

From 1984–1985, he was a YKK sponsored researcher at Tokyo Institute of Technology, where he conducted research into Built-In Self-Test (BIST) and knowledge based automatic test pattern generation (ATPG).

From 1986–1988 he was at GEC Research Ltd., U.K., where he was the manager of the VLSI Test Methodologies group. From 1988–1992 he was at Teradyne, where he was responsible for ATPG and design for testability (DFT). He is currently an engineering manager at CrossCheck Technology, where his research interests are in delay test generation, Iddq testing, test economics, BIST, DFT rule checking and test logic synthesis. He is the general chair of the 1994 IEEE VLSI Test Symposium.

Tushar Gheewala, CEO and chairman of CrossCheck Technology, Inc., founded the company in 1987. Today, CrossCheck supplies innovative test automation software solutions to the world's leading IC manufacturers and system developers.

Prior to founding CrossCheck, Gheewala headed the research facility for Sperry Corporation from 1985 to 1987.

Before joining Sperry, Gheewala was the director of packaging, testing and manufacturing for Gigabit Logic.

Prior to that, Gheewala was with IBM Research Center in Yorktown, New York.

While at IBM, Gheewala received several awards including the Outstanding Innovation and Invention Achievement awards. In 1981, he was named the Outstanding Young Electrical Engineer in the nation by Eta Kappa Nu and the Institute of Electrical Electronic Engineers (IEEE).

Gheewala attended the Indian Institute of Technology in Bombay. He completed his education with a master's degree and a Ph.D. in electrical engineering from Stanford University, specializing in semiconductor devices and processing.

His research interests are embedded test technologies, I_{DDQ} testing, and debug and diagnostics.

JOURNAL OF ELECTRONIC TESTING: Theory and Applications, 5, 195–206 (1994)

High Level Test Economics Advisor (Hi-TEA)

MAGDY ABADIR, ASHISH PARIKH, LINDA BAL*, PETER SANDBORN, AND CYNTHIA MURPHY
Microelectronics and Computer Technology Corporation, 3500 West Balcones Center Drive, Austin, Texas 78759

Abstract. To produce high-quality and cost-effective multichip systems, they must be designed with test and fault diagnosis as critical design requirements. However, deciding on where and when to test and whether to apply Design For Test DFT) and Built-In Self-Test (BIST) at the IC, multichip module (MCM) or board level requires considerable study and evaluation to determine the economics of the various solutions and the payback. In this article we describe a tool called High Level Test Economics Advisor (Hi-TEA) that analyzes the economics of various test strategies for multichip designs at an early stage of the design cycle. The tool also allows the user to perform trade-off analysis on the impact of various cost, yield, or test effectiveness parameter on the final cost and quality of multichip designs. Experimental trade-off analysis data that were generated using the tool for some leading-edge multichip designs will also be presented.

Keywords: MCM testing strategies, multichip module, test economics.

1. Introduction

As systems have increased in complexity, generating effective and practical tests for them has emerged as a major problem, regarded by experts in the field as the largest obstacle preventing full exploitation of new technologies such as MultiChip Modules (MCMs). In a recent article in *EE Times*, test problems were attributed as the primary cause for a 6-month delay in the introduction of a new MCM-based SPARC design by a major IC company. The resulting profit losses from that delay can very well be in the many millions of dollars.

Most of the discussion in this article is focused on MCM testing issues. However, the ideas and the results presented are applicable to any other multichip system. Testing MCMs is a difficult task because they contain multiple high pin count chips connected together into one circuit with high density interconnections. MCMs (more than any other packaging approach) stress the limits of available test solutions at the chip, substrate and module levels.

Why are conventional test methods inadequate? Traditionally, bed-of-nails testers have been used to test printed wiring board assemblies. However, because of the high chip density and the small interconnect line dimensions of emerging MCM and other packaging technologies, this approach cannot be used to gain acces to internal lines. Even if it were possible, this type of tester is extremely costly. Also, external testers are rapidly approaching another physical limitation—the circuits under test are faster than the delay along the tester probe lines.

To produce high quality and cost effective MCMs, they must be designed with test and fault diagnosis as critical design requirements. Testability and diagnosability must be planned early in the design cycle; otherwise, they may prove quite costly or impossible to accommodate later. However, test is still treated as an afterthought in the design process and designers of such complicated systems have little or no MCM testability guidance at this time. This can be attributed to the lack of awareness of the importance of test and the lack of tools that can provide the designers with some advice and guidance.

The cost and the resultant quality of an MCM are dependent upon 1) the yield of the chips, 2) the number of chips in the module, 3) the yield of the interconnect structure, 4) the yield of the bonding and assembly processes, and 5) the effectiveness of the testing and rework process in detecting, isolating, and repairing those defects. Chip yield plays an important role. For

*Visiting researcher from Eastman Kodak Co.

example, for a 50-chip MCM, if the yield of the incoming chips is 95%, then the yield of the assembled MCM before testing will be only 7.7% and without considering any other source of yield loss. This means that the incoming yield of bare chips must be pushed to nearly 100% if the module yield is to be high enough to have a cost-effective MCM process. For example, while the yield of packaged ICs can be pushed to nearly 100% using sophisticated IC testers, the infrastucture required to produce Known Good Bare Die to support MCM manufacturing doesn't exist today.

An obvious approach to alleviating the need of sophisticated testers at all levels of integration is to incorporate the tester into the circuit under test itsefl, and hence the notions of design for test and built-in self-test. This eliminates the need for expensive testers and provides a mechanism for accessing and exercising internal design circuitry. BIST, by definition, can be executed at the normal speed of the design under test. However, DFT and BIST are not free, they require an investment in chip area, and in certain cases may themselves present additional delays.

Ideally, to solve the MCM testing problem, one may require that every chip incorporates DFT and BIST. However, understanding the economics of DFT and BIST is crucial to determine the proper amount of DFT and BIST to include at the IC and module levels. In most cases, only a subset of the chips need to incorporate testability features, while others may not require DFT features. This situation cannot be avoided because an MCM will very likely contain off-the-shelf or existing components that have varying degrees of testability. The problem then becomes that of exploiting the partial DFT features in testing the system as much as possible to still maintain a high level of test quality and diagnosis.

It is also important to note that from the IC manufacturer point of view, the economics of incorporating DFT into an IC design may not be favorable for all applications without considering the ramifications at the MCM, board and system levels.

This article describes a tool that analyzes and assesses the economics of various test methods at the IC and MCM levels, and explores the trade-offs between various test strategies as well as the various design-for-test options at the wafer, die, and MCM levels. The tool employs a hierarchical computational model for modeling the complete manufacturing and test process starting at the wafer and going all of the way to the MCM level. This model is the basis for performing trade-off analyses among the various test techniques by computing their impact on both the final cost and quality of an MCM.

2. Scope

The determination of an optimal test strategy must be evaluated concurrently for all stages of the manufacturing process. Numerous factors must be concurrently analyzed for selecting an optimal test methodology for achieving a known good device. Figure 1 illustrates some of the key factors that must be modeled to determine the benefits and evenutal cost-effectiveness of various test solutions at the wafer, die, and module levels.

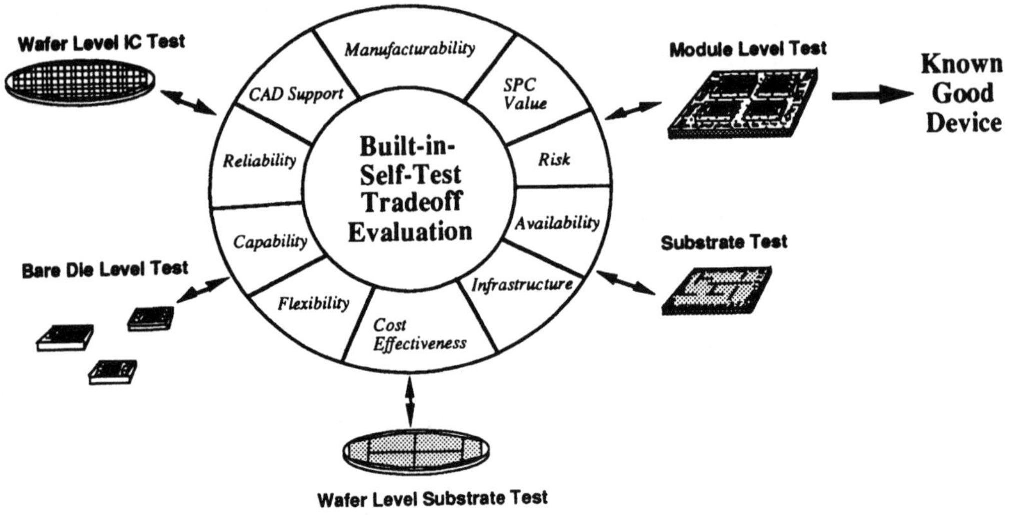

Fig. 1. Evaluation factors for test strategy selection.

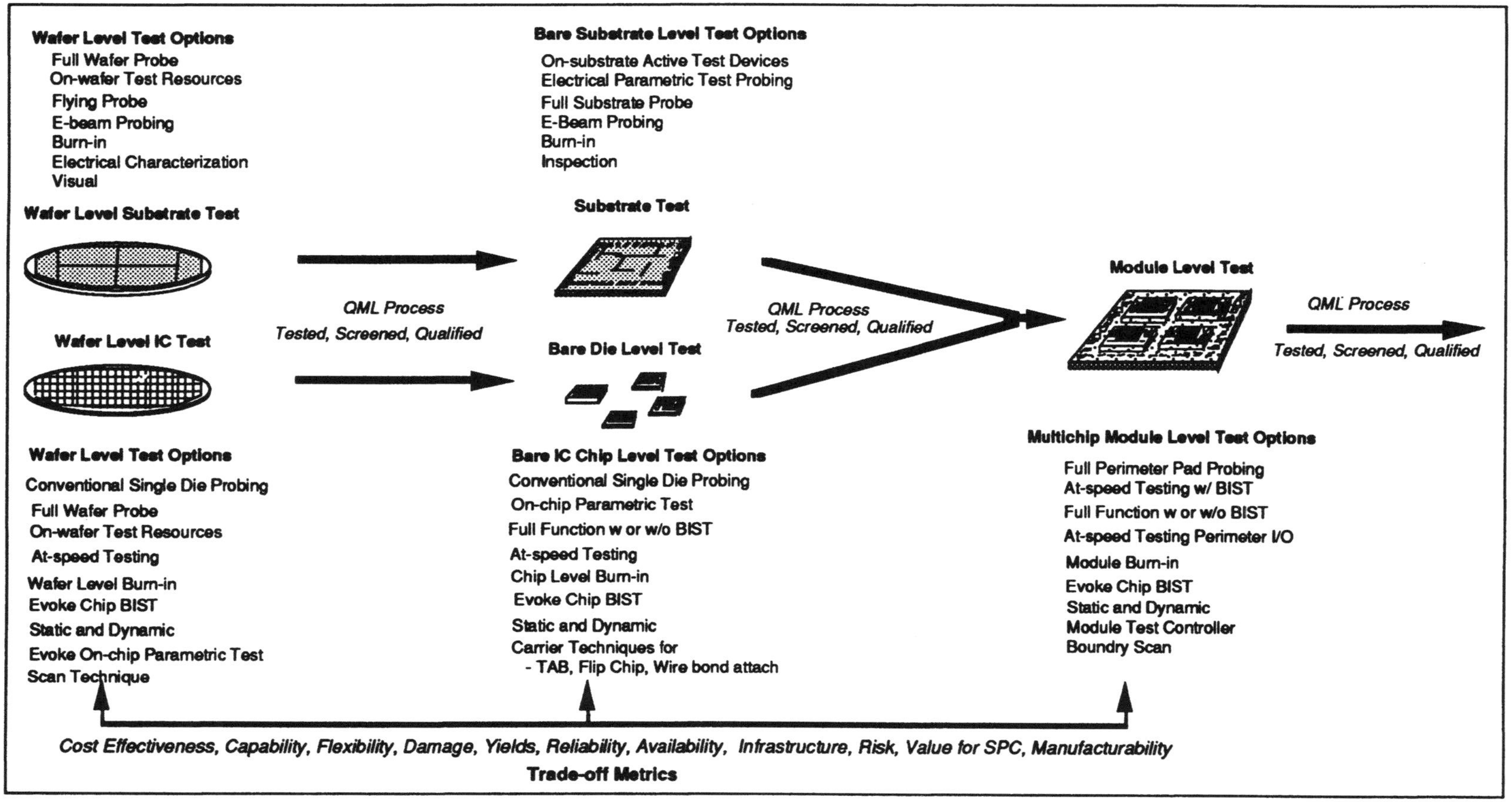

Fig. 2. Potential test solutions.

There are numerous test solutions which have their unique advantages and disadvantages. As an example, Figure 2 illustrates various test solutions which might be available through the complete process of producing a fully functional, fault-free, and reliable MCM.

3. Metrics for Optimization Analysis and Selection

To properly weigh the benefits of the approaches, which may include combinations of solutions, we need to compare the different approaches against a common set of metrics. Two primary quantitative metrics will be used as the basis for comparing the various test methods:

- The impact on the cost ($) of the MCM.
- The impact on the quality of the MCM measured by the defect level (in parts per million) in the produced MCMs.

The selection of cost as one of our comparison metrics reflects the important role that test plays in determining the final cost of an MCM. It also emphasizes the importance of finding cost-effective test solutions and not just test solutions at any cost.

The second performance metric serves as a balance against recommending cheap solutions with inferior quality. We will use the defect level (or equivalently, the outgoing yield) of the produced MCM as a measure of quality. Since all alternative test methods should not change the product functionality substantially (some DFT methods result in some performance degradation), cost and quality are the two prime characteristics for evaluating any product.

In addition to cost and quality, there are several other evaluation criteria that are hard to measure in a quantitative fashion.

- Capability, extendibility, and limitations of test strategy with respect to various design complexities, such as circuit size, pin count, technology variations, maximum speed, etc. Certain methods may be eliminated from consideration if they fail to meet state-of-the-art complexities, or if they possess certain limitations that would make them obsolete in the near future as complexities continue to increase. For example, certain probing methods might have an upper limit of 100 MHz, which is acceptable today, but would make that method obsolete in the very near future.
- Adverse impact of the method on the reliability of the product. For example, certain die test methods may cause damage to the IC pads or they may cause thermal overstress. Another example is MCM test strategies that rely on having a high number of rework cycles. The reliability of the product resulting from these methods might be significantly reduced, and hence they should be avoided if possible.
- Usefulness of the method at higher levels of integration and for field test, diagnosis, and maintenance.
- Electrical performance degradation caused by the method. Most DFT methods introduce delays into the design. However, proper design of DFT features could avoid affecting critical path and hence reduce the overall performance impact of DFT.
- Impact of the method on manufacturability. This covers issues such as:

  How easy is it to adapt this method into existing manufacturing processes?

  How big of an infrastructure is required to support the new method?

  How expensive is the equipment required to support this test method?

4. Test Economics Analysis Model

To perform an adequate testability trade-off analysis, we need to examine and understand the relationship between the cost parameters, the yield of the various components/steps, and various test strategy parameters. Some of these key parameters are listed below:

Cost Parameters

- Wafer cost
- Wafer probe cost
- Wafer burn-in cost
- Die cost
- Die burn-in cost
- Die test cost
- Substrate cost
- MCM assembly cost
- MCM test cost
- MCM diagnosis and repair cost

Yield Parameters

- Wafer defect level
- Wafer yield
- Wafer burn-in mortality rate (the fraction of die that fails during burn-in)
- Die burn-in mortality rate
- Substrate and interconnections yield
- MCM assembly yield

- MCM rework yield
- MCM burn-in mortality rate

Test Parameters

- Wafer-probe fault coverage
  Fault coverage is the ratio of the faults detected by the test to total faults possible. It may be associated with one specific type of fault, or it might be an average representation for all types of faults.
- Die-test fault coverage
- MCM-test fault coverage
- MCM-rework scrap rate
  The fraction of defective MCMs that cannot be repaired and are scrapped.
- Maximum number of repairs per MCM
  The maximum number of allowed repair cycles per MCM. An MCM that still appears defective after exceeding the limit of repairs is scrapped.

Various test methods have a different set of values for some of the parameters listed above. Obviously, some of these methods, e.g., a wafer-level at-speed test method, or using a chip carrier to do die burn-in, etc., impacts only a subset of the parameters, while other methods, such as incorporating BIST into an IC design, may have an impact on most parameters. To analyze the relationship among all the above parameters and how they impact the final cost and quality of the MCM, we developed a simple generic assembly, test, and rework simulation models. Figure 3 illustrates how the three basic process models are used to create a simple manufacturing flow.

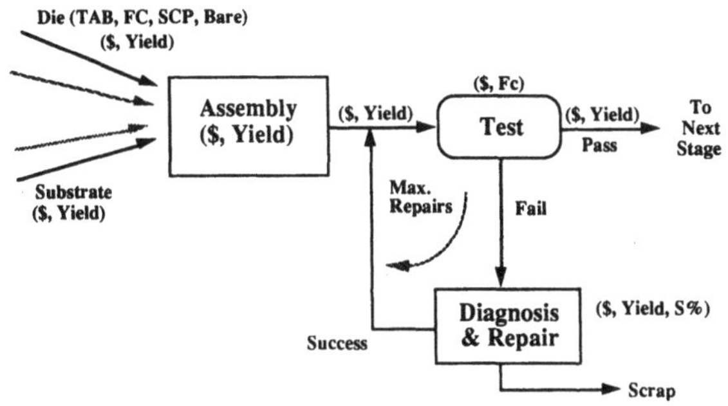

Fig. 3. A simple MCM manufacturing and test model.

Each one of the boxes in Figure 3 represents a certain type of process (assembly, test, and rework). The assembly box takes one or more inputs that corresponds to the various types of die to be assembled and the substrate itself. Each of the inputs is characterized by a cost, yield, and a count indicating how many times it is used (e.g., 10 SRAM chips). The assembly process itself has a cost and a yield factor, as does the pretested MCM that is produced. That module is then tested with a process that has a cost and a certain degree of fault coverage. The modules that pass the test are assumed good and are passed to the next level of assembly. However, since the testing process cannot detect all possible faults, some defective modules may escape and reduce the quality of the output modules.

The modules that fail the test are then diagnosed for possible repair—some modules may not be repairable and are scrapped. The repairable modules are repaired and are subjected again to the testing process. The diagnosis and rework process has an average cost and also has a yield (i.e., not all repaired modules are good due to rework assembly defects, new components defects, accidental defects, or because of miss diagnosis). A module is scrapped if it fails to pass the test after being repaired a maximum allowable number of times.

The simple model of Figure 3 does not have an explicit MCM burn-in processing step. Such a step can be added at the end an be followed by another test and repair cycle, or it can be inserted after assembly and before test. A burn-in step would obviously add to the cost, results in a certain mortality rate, and might reduce yield. However, it will improve the reliability of the subjected product.

5. Example: Chip Yield vs. MCM Cost and MCM Defect Level

To illustrate the usefulness of the simple model described in the previous section, consider the following example. An MCM is to be built containing 50 identical chips. These chips can be procured (purchased or made internally) at different prices and different quality levels as shown in Table 1 below. The table also shows the corresponding MCM cost and MCM defect level as computed by Hi-TEA. The following assumptions were used for the various parameters: assembly cost = $200, substrate cost = $200, test cost = $150, repair cost = $100, test coverage = 99%, 1 repair allowed, repair scrap rate = 5%, and repair yield = Min (95%, chip yield). Both the substrate and the assembly were assumed to have a perfect yield in this experiment. Hence, the chips were the only source of defects in this hypothetical experiment.

The results shown in Table 1 illustrate that although we more than quadrupled the chip-set cost (from $100

Table 1. Chip yield vs. MCM cot and qualtiy.

Cost	Yield	MCM Defect Level (After Test)	MCM Cost
$2	80%	105599 ppm	$1101
$3	85%	78073	$1120
$4	90%	51345	$1132
$5	94%	30493	$1135
$6	97%	15143	$1119
$7	99%	5042	$1034
$8	99.5	2533	$1025
$9	99.9	530	$1017

to $450), the cost of the MCM improved by about 10% while the quality of the outgoing MCM improved by almost three orders of magnitude! The first case (the 80% case) results in shipping more than 10% defective parts, while the last case results in about 0.05% defective shipped parts. Our experience with Hi-TEA has shown that either a bell-shaped curve, or in general a two minima curve relates MCM cost to chip yield. One minima (one end of the bell curve) corresponds to a high quality solution using high quality chips, while the other minima, corresponds to a low quality solution that uses cheap and low quality chips.

Having chips with various cost/quality metrics as explained above can be directly related to the test strategy used for testing the chip at the wafer level and as a bare die, the amount of testing, and whether DFT techniques were employed or not. The above experiment also points to an interesting problem at the business level. From a semiconductor company point of view, the goal is to produce good quality chips at the lowest possible price. Adding DFT structures may increase the cost of the chips. However, from a system manufacturer point of view, the less expensive chips can result in a more expensive MCMs, boards, or systems. Thus, it is clear that system houses are the ones who should demand chips with high quality and which are easy to test and be willing to pay for it.

The problem is further complicated by the fact that many chips are commodity items (off-the-shelf), and hence are used in a variety of applications that do not demand sophisticated DFT techniques.

An interesting question that is often raised: What happens if we increase the fault coverage of the test step after assembly (implies an added cost as well) to compensate for the poor quality of the parts? To answer this question we ran another experiment for three cases using the same data as before except as indicated below:

1. MCM Test Coverage = 95%, Test Cost = $100
2. MCM Test Coverage = 99%, Test Cost = $150
3. MCM Test Coverage = 99.9%, Test Cost = $200

The results are listed below in Table 2.

The results indicate that the quality level of the outgoing MCM improved by about a factor of 5 going from 95% to 99%, then it improved by another factor of 10 going from 99% to 99.9%. For a target quality level, say 2,500–3,000 ppm for example, we can use the $9 chips with case 1 MCM testing strategy which result in a $965 MCM. Alternatively, we can use the $8 chips with case 2 MCM test strategy and get an MCM that costs $1,025. Finally, we can use the $5 chip with case 3 MCM test strategy and get an MCM that costs $1,213. All three solutions result in an MCM with roughly the same quality. The answer to the question we posed earlier is obvious here: It is more economic to start with high quality parts than to rely on higher coverage testing later on.

Table 2. MCM cost and MCM defect level for three test strategy cases.

		MCM Test Coverage and Test Cost					
		95%, $100		99%, $150		99.9%, $200	
Chip Cost ($)	Chip Yield (%)	Defect Level (ppm)	Cost ($)	Defect Level (ppm)	Cost ($)	Defect Level (ppm)	Cost ($)
2	80	427652	785	105599	1101	11098	1316
3	85	333987	860	78073	1120	8095	1274
4	90	231681	937	51345	1132	5257	1239
5	94	143446	1007	30493	1135	3091	1213
6	97	73460	1026	15143	1119	1524	1164
7	99	24959	966	5042	1034	505	1078
8	99.5	12601	966	2533	1025	253	1072
9	99.9	2647	965	530	1017	53	1066

A manufacturing manager who was comparing the entries of the last row in the above table indicates that from an economic standpoint, most manufacturers will be willing to pay the additional $101 to reduce the defect level from 2,647 ppm to 53 ppm. This may not only reduce test and repair costs at higher levels of integration, but more importantly it reduces field returns and the potential loss of customer satisfaction and loss of future sales.

6. Hi-TEA Model Capabilities

Hi-TEA has been prototyped using C++ on a UNIX platform. The effort is a part of the Multichip Systems Design Advisor (MSDA) project. The MSDA project is a consortia effort to develop and facilitate the commercialization of CAD tools for conceptual design of multichip systems. A new effort is currently under way to enhance and expand the capability of the tool working with Martin Marietta with government funding under the RASSP program. Analysis data from other MCC projects such as the Commercial Open Systems Project are being used to populate and verify the Hi-TEA model.

The simple modeling elements described above are used to model bigger parts of the manufacturing flow. The user can quickly create a customized manufacturing flow by sequencing the appropriate type of processes, and then use the resulting model to evaluate the cost and quality metrics associated with his design. For example, the model in Figure 4 starts at the wafer level processing and test steps, followed by die burn-in and test, then MCM assembly, test, and repair, and finally MCM burn-in, final test, and repair. Each of these steps is characterized by various cost, yield, and fault coverage parameters.

Hi-TEA supports a two level Macro/Micro economics model hierarchy. At the Macro level, a manufacturing flow is modeled using a number of processing steps as shown in Figures 3 and 4. Each one of these processing steps, e.g., MCM assembly, can be broken further down at the micro level into a number of steps. Figure 5 illustrates this concept. The figure illustrates a manufacturing flow of a hybrid MCM using both flip-chip and surface mount technology (SMT). The assembly is broken into two macro stages, each followed by a test and a rework cycle. Each of the macro boxes in the figure can be characterized by a number of global parameters (e.g., cost, yield, time, etc.) Alternatively, each of these processes can be modeled as a number of micro steps as shown in the figure. At the micro level the models use parameters such as equipment cost, labor rates, depreciation base, tooling cost, material cost, etc.

Figure 6 illustrates graphically how the MCM assembly step of Figure 4 can be defined as a sequence of micro processing steps. The steps are either selected from a library of steps as shown in the figure, or they could be defined by the user. The figure also shows the detailed parameters required to define each of the micro steps. Not all of these parameters are applicable to each step.

Hi-TEA allows the user to model some of the steps at the micro level while retaining the macro level abstraction of the other steps. Obviously, this provides the user with great flexibility and combines the accuracy of detailed micro cost models with the ease of macro modeling and the ease of performing quick what-if trade-off analysis on macro models.

7. A Testability Trade-Off Case Study Example

An application specific testability trade-off analysis was performed in the commercial open system II corsortia project at MCC. The analysis was done for a leading-edge MCM test vehicle containing a large CPU, a large co-processor and controller chip, and 10 4-Mega bit SRAM chips. The target speed of the MCM leapfrogged all known workstation RISC processors speed.

A Testability Base Case was defined to serve as the basis for comparing the various test strategies. First, we assumed that the chips will be tested at the wafer level, using traditional wafer probing methods. Wafer defect level was assumed to be 3 defects per square inch for the CPU and co-processors (CMOS), and 6 defects per square inch for the SRAM wafers (BiCMOS). The wafer is the diced, and the die that passes the wafer test is attached to temporary chip carriers. The chips are burned-in and then electrically tested. Ten percent of the die were assumed to fail during burn-in. The chip carriers are removed for reuse. The die that passed previous testing steps are used for the MCM assembly.

The substrate with the interconnections is assumed to have been pretested accurately prior to MCM assembly. The cost of the substrate depends on its type and its size. A detailed model for that was used. Also MCM assembly was modeled at the micro-level.

The MCM is first tested for assembly defects and then tested functionally. Two of the twelve die had boundary scan. Additional test points are added to help in testing the MCM assembly defects. Very little DFT

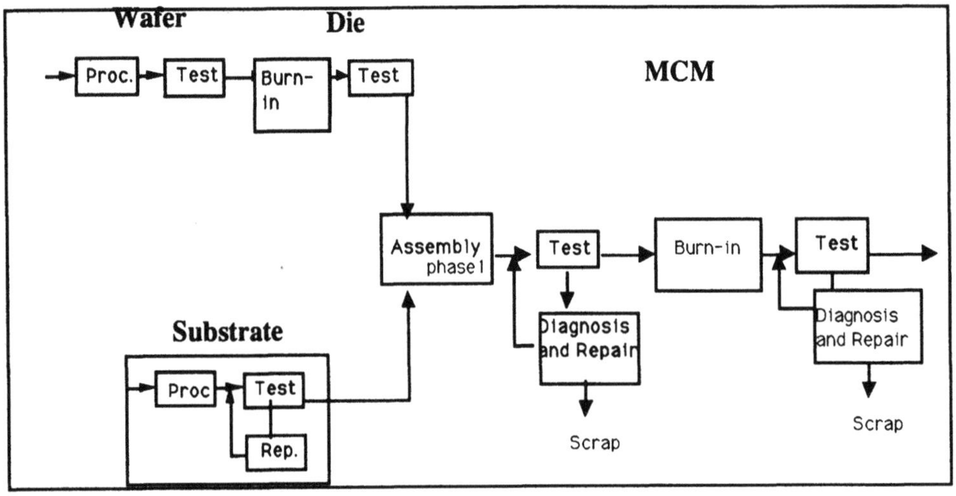

Fig. 4. An extended test and manufacturing process flow model for MCMs.

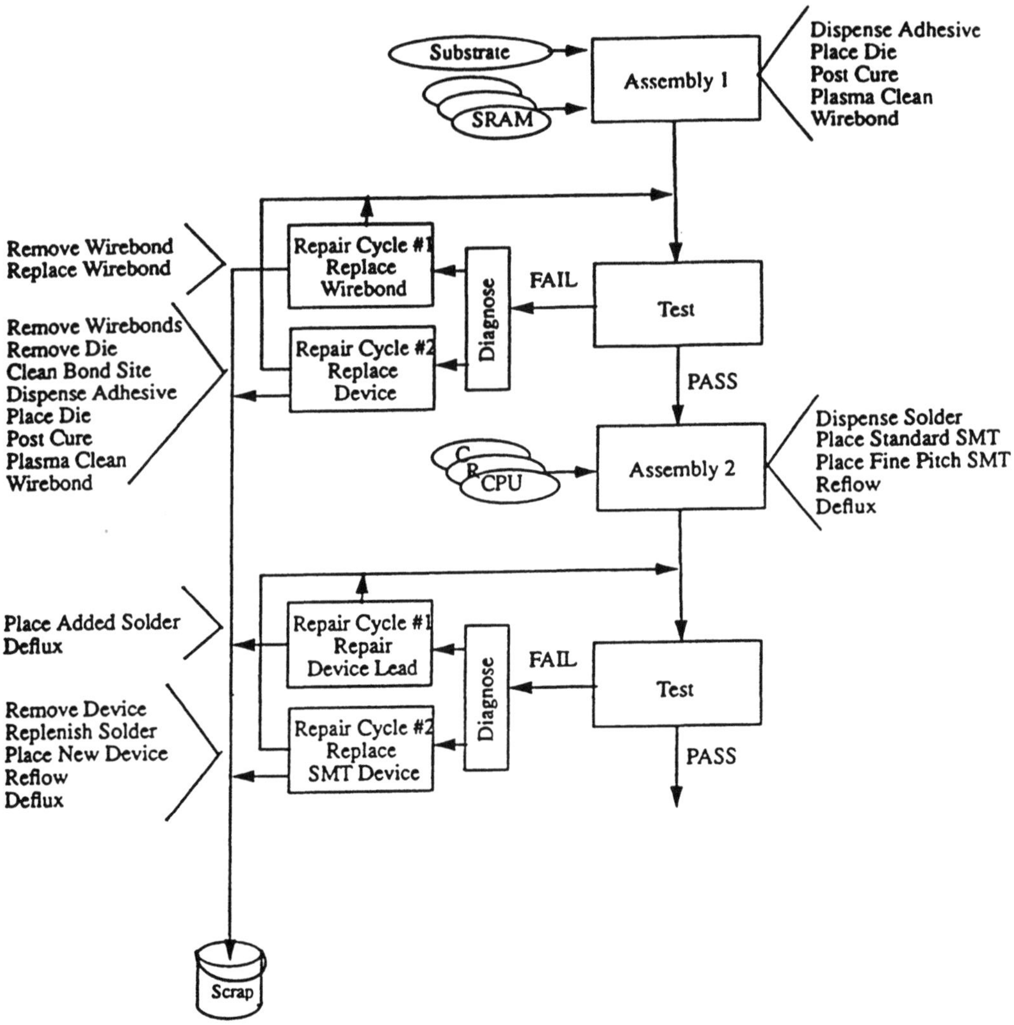

Fig. 5. Model abstraction in Hi-TEA.

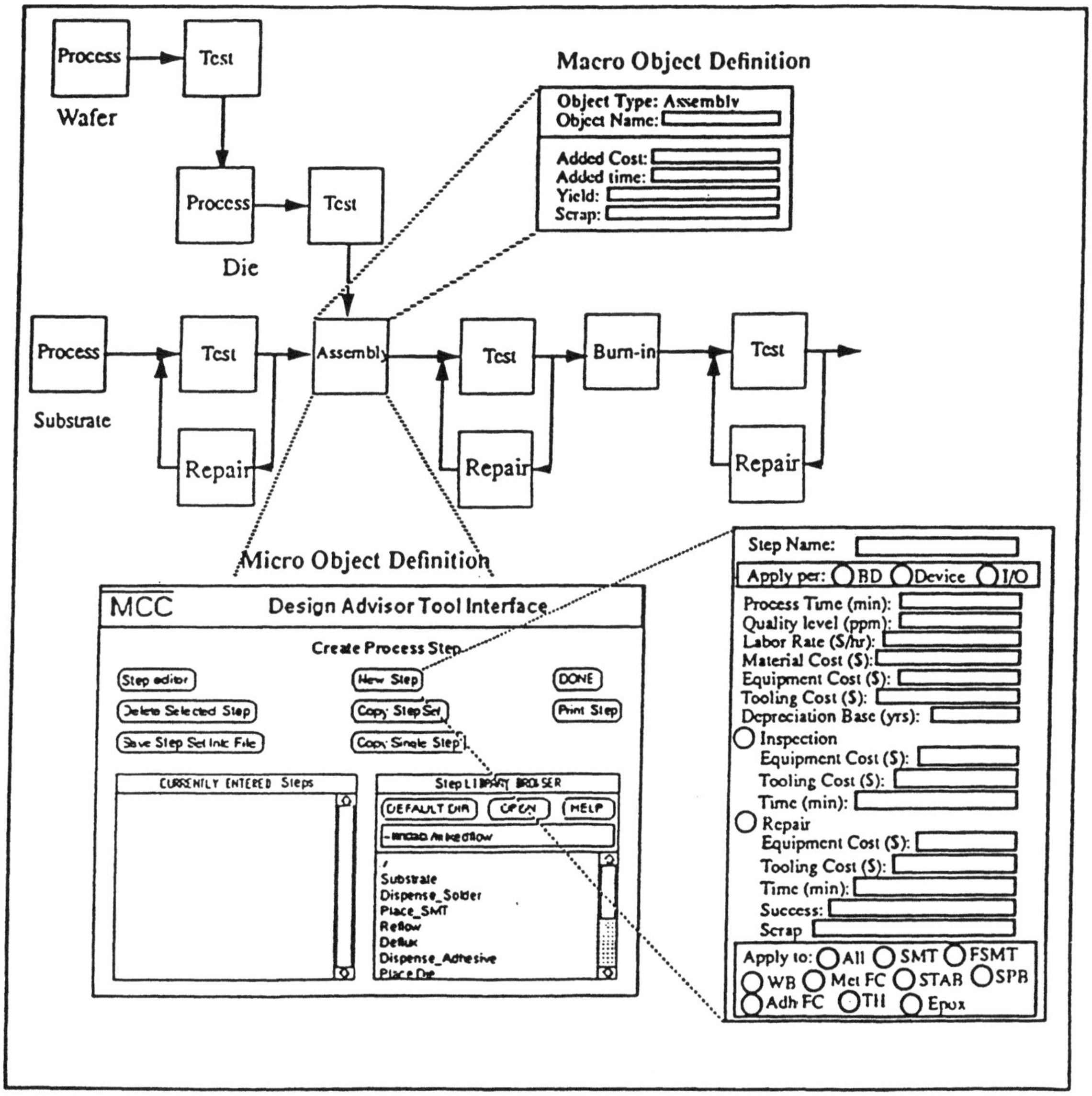

Fig. 6. An example macro/micro economic model.

and BIST is available inside the chips for use during MCM testing (except for boundary scan). A functional testing environment that mimics the normal use environment (workstation) of the MCM was assumed. MCMs that fail the test are diagnosed for possible repair. The repair may require replacing one or more of the chips. After rework, the MCM is retested. If it fails again, it is sent back to the rework station. MCMs that pass the test are then prepared for the next level of assembly. Our analysis concentrated on the cost and quality of those MCMs that pass the test. The first two columns of Table 4 lists the key parameters used for the testability base case.

In addition to the base case, we also modeled a number of alternative testing strategy cases. These cases are summarized below and the key parameters used are listed in Table 3.

1. Low coverge die test case: the fault coverage of die test was decreased to 95% instead of 99% while reducing the die test time in half.
2. Full BIST case: BIST is incorporated into all of the die and was used to facilitate testing at all levels. Of course the impact of BIST on die area and ultimately on wafer yield, die cost, etc. was taken into account.

Table 3. Test strategy parameters/assumptions.

Description	Base Case	Case 2 Partial BIST (into CPU and Co-Proc.)	Case 3 Full BIST (into all Chips)	Case 4 Addition of a Test Control Chip	Case 5 Partial Assembly, Test and Repair Phase 1	Phase 2
Overhead	None	5% chip area	5% for CPU and Co-proc 2% for SRAMs	An extra chip that costs twice as much as SRAMs	None	None
Wafer test coverage	80%	99%	99%	80%	80%	80%
Wafer test cost	50¢,$2,$3	50¢,$1,$1	25¢,$1,$1	50¢,$2,$3	50¢,$2,$3	50¢,$2,$3
Die test coverage	99%					
Die test cost		Reduced by 50%	Reduced by 50%			
MCM test cost	$75	$45	$30	$45	$25	$25
MCM test coverage	95%	99%	99%	99%	99%	95%
Repair cost	$200	$150	$150	$150	$75	$150
Repair yield	80%	85%	85%	85%	90%	90%
Repair scrap rate	20%	10%	10%	10%	10%	10%

Table 4. Die cost and yield before MCM assembly.

Description	Cost $	Yield %
CPU	198.52	99.56
Co-processor	80.77	99.66
SRAM	29.63	99.72

3. Partial BIST case: Only the large CPU and controller chips incorporated BIST, while the SRAMs were unchanged.
4. Test controller chip case: A test controller chip was added to the MCM design to provide various monitoring and control features to the MCM to facilitate MCM test and diagnosis. In this case, we analyzed the incorporation of a test controller chip into the MCM. This solution is suitable in cases where the MCM manufacturer has no control over the chips. The test controller is designed to provide access to the key internal buses of the module as well as built-in test resources that can be used to control and observe the various design components. The test controller also acts as the primary test interface of the module to the external board (other test controllers, testers, etc.) via a standard test bus interface.
5. Partial MCM assembly case: In this case, we analyzed the use of a two-phase assembly and test process for the MCM. During the first phase, the SRAMs are attached to the MCM, tested, and repaired (if necessary). Then in the second phase, the two large chips are attached and the full module is tested and repaired if needed. This partial assembly and test strategy make it possible to efficiently test the SRAM chips in parallel. Special fixtures or removable interconnects could be designed to provide access to the data and address buses during the first phase testing process. Once the cache integrity is verified, the second assembly and test phase can proceed. The ability to diagnose the cause of failures during both phases of testing is higher in this case, as a natural implication of the partitioning that took place.

7.1. Results

A Hi-TEA model similar in structure to the one in Figure 4 (without MCM burn-in) was used to perform the trade-off analysis computation. Several parts of the model, such as, die processing, die test and burn-in, and MCM assembly, were modeled at the micro level, while the rest of the steps were modeled at the macro level. The table below illustrates die cost and yield for all the chips before MCM assembly as they were computed for the base case.

Figure 7 summarizes the results of our trade-off analysis study by comparing the cost and defect level of the MCM for all the test strategy cases described above against each other. MCM test and rework contributed about 20% to 35% to the final cost of the MCM.

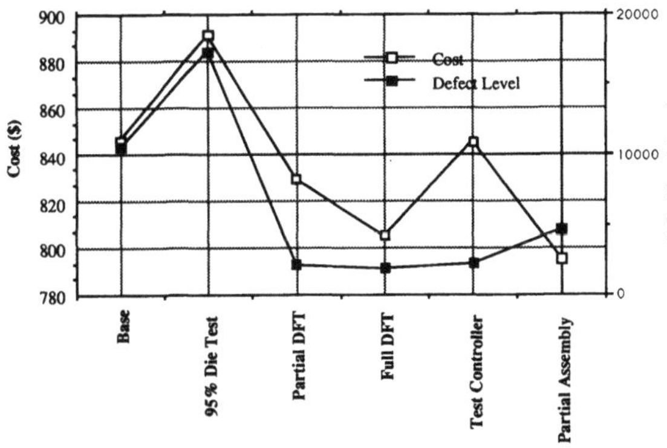

Fig. 7. Test strategy trade-offs example.

The low die-test fault coverage case (case 1) resulted in an increase in MCM cost of about 5% (although the die test cost was reduced by half). The defect level of the outgoing MCM increased by about 70%! This clearly indicates that high coverage die test is essential to provide known good die, and is also economically justifiable.

As the figure indicates, the two DFT cases at the chip level and the partial assembly case result in significant cost savings as well as quality improvements. In both DFT/BIST insertion cases the initial cost of adding DFT was economically justified, because it reduces MCM cost as well as significantly improves MCM quality. A reduction in cost of about 2 to 3% resulted in the partial BIST case, while savings of about 5% were realized for the full BIST case. The full BIST case has the lowest defect level, but the partial assembly case has the lowest cost. The impressive result was the reduction in the outgoing defect level of almost 75% in cases 2–5 as compared to the base case.

Incorporating BIST into the CPU chip resulted in about a 15% cost increase at the die level (it increased from $206 to $237). Thus, BIST incorporation may not be economically justifiable at the die level alone. However, once the MCM impact was analyzed, it became obvious that using BIST saves money as well as improves quality. Continuing the analysis to higher levels of integration (e.g., motherboard, systems, etc.) will result in more cost savings and more quality improvements.

To study the impact of the wafer defect density we ran several experiments with different defect levels. The cost of the MCM more than doubles if the defect level increase to 6 and 10 defects per square inch for the CPU and SRAMs, respectively (originally assumed to be 3 and 6).

The previous experiments assumed a particular bonding method was used to bond the chips to the MCM substrate. Figure 8 compares the results of using a different bonding method that requires fewer processing steps, and hence it is less expensive. However, the second method has a higher defect level (100 ppm instead of 50 ppm per bond). The results indicate that for this particular application, the second bonding method reduced the MCM cost only when full DFT/BIST or the partial assembly strategy were used. The first method had lower cost in all other test strategies. The MCM defect level increased by about 25% when the second bonding method was used.

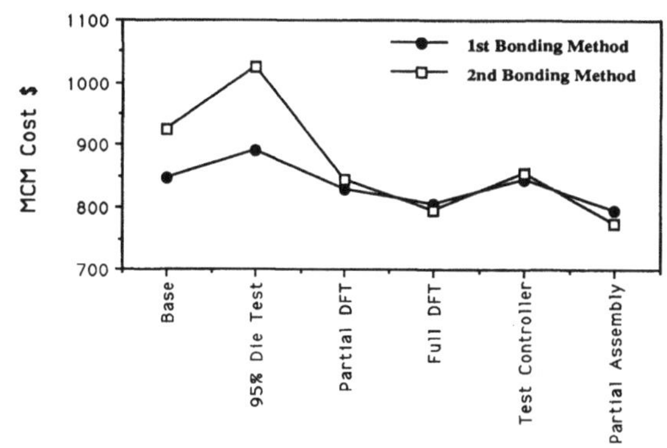

Fig. 8. Cost trade-offs among two different bonding approaches.

8. Conclusion

In this article we presented a tool called Hi-TEA for modeling the economics of manufacturing and test strategies associated with multichip systems in general, and of MCMs in particular. The tool features a two level macro/micro economics modeling capability. We also presented various experimental resuts obtained using the tool. The results clearly indicate that incorporating DFT and BIST with various degrees at the chip or MCM levels is economically justifiable and results in cost reduction as well as quality improvement. The results also indicate the importance of using known good die. We also illustrated how the tool can be used to evaluate various manufacturing flows (e.g., two phase assembly, test and repair).

References

1. *Economics of Design and Test for Electronic Circuits and Systems*, A.P. Ambler, M. Abadir, and S. Sastry, (eds.), Ellis Horwood Publisher, 1992, contains extended versions of papers presented at the First International Workshop on the Economics of Design and Test, Austin, Texas.
2. T.W. Williams and N.C. Brown, "Defect Level as a Function of Fault Coverage," *IEEE Trans. on Computers*, vol. C030, pp. 987–988, Dec. 1981.
3. C. Dislis, J. Dick, and A.P. Ambler, "An Economics Based Test Strategy Planner for VLSI Design," *Proceedings of the 2nd European Test Conference*, 1991.
4. C. Dislis, A.P. Ambler, and J. Dick, "Economic Effects in Design and Test," *IEEE Design and Test of Computers*, vol. 6, Dec. 1991.

Magdy S. Abadir was born in Cairo, Egypt, in 1956. He received the B.S. degree with honors in Computer Science from the University of Alexandria, Egypt in 1978, the M.S. degree in Computer Science from the University of Saskatchewan, Saskatoon, Canada, in 1981, and the Ph.D. degree in Electrical Engineering from the University of Southern California, Los Angeles, in 1986. Prior to joining the Microelectronics and Computer Technology Corporation (MCC) in May 1986, he was a research associate at the University of Southern California. He is currently a senior member of the technical staff in the High Value Electronic Division at MCC where he is involved in a number of projects associated with testing of multichip modules, test economics, Design For Test Synthesis, and the development of computer-aided software tools to support these various test activities.

Dr. Abadir has cofounded and chaired a series of International Workshops on the economics of design, test and manufacturing. He is a coeditor of a book entitled "Economics of Design and Test of Digital Circuits and Systems," Ellis Horwood, 1992. He has published over 45 technical journal and conference papers in the areas of test economics, design for test, computer-aided design, high-level test generation, design verification and layout verification. His current research interests include test economics, design for testability synthesis, hierarchical test generation, trade-off analysis, and optimization tools.

Ashish R. Parikh received the B.S. degree in Electrical Engineering from the University of Texas at Austin in 1991. He is currently working on his M.S. degree in Computer Engineering at the UT Austin.

He joined the Microelectronics and Computer Technology Corporation (MCC) in 1992 as a Research Assistant. He is presently an Associate Member of Technical Staff in MCCs High Value Electronics Division. He is a programmer on the Multichip Systems Design Advisor project. His current research interests include high-level testability analysis, test economics, software engineering, and object-oriented software systems.

Linda Bal of Eastman Kodak joined MCC's High Value Electronics Division as a Visiting Researcher in June of 1991. She is a member of the Multichip System Design Advisor Project team. At Eastman Kodak, Ms. Bal is a Senior Development Engineer in the Research and Engineering Department of the Electronic Products Business Unit. Since joining that group in July of 1990, she has designed Low Temperature Cofired Ceramic and Thick Film Multi-layer Hybrid Packages.

Prior to joining packaging engineering, Ms. Bal spent two and a half years in the Electonic Products Test Engineering department at Kodak where she developed and subcontracted in-circuit test programs. Prior to test engineering, Ms. Bal was a manufacturing and development engineer in Kodak's Mass Memory Divison for Magnetic Peripheral Systems Products.

Ms. Bal received a B.S. in Electrical Engineering from Purdue University in 1985. While at Purdue, she completed the Cooperative Education Program by working for IBM in their San Jose Magnetic Recording Development Lab.

Peter A. Sandborn received the B.S. degree in engineering physics from the University of Colorado, Boulder, in 1982, and the M.S. degree in electrical science and the Ph.D. degree in electrical engineering from the University of Michigan, Ann Arbor, in 1983 and 1987, respectively. His dissertation research concerned the physics and modeling of GaAs MESFETs.

He joined the Microelectronics and Computer Technology Corporation (MCC) in 1987 and is presently a Senior Member of the Technical Staff in MCC's High Value Electronics Divison and the technical leader of the Multichip Systems Design Advisor Project. Prior to joining the High Value Electronics Division in November 1988 he was in the Analog Technology Laboratory of the Computer Aided Design Program at MCC. While in the Analog Technology Laboratory he worked on the application of non-stationary transport models to modeling submicron silicon and GaAs device structures and developed a large-signal time-domain transient device simulation platform oriented toward the development of equivalent circuit models for transistor structures. Since joining the High Value Electronics Division he has worked on power distribution problems in multichip packaging systems, CAD software for packaging and interconnect applications and is an expert in the design of multi-chip systems. Dr. Sandborn is the principle developer of the Multichip Systems Design Advisor trade-off analysis software.

Dr. Sandborn is the author of a book entitled "Conceptual Design of Multichip Modules and Systems," Kluwer Academic Publishers, 1994, and is the principle author or co-author of over 30 technical publications in the areas of semiconductor device simulation and packaging and interconnect.

Cynthia F. Murphy received the B.S. degree from the College of William and Mary in 1977 and the M.S. from the University of North Carolina, Chapel Hill, in 1982. She joined the Microelectronics and Computer Technology Corporation (MCC) in 1988 and is presently a Member of the Technical Staff in MCC's High Value Electronics Division and a technical leader for the Known Good Die Project. Prior to joining MCC she was a process and product engineer with the Semiconductor Products Sector of Motorola, Inc. While with Motorola, she was involved with process and yield analysis of high-end CMOS devices. Since joining MCC she has worked process characterization, yield and reliability analysis, and technology assessment of a number of advanced packaging technologies, As part of the technology assessment work, she has developed or assisted in the development of cost models for incoming die, assembly, and test for multichip modules. She has also developed a cost model for test and burn-in of known good die for both single die and wafer-level applications.

JOURNAL OF ELECTRONIC TESTING: Theory and Applications, 5, 207–218 (1994)

Multichip Systems Trade-Off Analysis Tool

PETER A. SANDBORN, RAJARSHI GHOSH, KEN DRAKE, MAGDY ABADIR, LINDA BAL* AND ASHISH PARIKH

Microelectronics and Computer Technology Corporation (MCC), 3500 West Balcones Center Drive, Austin, TX 78759

Abstract. This article discusses the conceptual/specification phase of multichip system design. The automation of conceptual design practices will decrease design and manufacturing risks by allowing system designers to view the entire performance design space early in the design cycle prior to the initiation of traditional physical design activities.

This article describes a software tool that performs interdisciplinary trade-off analysis and partitioning for multichip systems including multichip modules (MCMs) and traditional assembly approaches. The tool concurrently computes physical, electrical, thermal, reliability, testability, and cost performance metrics for multichip systems. This tool allows a large fraction of possible design, technology, material, test, and manufacturing approaches to be explored before decisions are made and resources committed.

An example set of trade-off results are presented which examine the trade-off between peripheral I/O format die and area array I/O format die in an MCM, as a function of partitioning a fixed functionality into a variable number of die.

Keywords: Conceptual design, flip chip bonding, multichip module (MCM), packaging and interconnect, trade-off analysis.

1. Introduction

The need for increasing system performance has brought about the requirement for advanced packaging and interconnection approaches that do not compromise the performance of advanced chips. This requirement dictates the need for advanced computer aided design (CAD) directed at packaging and interconnect. One of the serious shortcomings in the present CAD tool set for packaging and interconnect design is in the area of conceptual design support; e.g., formal trade-off analysis and partitioning. Traditionally, packaging and interconnect CAD tools have evolved from more mature tools used in IC design activities. This scenario is changing as designers realize that IC tools cannot always be evolved to support the unique problems posed in the multichip systems design space. Design of complex modules and systems is characterized by a greater diversity of system and subsystem components (and assembly approaches) than chip-level design, requiring an interdisciplinary analysis, which is highly dependent upon physical and functional partitioning. Trade-off analysis for these systems requires expertise in electrical, thermal, mechanical, manufacturing, cost analysis, testing, and reliability. System packaging and interconnect design is further complicated by the presence of a large and continuously changing set of competing technologies, materials, and process alternatives.

For most system designers, the goal is to find a design that balances performance with ease of manufacture and support, while minimizing cost. Early in the design cycle, trade-off decisions have to be made intelligently and quickly before major investments are committed or detailed design work begins. Critical decisions made within the first 10% of the total design cycle time will ultimately define up to 80% of the final product cost.

Besides time and budget constraints, system packaging trade-off decisions are hampered by the lack of adequate and accessible alternative technology data. This situation is an especially serious problem in the packaging and interconnect field since the number of technologies, processes, materials, and approaches is substantial, and selecting optimums is arduous and non-trivial if one truly wants a balance in cost and performance. Alternatives include interconnects (printed

*Visiting researcher at MCC from Eastman Kodak Company.

circuit boards, ceramic, thin film, etc.), bonding techniques (wirebond, TAB, flip chip), test techniques, and manufacturing methods. Few designers can comprehend all the interdependencies and ramifications the technologies and materials chosen may have on a particular design's performance, cost, and manufacturability.

Trade-off analysis activities could be carried out using detailed simulation tools; however, because not enough information about the system is available during the conceptual design phase, the usability of detailed simulators is limited. In addition, the lack of a well-defined theory relating behavior and structure at the system-level often makes approaches based on detailed computations less favorable than knowledge-base supported estimation methods for conceptual system design and synthesis. Estimation-based conceptual trade-off tools are not intended to replace detailed simulation, but to provide support for early (conceptual) specification design when simulation may not be practical.

The next section of this article outlines the role of conceptual design and trade-off analysis in a concurrent design environment, and describes the trade-off analysis tool developed by the authors. Section 3 describes a portion of a trade-off study performed for an MCM containing peripheral or area array bonded components.

2. The Packaging Design Environment

Many studies have appeared that provide comparisons of packaging technologies, [1], [2] and many others. While these studies contain a great deal of useful information, attempting to draw general conclusions about the applicability of one technology over another from them is not straightforward. The problem is rarely confined to just the selection of packaging technologies for a module or just the selection of a chip technology or system architecture. Effective trade-off analysis includes chip design through whole system design with all phases of the design process treated concurrently. A design/life-cycle environment vision which encompasses these concepts is shown in Figure 1.

Figure 1 is a datacentric representation of the concurrent design and life-cycle of a complex electronic system. The designer begins at the center with very little data about the system; the concentric rings represent

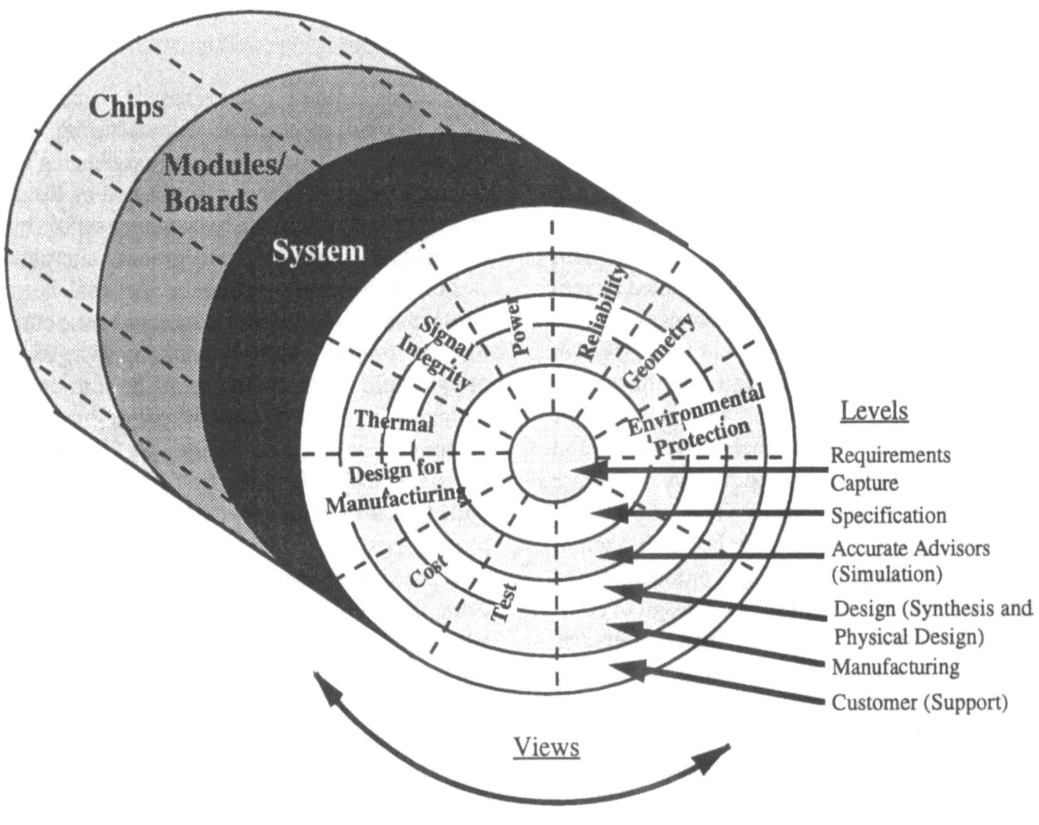

Fig. 1. Datacentric model for the concurrent design and life-cycle of a complex electronic system.

the *levels* of activity in the design, production, and life-cycles (i.e., the design progression and increased system knowledge). The specification level formulates an executable system specification from the requirements and is the focus of conceptual design. The rings in the model are divided into sections representing the performance oriented *views* which must be considered in the design/life-cycle.[1] The pie shaped elements cover all the rings, indicating that each subject should be considered at all levels of design and production.

2.1. System Synthesis

Design synthesis is the process of producing an artifact that satisfies some high level behavioral and structural specification. Synthesis activities can be placed in perspective using a Y-chart [3]. The Y-chart was originally intended to demonstrate the behavior of VLSI component design tools but can be extended to illustrate system level tasks as shown in Figure 2 [4]. The axes represent three design perspectives: 1) Functional (or behavioral)—the input/output of the system; 2) Structural—the interconnection of components to realize the required functionality; and, 3) Physical—the physical manifestation of the components and interconnections used to realize the required functionality. Along each axis, multiple levels of abstraction, or levels of detail, are represented.

In general, synthesis activities map perspectives from one axes to another in a counter clockwise direction. Synthesis activities near the center of the Y-chart are generally better defined and more mature than those farther from the center. The synthesis process treated in this article is the selection of packaging and interconnection technologies from a performance specification and a component set.

2.2. A Trade-Off Analysis Software Tool

A software tool for enhancing the manufacturability and decreasing the design risk associated with the selection of packaging technologies for integrated circuits has been developed. MSDA (Multichip Systems Design Advisor)[2] is designed to concurrently compute physical (size, weight, interconnect routing requirements), electrical (delays, noise), thermal (internal and external thermal resistances), reliability, testability, and cost performance metrics for multichip systems. Technologies

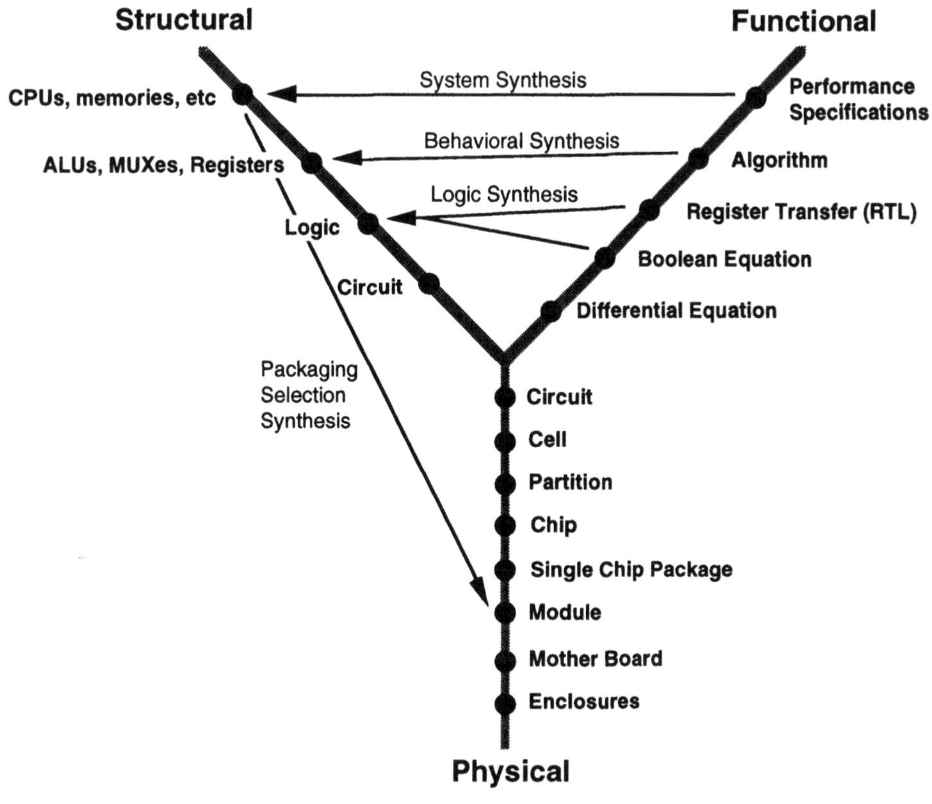

Fig. 2. Synthesis tasks represented on a Y-chart. Common synthesis activities are plotted as mappings on the chart.

treated by MSDA include traditional and fine-line printed circuit boards, low temperature co-fired ceramic, and thin film. Multiple design rule sets for each of these technologies are included in the tool; however, users may easily customize the properties or add their own proprietary technologies. Component assembly approaches include wirebonding, TAB, flip chip, and single chip packages. Materials are also available for bare die attach, encapsulation, heat exchanger attach, and for defining the bonding and interconnect technologies.

Figure 3 shows a block diagram of the MSDA tool. The tool begins by capturing design and performance budgets and constraints. Next, the components (active and passive) to be packaged are described. Component descriptions include the entry of physical (dimensions), material, I/O (number of signal, power, ground, unused), electrical (bias level, logic swing, simple driver and receiver descriptions, or driver and receiver SPICE models), thermal (allowable junction temperature, power dissipation), and manufacturing (cost, tested yield, burned-in yield, test and burn-in costs). All of this information is included in the "chip" class. The chip class also contains details of the type of bonding, die attach material, encapsulation, and single chip package (if there is one) associated with a particular component. In addition to these properties, the characteristics of heat spreaders and thermal vias are captured as properties of each component rather than the module or board so that they can be easily manipulated on a per component bias.

After the components are described to the tool, they are placed into "partitions." Partitions consist of any subset of the defined components grouped into a few chip pacakge, MCM, or printed circuit board. It is through the creation of partitions that complex systems can be built using the tool. The characteristics of partitions include interconnect technologies, heat exchangers, and connectors. A netlist corresponding to the partition under design, a group of partitions, or a design larger than that under consideration is an optional input to the trade-off tool; however, often conceptual design takes place prior to the synthesis of a netlist. The partitions are designed to be hierarchical and organized in a tree structure so that they may be placed inside of each other to build systems and study trade-offs associated with alternative distributions of components among system interconnects. The hierarchical nature of partitions is possible because an integrated circuit and a completed module share many identical physical properties and can often be treated in the design of a system in a similar manner.

Besides providing concurrent performance estimation through closed form algorithms, the MSDA tool supplements its internal analysis by linking to detailed simulation through model building and simulator management activities. Detailed simulation links are presently used as part of the electrical performance analysis and reliability evaluation in the tool.

The analysis of critical net delays is one example of the fusing of estimation level analysis and simulation, and the effective use of simulators in conceptual design activities. The evaluation of critical net delays at the conceptual design level consists of the formulation of a "packaging delay" that includes time-of-flight, RC charging, and contributions from reflections. Packaging delays plus noise contributions, along with the delays associated with gates, receivers, and memory access times, determine the maximum operating frequency of multichip systems.

The process by which critical net delays are evaluated in the trade-off tool is shown in Figure 4. To evaluate delays in the MSDA tool, the designer draws a critical connection on a placement editor used to capture conceptual level layout information. The net may have multiple loads, bends, vias, and stubs. Next, the net path and connections are processed by the tool to form a model of each net segment for use in a detailed electromagnetic simulator. The simulator computes inductance and capacitance matrices for the segment, which are post processed by MSDA into SPICE subcircuits. After all the segments are modeled, the subcircuits are combined along with driver and receiver models (either input with the component descriptions in subcircuit form or defined using Thevenin equivalent circuits from user supplied characteristics) and the whole net is analyzed using SPICE. All of the simulation management and post processing is handled by the trade-off tool and performed concurrently with size, routing, cost, reliability, testability, and thermal analyses.

While a number of efforts have been directed toward the conceptual design and specification of chips [5], [6], little work has been directed at the broader module/board and system level problem. Tools have been developed that give advice on specific "views" such as reliability, manufacturability, or power distribution (see [7]-[9] for examples). But only one other effort is known to be directed at concurrent interdisciplinary module/board level design [10].

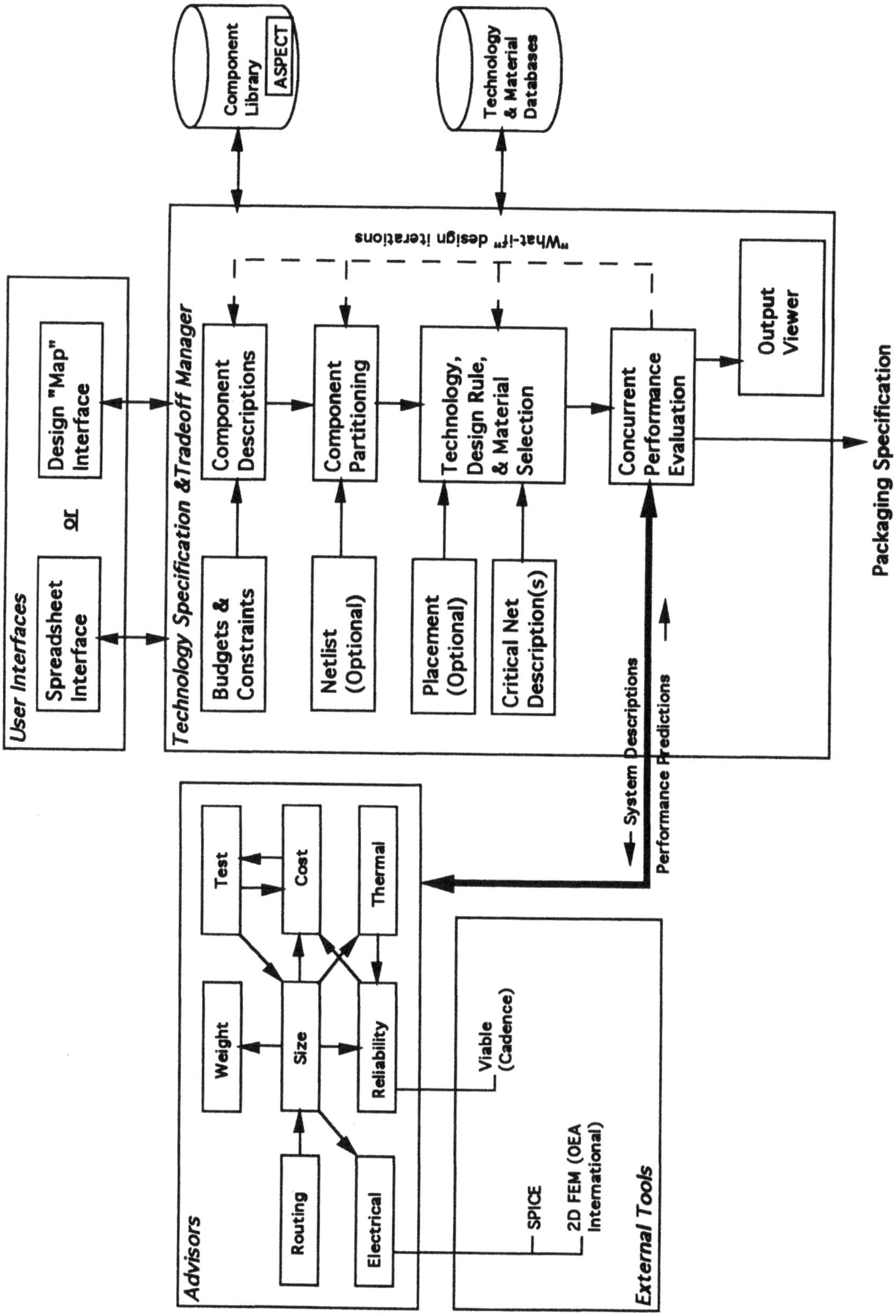

Fig. 3. Block diagram of the Multichip Systems Design Advisor (MSDA) trade-off analysis tool.

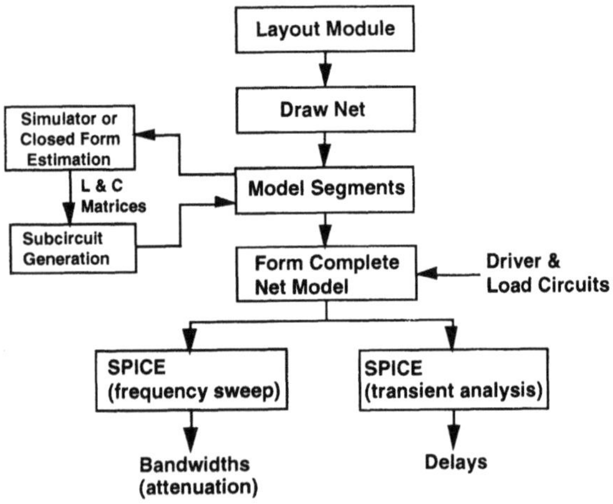

Fig. 4. Critical net delay and bandwidth (attenuation) evaluation. A segment is defined in the model as the length of line between two nodes (i.e., no internal bends, taps, or vias).

3. Example Results

Trade-off studies for numerous multichip systems performed using the tool described in the last section have been published previously [11]–[13]. The trade-off analysis presented in this section examines the trade-off between peripheral I/O format die (for wirebonding, tape automated bonding, or peripheral flip chip bonding) and area array I/O format die (for flip chip bonding), as a function of partitioning a fixed functionality into a variable number of die. The comparison in this study has been made in the context of a multichip module (MCM). The results presented in this article focus on a comparison of cost (including module assembly, test, and rework) to assess the overall applicability of one bonding format over the other. Additional trade-offs (electrical, thermal, and size) for this comparison of peripheral and area array bonding are presented in detail in [14].

The partitioning of functionality into chips is a fundamental design concern for high performance systems. The challenge is to determine the optimal number of chips for a given functionality, based on a cost/performance trade-off. Putting a large amount of functionality into a single chip may provide electrical performance and system size advantages, but often results in large die with low yields and high costs. Alternatively, realizing the same functionality using a large number of small die, results in less costly die, at the expense of a larger system size and possible performance degradation. In addition to cost and performance issues, the testability of a single large die is often more complex than that of several small die that perform the same function.

The optimal integration density (number of logic gates or memory bits per chip) depends on a large number of tightly coupled design goals, including: cycle time, noise, power dissipation, cost (yield), physical size, and testability. Mainframe computers have traditionally been built using a large number of bipolar chips containing a relatively small number of gates that consume a large amount of power. Personal computers are built using highly integrated (usually single) CPU systems with much smaller power dissipations, lower costs, and poorer performance than high-end systems. These market segments are making functional partitioning trade-offs.

In light of these trade-offs, the objective of this study is to determine the optimal number of chips necessary for partitioning a given functionality into, as a function of peripheral and area array die formats. In this article we are concerned with multichip modules (MCMs). Direct chip attach methods used in MCMs include peripheral bonding approaches such as wirebonding and tape automated bonding (TAB), and area array approaches such as flip chip or array TAB (ATAB). Peripheral approaches require all of the die I/O to be in a single (or at most double) row around the perimeter of the die. Area array approaches distribute the die I/O over the entire face of the die. In general, flip chip approaches result in smaller systems (since die can be placed closer together) and higher performance systems (since the effective inductance associated with the chip bonds is reduced and line drivers can potentially be smaller). On the other hand, area array bonding requires extensive die preparation (today, this is only economical before wafers are diced) and the infrastructure to support peripheral bonding is considerably greater than that for area array (i.e., very few IC manufacturers are producing area array format die).

In order to assess the impact of peripheral versus area array bonding on a system, a concurrent analysis of system size, performance, and cost including chip cost, test, and rework must be performed. Analyzing system costs while neglecting the chips will not generally lead to a correct relative cost comparison between packaging alternatives. Similarly, analyzing chip costs to meet a specified functionality without considering the cost of packaging and the costs associated with test and rework can be equally misleading; i.e., a selected functional partitioning may appear economical at the chip level but could result in a more expensive system once test and rework costs are accounted for.

3.1. Analysis Approach

The analysis approach employed for this study makes use of a combination of new and existing models and algorithms implemented within the MSDA tool. Figure 5 shows the process used to determine the overall module cost for various partitionings and bonding approaches. The die characteristics were predicted using the models from SUSPENS [15]. Our implementation of these models accepts the total number of gates and chips into which to equally divide the gates as inputs (the gates are assumed to be indistinguishable). The model computes the number of signal and control I/O from Rent's rule and the die core size using an estimation of the average wire length which assumes a die core size that is interconnect capacity limited. The number of power and ground I/O that a chip requires is application specific; however, for this study we have assumed constant values for the signal I/O to ground I/O ratio: 4.2 for peripheral bonding cases (DEC Alpha chip [16]), and 6.0 for area array cases (an equal number of power and ground I/O are assumed). A larger signal to ground ratio is allowed for area array bonding due to the lower effective inductance associated with flip chip bonding (due to shorter bond lengths). With the die core size and the total number of I/O determined, the final die size can be computed. The size estimations for peripheral and area array die are similar to the formulations developed in [17]. For a peripherally bonded die, the area is given by the maximum of two limitations:

$$\textit{peripheral area } 1 = (2length_{pad} + pitch_{pad} \lfloor N_p/4 \rfloor)^2 \quad (1a)$$

$$\textit{peripheral area } 2 = N_p width_{pad} length_{pad} + (1 + kN_p) area_{core} \quad (1b)$$

where

- $length_{pad}$ = length of a peripheral bond pad on the die
- $width_{pad}$ = width of a peripheral bond pad on the die
- $pitch_{pad}$ = minimum center-to-center pitch of peripheral bond pads on the die
- N_p = total number of die I/O
- k = the fractional increase in the core die area necessary to accommodate redistribution of one I/O to the periphery of the die
- $area_{core}$ = core die area.

Equation (1a) is the I/O limited chip area and Equation (1b) is the peripheral redistribution limited area, both assume a single row of bond pads. For an area array bonded die, the die area is given by the maximum of two limitations:

$$\textit{array area } 1 = (pitch_{pad} \lceil \sqrt{N_p} \rceil)^2 \quad (2a)$$

$$\textit{array area } 2 = N_p length_{pad} width_{pad} + area_{core} \quad (2b)$$

where

- $length_{pad}$ = length of an area array bond pad on the die
- $width_{pad}$ = width of an area array bond pad on the die
- $pitch_{pad}$ = minimum center-to-center pitch of area array bond pads on the die.

Equation (2a) is the I/O limited chip area and Equation (2b) is the bond pad area limitation, assuming that active circuitry cannot be placed under the bond pads. Subtracting Equation (2b) from Equation (1b) yields the size gradient results derived in [17], with the exception of the "outer core" redistribution term, which is accounted for in our derivation by the Equation (1a) limitation.

After the final die size is determined, the cost of the die is computed. The yield of die on the wafer was computed using Murphy's yield law [18] with a fixed defect density,

$$yield_{wd} = \left(\frac{1 - e^{(-area_{die} D)}}{area_{die} D} \right)^2 \quad (3)$$

where D is the average defect density (defects per unit area) on the wafer. The number-up on the wafer was computed assuming a fixed minimum spacing between die (50 mils). In order to compute the die yield and cost at the beginning of the assembly process, the methodology shown in Figure 6 was followed.

The die yield after wafer test is computed from the test coverage (fraction of the defects identified in the test) and the actual yield of the die on the wafer,

$$yield_{die} = yield_{wd}^{(1 - \textit{test coverage})} \quad (4)$$

Equation (4) is based on the Williams and Brown model [19]. After sawing, the known defective die are scrapped and the rest are sent on to burn-in. The fraction of die which continue to burn-in and test is given by,

$$\textit{pass fraction} = yield_{wd}^{\textit{test coverage}} \quad (5)$$

The burn-in step is characterized by a cost and a fixed fraction of die which fail during burn-in. The die

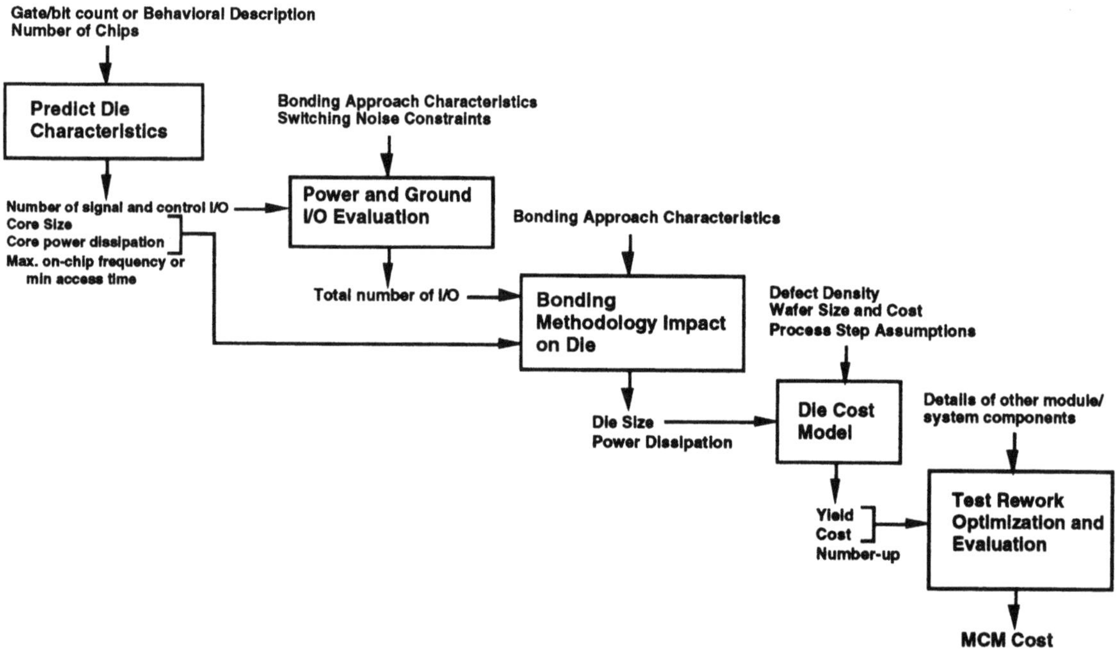

Fig. 5. The methodology used to determine the Multichip Module (MCM) cost for various partitioning and bonding approaches.

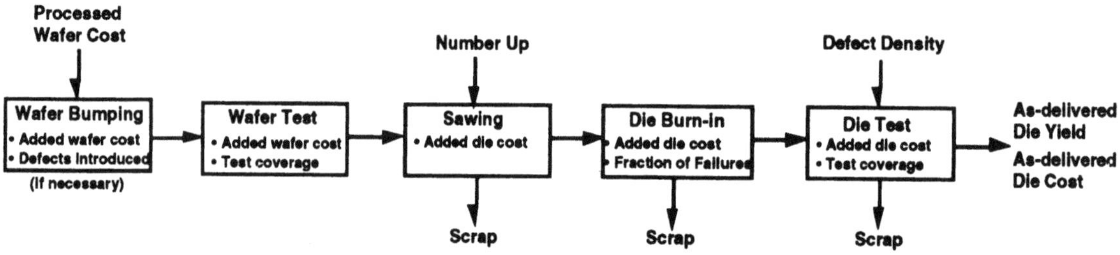

Fig. 6. Simple die cost model.

test step is treated using Equations (4) and (5) and replacing $yield_{wd}$ with the incoming die yield. The wafer or die cost is computed at the end of each step shown in Figure 6 using the relation,

$$cost_{cumulative} = \frac{cost_{previous\ steps} + cost_{step}}{pass\ fraction} \quad (6)$$

The characteristics of the rest of the module (size, type of substrate, wiring capacity, etc.) are estimated using the trade-off analysis tool described in Section 2.2 of this article. Included in this analysis is a detailed estimation of module costs, along with the cost of assembly, test, and repair as discussed in [20].

3.2. Results

MCM cost results for a conventional chip in which there is negligible unused die area are presented in this section. In all cases the chip (or gate set) that was studied was assumed to be part of a processor module that included an additional large chip (an ASIC of some type) and ten SRAMs. The characteristics of the chips in the module are given in Table 1.

To begin the analysis, we fix the number of signal I/O of the CPU chip to 527 and adjust Rent's rule so that there is no empty area inside the peripherally bonded chip. The number of gates in the chip in this case is approximately 2,388,000 (0.7 μm design rules). The results for this case are then generated by fixing the number of gates in the CPU at the above value and varying the number of chips used to realize them.

The results of this analysis are shown in Figures 7 through 11. Figure 7 shows how the number of signal and total I/O per chip varies as the given functionality is divided into various numbers of die. The number of signal I/O is the same in the peripheral and area array

Table 1. Characteristics of the test module used for the example in this article. The yield corresponds to the yield of the die at the start of assembly (i.e., at the end of the process shown in Figure 6). The third chip (CPU) is the functionality which is the subject of the partitioning exercise, the numbers in parenthesis indicate its characteristics in single chip form, where p = peripheral and aa = area array. The critical assumptions used to generate the results in this article are included in Table 2.

Chip	Quantity	Number of Signal and Control I/O	Total Number of I/O	Die Size (mils)	Power Dissipation (W)	Yield (fraction)	Cost ($)
SRAM	10	62	100	470 × 300	1.5	0.9972	29.63
ASIC	1	484	750	600 × 600	50	0.9967	80.77
CPU	Variable (1)	(527)	(781 p) (701 aa)	(790 × 790 p) (744 × 744 aa)	(58.4)	(0.9968 p) (0.9970 aa)	(169 p) (176 aa)

Table 2. Critical analysis assumptions used to generate the trade-off results presented in this article.

Property	Value
Signal/Ground (peripheral)	4.2
Signal/Ground (area array)	6.0
Bond Pad Size (peripheral)	2.5 × 3 mils
Bond Pad Size (area array)	5 × 5 mils
Min. Bond Pad Pitch (peripheral)	4 mils
Min. Bond Pad Pitch (area array)	10 mils
Wafer Defect Density	3 defects/square inch
Processed Wafer Cost	$800
Wafer Diameter	6 inches
Unusable Wafer Boarder	0.4 inches
Minimum Space Between Die on Wafer	50 mils
Wafer Bumping Cost (per wafer)	$200
Defects Added by Wafer Bumping	0.2 defects/square inch
Wafer Test Cost	$3.00 per die
Wafer Test Coverage	80%
Die Burn-in Cost	$2.60 per die
Induced Failures at Burn-in	10%
Die Test Cost	$3.00 per die
Die Test Coverage	99%
Substrate Yield	99.999%
Assembly Yield	99.995%
Repair Yield	80%
Module Test Cost	$75.00 per module
Module Test Coverage	95%
Die Logic Type	CMOS
Min. Feature Size on Die	0.7 μm
Quantity	50,000 modules

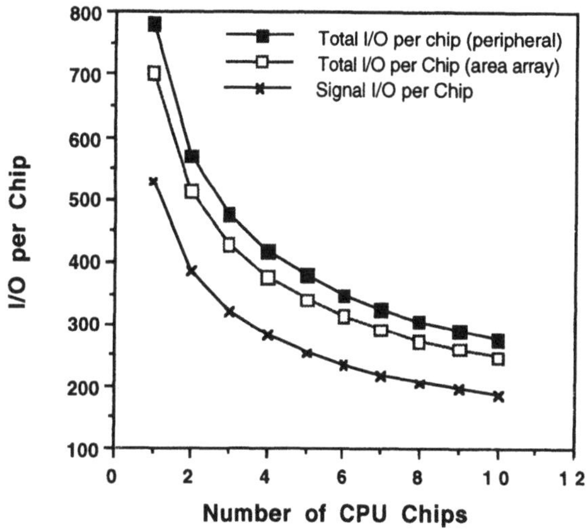

Fig. 7. I/O per CPU chip as a function of the number of chips the CPU is divided into. Signal I/O refers to all I/O which are not powers or grounds.

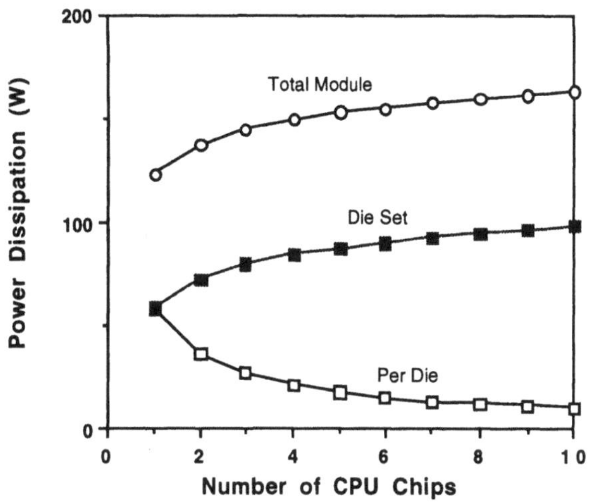

Fig. 8. Power dissipation per CPU chip and total power dissipation for the set of chips the CPU is divided into (the power dissipations of all other components in the MCM are included in the Total Module curve).

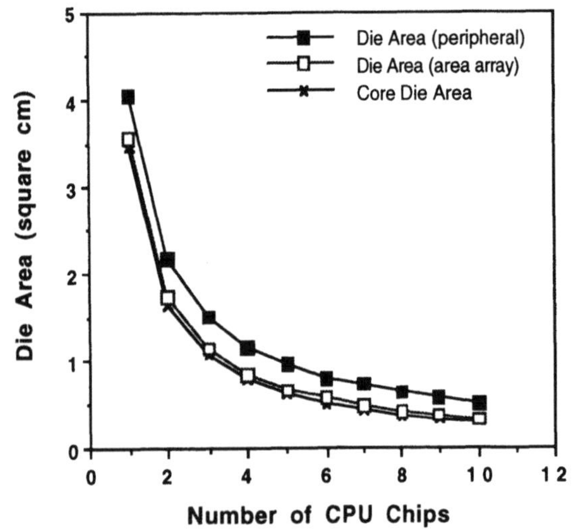

Fig. 9. Single CPU die area and core die area as a function of the number of chips the CPU is divided into.

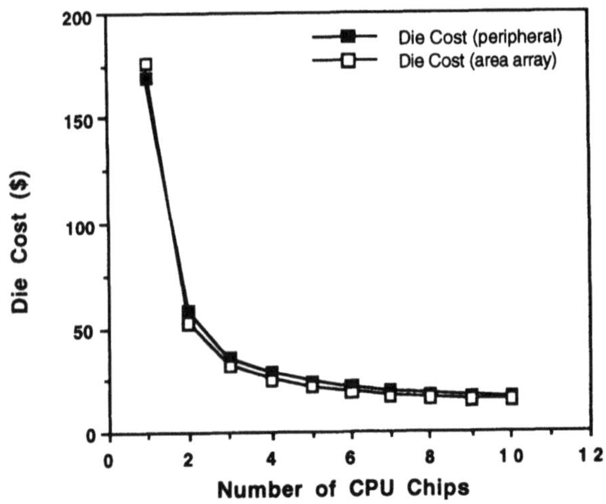

Fig. 10. Single die cost as a function of the number of chips the CPU is divided into.

cases and the number of total I/O in the area array case is less than the peripheral case (as previously discussed). Figure 8 shows the power dissipation per chip (Per Die) and the total power dissipation of all the chips used to realize the CPU (Die Set). A chip's power dissipation is proportional to its number of gates, so power dissipation drops as the number of gates drops. The power dissipation per gate, however, increases as the size of the die decreases, since the average dynamic power dissipation is proportional to on-chip frequency, and the maximum on-chip frequency is increasing due to smaller die areas. The result is a net increase in the power dissipation of the system as the number of chips increases (assuming that we will want to run the chips faster).

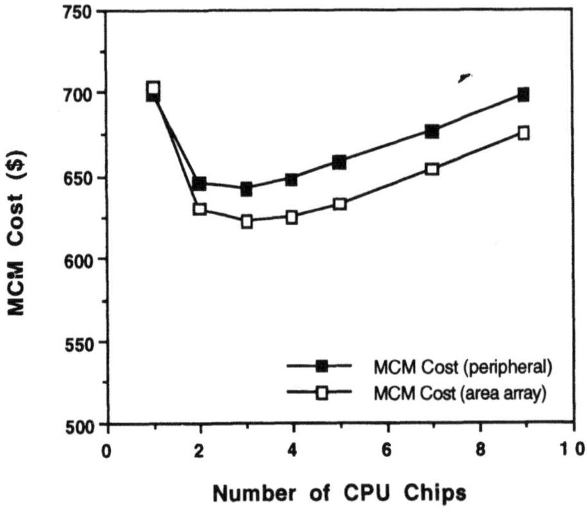

Fig. 11. Completed MCM cost. The MCM cost includes test and repair and the costs associated with all components (see Table 1).

Figure 9 shows the die area variation. Both the peripheral format chip and the area array format chip track the core area. The area array chip is smaller than the peripheral chip due to the extra area required for the redistribution of connections from inside the core of the die to its periphery. As the I/O per area decreases, the percentage of the die area necessary for redistribution decreases and the peripheral and area array format chips limit to the same size.

Figures 10 and 11 summarize the cost results. Figure 10 shows that the difference in cost per die for peripheral and area array format die is minimal. When only one die is used, the peripheral die cost is actually less than the area array die cost because the cost of bumping the wafer containing the area array die is amortized over relatively few working die per wafer (due to larger die size and therefore fewer die/wafer). As the number of working die per wafer increases, the cost savings for having slightly smaller area array die than peripheral die, more than offsets the bumping cost incurred by each die. For this study, our primary interest is not the cost of a single die but the cost of the completed multichip module and when assembly, test, and repair effects are considered.

Figure 11 shows the MCM cost as a function of the number of chips into which the functionality is divided. In all cases an 11 layer (3 mil lines and spaces) low temperature cofired ceramic (LTCC) interconnect was assumed. For this application, the cost of the chips is dominant ($473–$550), the cost of the LTCC interconnect is approximately $45 in all cases, and the rest of the MCM cost in Figure 11 is assembly, repair, and testing. The figure shows that for this application, the optimal number of chips into which to divide the functionality, from a cost standpoint is three. With the exception of using one chip, the area array approach results in lower MCM costs for all partitionings than the peripheral case. The results in Figure 11 are application specific. An application where the die do not contribute as great a fraction of the total module cost could display different characteristics.

Cost may be the greatest driving factor in many MCM designs, but it cannot be considered in isolation. We must also consider the size, electrical, and thermal performance of the resulting MCM. Several quantities used in this study have been treated as constants, when in reality, they are a function of the die size and its I/O count. These quantities include the wafer test cost (assumed to be a function of the number of I/O), wafer test coverage, die test cost, assembly yield, and repair yield. Exensions to this work should concentrate on formulating functional relations for these quantities in order to obtain more accurate comparison results.

4. Summary

This article described a software tool that performs trade-off analysis for multichip modules and systems. The tool operates at the conceptual design level prior to traditional simulation and physical design activities. The present version of the tool predicts the physical, electrical, thermal, reliability, test, and cost performance of a multichip system through a combination of estimation and simulation techniques. The use of trade-off analysis can reduce design time significantly and increase the quality of complex systems.

Notes

1. Alternatively the model could be divided into more general system views such as: Architecture/Algorithm, Physical, Software, Performance, Test/Verification, Manufacturing/Cost, and Documentation.
2. Also previously referred to as SPEC (System Performance Evaluation and Comparison).

References

1. C.A. Neugebauer and R.O. Carlson, "Comparison of Wafer Scale Integration with VLSI Packaging Approaches," *IEEE Transactions on Components, Hybrids, and Manufacturing Technology*, CHMT-10, pp. 184–189, June 1987.
2. V.K. Nagesh, D. Miller, and L. Moresco, "A Comparative Study of Interconnect Technologies," *Proc. Ninth Annual International Electronics Packaging Symposium*, pp. 433–443, Sept. 1989.
3. D.D. Gajski and R.H. Kuhn, "Guest Editors' Introduction: New VLSI Tools," *Computer*, 16, pp. 11–14, Dec. 1983.
4. W.P. Birmingham, A.P. Gupta, and D.P. Siewiorek, *Automating the Design of Computer Systems, The MICON Project*, Jones and Bartlett Publishers, Boston, MA, 1992.
5. A.M. Dewey and S.W. Director, *Principles of VLSI System Planning: A Framework of Conceptual Design*, Kluwer Academic Publishers, Boston, MA, 1990.
6. J. Bortolazzi and K.D. Mueller-Glaser, "Towards Computer Aided Specification of Analog Components," *Proc. IEEE Custom Integrated Circuits Conference*, pp. 10.7.1–10.7.4, May 1990.
7. P. Edmond, A.P. Gupta, D.P. Siewiorek, and A.A. Brennan, "ASSURE: Automated Design for Dependability," *Proc. of the 27th Design Automation Conference*, pp. 555–560, June 1990.
8. D.L. Hilbert, "The Integrated Product Engineering Expert System (IPEX) a Design/Producibility Advisor," *Proc. of NEPCON-West*, pp. 157–163, Feb. 1992.
9. P.A. Sandborn and H. Hashemi, "A Design Advisor and Model Building Tool for the Analysis of Switching Noise is Multichip Modules," *Proc. International Symposium on Microelectronics (ISHM)*, pp. 652–657, Oct. 1990.
10. J.P. Krusius, "System Interconnection of High Density Multichip Modules," *Proc. International Conference on Advances in Interconnects and Packaging*, SPIE vol. 1390, pp. 261–270, Nov. 1990.
11. P.A. Sandborn, "A Software Tool for Technology Trade-off Evaluation in Multichip Packaging," *Proc. Eleventh International Electronics Manufacturing Technology Symposium*, pp. 337–341, Sept. 1991.
12. P.A. Sandborn, "Technology Application trade-off Studies in Multichip Systems," *Proc. 1st International Conference on Multichip Modules*, pp. 150–158, April 1992.
13. P.A. Sandborn, H. Hashemi, and L. Bal, "Design of MCMs for Insertion into Standard Surface Mount Packages," *Proc. NEPCON-West*, pp. 651–660, Feb. 1993.
14. P.A. Sandborn, M. Abadir, and C. Murphy, "A Partitioning Advisor for Studying the Trade-off Between Peripheral and Area Array Bonding of Components in Multichip Modules," *Proc. Fifteenth International Electronics Manufacturing Technology Symposium*, pp. 271–276, Oct. 1993.
15. H.B. Bakoglu, *Circuits, Interconnections, and Packaging for VLSI*, Addison-Wesley Publishing Company, Reading, MA 1990.
16. D.W. Dobberpuhl, et al., "A 200-MHz 64-b Dual-issue CMOS Microprocessor," *IEEE J. of Solid-State Circuits*, vol. 27, pp. 1555–1567, Nov. 1992.
17. P.H. Dehkordi and D.W. Bouldin, "Design for Packagability: the Impact of Bonding Technology on the Size and Layout of VLSI Dies," *Proc. of the IEEE Multichip Module Conference*, Santa Cruz, CA, pp. 153–159, March 1993.
18. B.T. Murphy, "Cost-size Optima of Monolithic Integrated Circuits," *Proc. of the IEEE*, vol. 52, pp. 1537–1545, Dec. 1964.
19. T.W. Williams and N.C. Brown, "Defect Level as a Function of Fault Coverage," *IEEE Trans. on Computers*, C-30, pp. 987–988, Dec. 1981.
20. M. Abadir, A. Parikh, L. Bal, C. Murphy, and P. Sandborn, "High Level Test Economics Advisor (Hi-TEA)," *Journal of Electronic Testing*, 5, pp. 195–206, 1994.

Peter A. Sandborn received the B.S. degree in engineering physics from the University of Colorado, Boulder, in 1982, and the M.S. degree in electrical science and the Ph.D. degree in electrical engineering from the University of Michigan, Ann Arbor, in 1983 and 1987, respectively. His dissertation research concerned the physics and modeling of GaAs MESFETs.

He joined the Microelectronics and Computer Technology Corporation (MCC) in 1987 and is presently a Senior Member of the Technical Staff in MCC's High Value Electronics Divison and the technical leader of the Multichip Systems Design Advisor Project. Prior to joining the High Value Electronics Division in November 1988 he was in the Analog Technology Laboratory of the Computer Aided Design Program at MCC. While in the Analog Technology Laboratory he worked on the application of non-stationary transport models to modeling submicron silicon and GaAs device structures and developed a large-signal time-domain transient device simulation platform oriented toward the development of equivalent circuit models for transistor structures. Since joining the High Value Electronics Division he has worked on power distribution problems in multichip packaging systems, CAD software for packaging and interconnect applications. Dr. Sandborn is the principle developer of the Multichip Systems Design Advisor trade-off analysis software.

Dr. Sandborn is the author of a book entitled "Conceptual Design of Multichip Modules and Systems," Kluwer Academic Publishers, 1994, and is the principle author or co-author of over 30 technical publications in the areas of semiconductor device simulation and packaging and interconnect.

Rajarshi Ghosh received his MS (EE) from the University of Texas at Austin and his B.E. from the University of Calcutta.

Mr. Rajarshi Ghosh is a Member of the Technical Staff with MCC's High Value Electronics (HVE) Division. He has worked at MCC since 1988. He has been working for the Multichip Systems Design Advisor (MSDA) Project since its inception in 1990 as the project's software developer. He is responsible for MSDA's software architecture and user interface. Prior to this assignment he worked for the Quick Turnaround Interconnect (QTAI) Project as a software developer. His contributions to that were in developing the Graphical User Interface for the QTAI software and authoring a design-data conversion and manipulation program.

Ken Drake received his Ph.D. in EE from the University of Michigan, an M.S. in EE and an M.S. in BioE from the University of Michigan and University of Miami, respectively, and his B.S. from Oakland University.

Dr. Drake is a Project and Business Development Manager at the Microelectronics and Computer Technology Corporation (MCC), Austin, TX. Since May 1990, he has successfully defined, launched, and managed both commerical- and government-sponsored programs involved in electronic design CAD software development and industry alliances of EDA vendors and electronic manufacturers. With over 18 years of experience, Dr. Drake is a seasoned personnel, financial, and technical manager of small to medium size publications in the area of advanced electronic packaging, electronic design, and microsensor technologies.

Prior to MCC, Dr. Drake was the Head, Electronic Devices and Packaging Technology Department at ERIM, an R&D institute specializing in image processing hardware and software development for DoD and commercial applications. Before ERIM, Dr. Drake was the President of Integrated Microsystems Inc., which specialized in silicon microsensor technology. Prior to Integrated Microsystems, Inc., he was a senior research engineer with Cordis-Dow Corporation where his specialty was clinical R&D of biomedical devices. Dr. Drake served as a Naval officer for four years and has remained active in the Naval Reserve program with the current rank of Commander.

Magdy S. Abadir was born in Cairo, Egypt, in 1956. He received the B.S. degree with honors in Computer Science from the University of Alexandria, Egypt in 1978, the M.S. degree in Computer Science from the University of Saskatchewan, Saskatoon, Canada, in 1981, and the Ph.D. degree in Electrical Engineering from the University of Southern California, Los Angeles, in 1986. Prior to joining the Microelectronics and Computer Technology Corporation (MCC) in May 1986, he was a research associate at the University of Southern California. He is currently a senior member of the technical staff in the High Value Electronics Division at MCC where he is involved in a number of projects associated with testing of multichip modules, test economics, Design For Test Synthesis, and the development of computer-aided software tools to support these various test activities.

Dr. Abadir has cofounded and chaired a series of International Workshops on the economics of design, test and manufacturing. He is a coeditor of a book entitled "Economics of Design and Test of Digital Circuits and Systems," Ellis Horwood, 1992. He has published over 45 technical journal and conference papers in the areas of test economics, design for test, computer-aided design, high-level test generation, design verification and layout verification. His current research interests include test economics, design for testability synthesis, hierarchical test generation, trade-off analysis, and optimization tools.

Linda Bal of Eastman Kodak joined MCC's High Value Electronics Division as a Visiting Researcher in June of 1991. She is a member of the Multichip System Design Advisor Project team. At Eastman Kodak, Ms. Bal is a Senior Development Engineer in the Research and Engineering Department of the Electronic Products Business Unit. Since joining that group in July of 1990, she has designed Low Temperature Cofired Ceramic and Thick Film Multi-layer Hybrid Packages.

Prior to joining packaging engineering, Ms. Bal spent two and a half years in the Electonic Products Test Engineering department at Kodak where she developed and subcontracted in-circuit test programs. Prior to test engineering, Ms. Bal was a manufacturing and development engineer in Kodak's Mass Memory Divison for Magnetic Peripheral Systems Products.

Ms. Bal received a B.S. in Electrical Engineering from Purdue University in 1985. While at Purdue, she completed the Cooperative Education Program by working for IBM in their San Jose Magnetic Recording Development Lab.

Ashish R. Parikh received the B.S. degree in Electrical Engineering from the University of Texas at Austin in 1991. He is currently working on his M.S. degree in Computer Engineering at the UT Austin.

He joined the Microelectronics and Computer Technology Corporation (MCC) in 1992 as a Research Assistant. He is presently an Associate Member of Technical Staff in MCCs High Value Electronics Division. He is a programmer on the Multichip Systems Design Advisor project. His current research interests include high-level testability analysis, test economics, software engineering, and object-oriented software systems.

JOURNAL OF ELECTRONIC TESTING: Theory and Applications, 5, 219–228 (1994)

Trade-off Analysis on Cost and Manufacturing Technology of an Electronic Product: Case Study

SHEKAR RAO, BERT HASKELL AND IAN YEE
Microelectronics and Computer Technology Corp., 12100 Technology Blvd., Austin, TX 78727

Abstract. A trade-off analysis on the cost and system packaging metrics of an electronic product aimed at the commercial/retail industry has been carried out. By comparing the system cost and packaging metrics with those of comparable consumer products, we have determined that there is opportunity for significant cost, size, and weight reduction of the overall electronics packaging system. These include the use of fine pitch IC packages, smaller discrete components, denser PCB wiring technology, double sided IC package surface mount, surface mount connectors, and improved plastics for the product housing. The analysis concluded that PCB area reduction of 40%, using a single PCB instead of three boards, reduction in board cost of over 50% and product weight reduction of over 28% are possible using available technologies.

Keywords: Benchmarking, redesign, trade-off, analysis of PCB (printed circuit boards).

1. Introduction

Performing a trade-off analysis on a product can be very beneficial for understanding and reducing manufacturing and final cost. At the very least, it helps to compare and hence benchmark the existing system metrics with those of other products currently available in the market and thus set the stage for incorporating more improvements with advanced technologies in newer models under development. Other advantages include understanding the evolution of technology and the current standards being used in similar products. MCC is currently analyzing the electronics packaging technology in consumer electronic products. We have accumulated extensive experience in the area of cost modeling and trade-off analysis on these products and has used the technologies to create a database and observe trends in different aspects of electronics packaging.

In this article, we have done a trade-off analysis on the electronics packaging technology used in a data input device for the retail industry (henceforth referred to as the *product X*). The product is used in commercial and retail outlets. In addition to the electronics packaging, we also examined the external housing and electromechanical assembly, and suggested ways in which cost and weight reduction could be achieved by using ligher materials found in other commercial products.

The main focus of the trade-off analysis is the electronics packaging—especially the printed circuit boards (PCBs), the discretes (small rectangular-shaped, passive chip surface-mounted devices resistor and capacitor components) and the connectors which link up the different PCBs used in the system. By comparing the present board metrics with those seen in other consumer and portable products like camcorders, notebook and palmtop computers, and cellular phones, we have identified potential savings in area and in cost if the metrics on the product were changed to match those of products with more advanced metrics.

2. Trade-off Analysis

Trade-off analysis enables the selection of appropriate packaging technologies to meet electrical, physical and economic requirements. After the definition of which technologies would be used to implement a particular product, calculations of major outputs, such as weight,

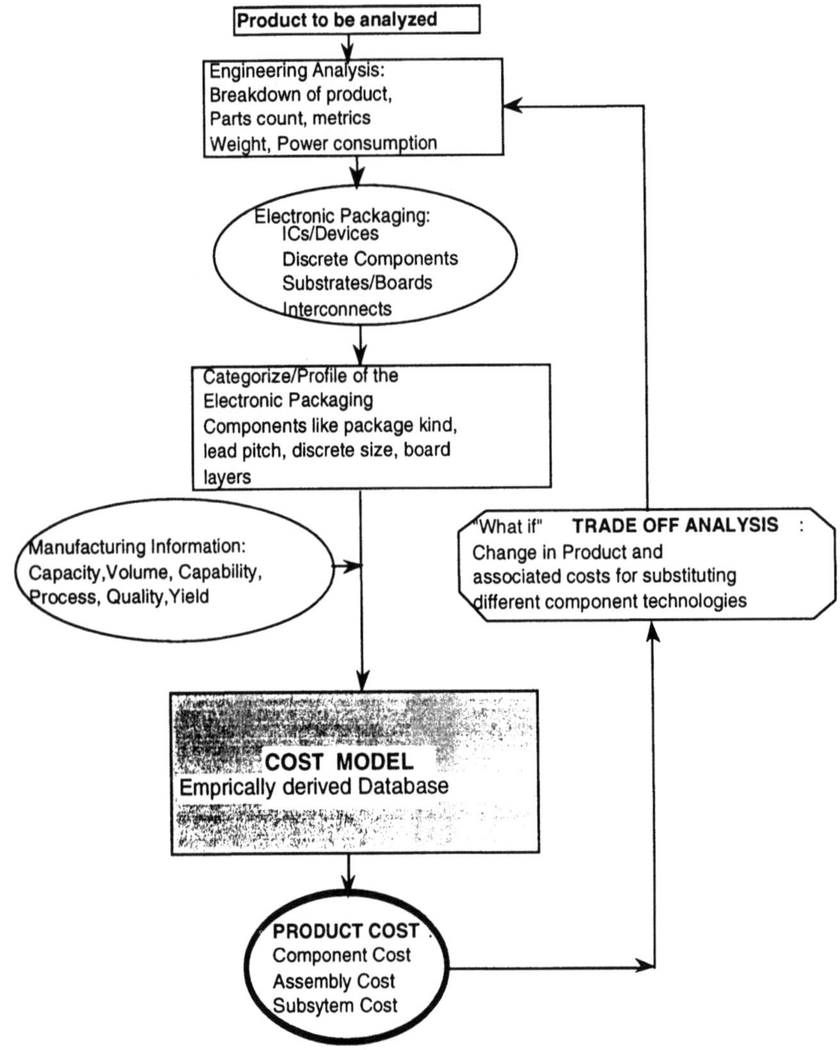

Fig. 1. Outline of Design/manufacturing/trade-off analysis tool.

volume, board area utilized, cost, and relative reliability index, could be summarized. This information can then provide a baseline for further revisions or iterations in order to optimize the technology selection for a given product. The key methodologies involved in doing a design/manufacturing trade-off analysis are outlined in Figure 1.

A number of trade-offs on the different components subsystems of the product under study are carried out. One example of this might be in choosing the optimum discrete component size to use in a palmtop computer. Given a list of chips to be used, the number and type of discretes, connection vehicles and format for partitioning the system, a user might find that using 0402 sized discretes may allow for the use of a smaller and lighter PCBs compared to the usage of 1208 discretes, but at an increased cost and reduced yield. Figures of merit would provide a semi-quantitative basis for comparison.

2.1. *Variables Used in Trade-off Analysis*

Table 1 lists the independent variables used in the cost model for the various components.

3. Product Trade-offs

3.1. *Product Description*

This product under analysis is small and compact and weighs less than two pounds. The electronics is contained

Table 1. Independent variables used in trade-off analysis.

Components	Independent Variables
Substrates	Substrate type: PCB, cofired ceramic, thin film Board size No. of layers Minimum line/space geometry No. and size of through-holes/vias Volume
ICs	Technology—Microprocessors, Logic, TTL, Memories Silicon area Number of leads Power dissipation
Discretes	Type Size
Assembly Cost	Technology: TAB, Wirebond, SMT No. of parts Type of parts Pitch
Connectors	Type of connectors (connection vehicle) Number of connections
Flex Circuits	Area Type: Cu/PI or PTF, 1 or 2 sided
LCD Panel	Type Area
Housing/Mechanical Parts	Number of parts Materials: plastic, metal, metallized plastic
Batteries	Type Capacity

on three PCB boards. The mother board (main PCB), a 6 layer PCB, contains most of the electronic control circuitry including the control ICs (Figure 2). The components are mounted using both SMT and through-hole. Two auxiliary boards are connected to the main PCB. Both of these boards are four layer PCBs. One of the auxiliary boards is an L-shaped power board and is connected to the mother board by means of a pin-and-socket connector while the other auxiliary board is connected by means of PTF (polymer thick film) and ZIF (zero insertion force) connectors. The L-shaped power board contains most of the power devices and circuitry that controls the electromechanical functions. The second auxiliary board, has two aluminum covers on either side—these serve to shield the components from electromagnetic interference (EMI) from nearby electro-optical devices.

3.2. Weight Analysis

The weight of the different components in product X is shown in Table 2. More than half of the weight is taken by the external housing. The electromechanical assembly accounts for slightly more than 25% of the rest of the weight of the product. The total system electronics comprising the three PCBs and some flex circuitry accounts for 17% of the total system weight. Table 2 shows a breakdown of the absolute and relative (%) weight of the different system components.

3.3. Benchmark of Product X Metrics

The connection density on the main PCB is 58 connection/sq. inch and a pin count of 4 pins/device. (Connection density is the total number of connections on the board divided by the PCB area while pin count is the total number of connections divided by the number of parts on the board.) The through-hole mounted components account for 6% of the total connections on the main board. On the power board, nearly 50% of the components are through-hole mounted with a connectin density of 25 connections/sq. inch and a pin count of 4 pins/device. The auxiliary board has a connection density of around 36 connections per sq. inch and pin count of 2 pins/device.

In Figure 3 we have plotted the connection density of the Main PCB of product X relative to typical main PCB densities of a range of consumer products. This graph shows the density performance of product X to be comparable to many existing consumer products. However, if the average pin (connection) count of the system is taken into account, the comparison is less favorable. The products with lower connection density than product X, such as notebooks and organizers, have high average connections counts typically around 9 to 10 connections per part. It is more appropriate to compare product X then to products such as camcorders, which have a similar average pin counts at around 3 to 4 pins per device. When compared to these camcorder products product X achieves less than half of the average connection density.

In Figure 4 we have plotted the estimated cost per connection for a number of products including the one under analysis. This is calculated by dividing the total estimated system electronics cost (including silicon, packaging, and assembly) by the number of connections in the system (IC, discretes, and connectors). The product X cost metric is 25% to 100% greater than

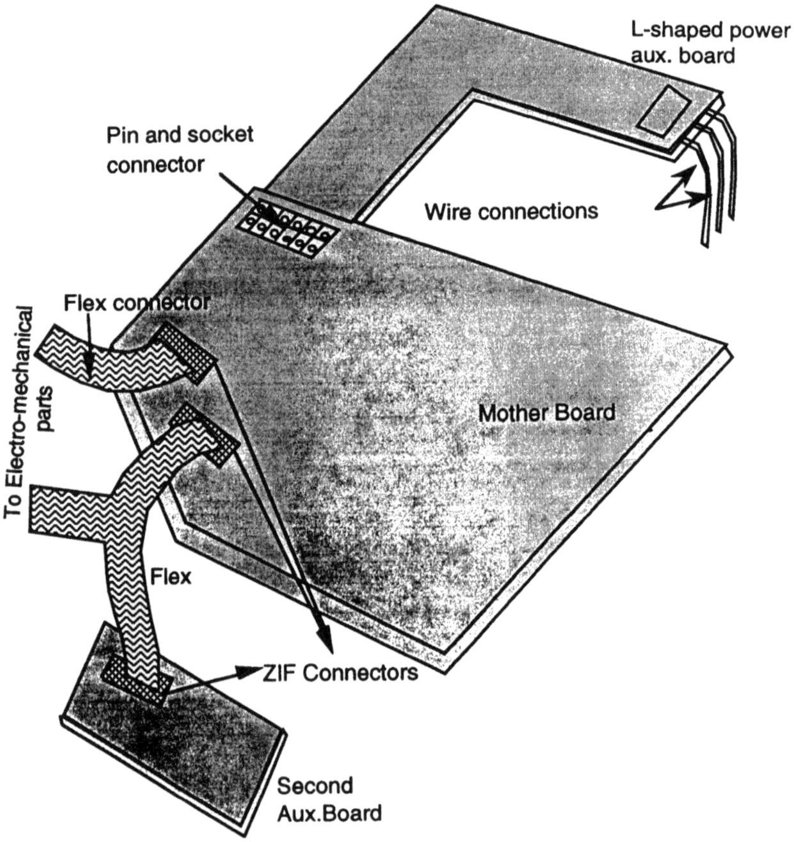

Fig. 2. Schematic of system electronics.

Table 2. Product X weight analysis.

Components	Absolute Weight (gms)	Relative Weight (%)
Housing	298.8	56.25
Main PCB	28	5.27
Power PCB	44.8	8.48
Auxiliary PCB	14	2.6
Electro-Mechanical assembly	145.6	27.4
Total	531.2	100

any of the camcorders. (Not actual manufacturers cost.) Based on this analysis, we conclude that the size of product X electronics as well as the component cost can be reduced by using electronics packaging technologies similar to those used in camcorders.

3.4. *Trade-off Analysis*

In this section we have made some first order estimates of how different packaging technologies might impact the characteristics of product X electronics.

The finest line/spacing dimension on the three boards of product X are 7/10 mils. The via pad diameter is 30 mils. The routing of conductors between through-holes is of the single conductor, low density kind. The total length of wiring required in all the three boards is 2795 inches. This is obtained by multiplying the wiring density estimates for each of the three boards by the area. A PCB cost model (developed internally at MCC) allowed us to estimate the cost of manufacturing a board using the above mentioned metrics and then to do a trade-off analysis by substituting technologies and metrics of other products in place of product X.

Let us first consider Table 3, which is a chart of the wiring density available from different PCB technologies as defined by their feature geometries. By multiply-

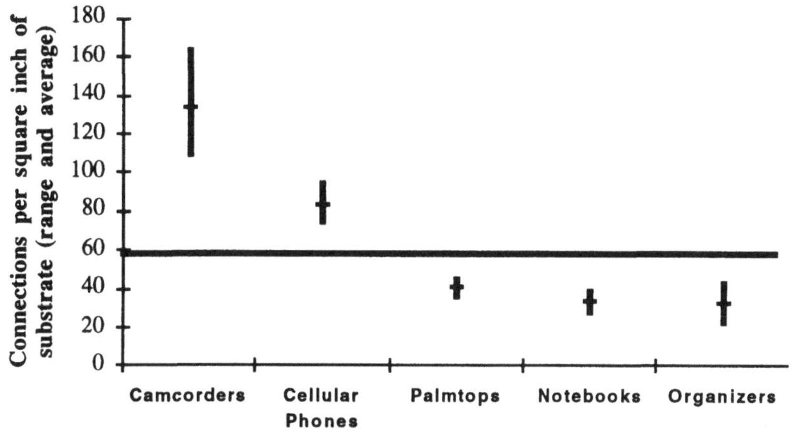

Fig. 3. Main PCB connection densities for different types of portable electronics products. (Dark continuous line is for product X under study.)

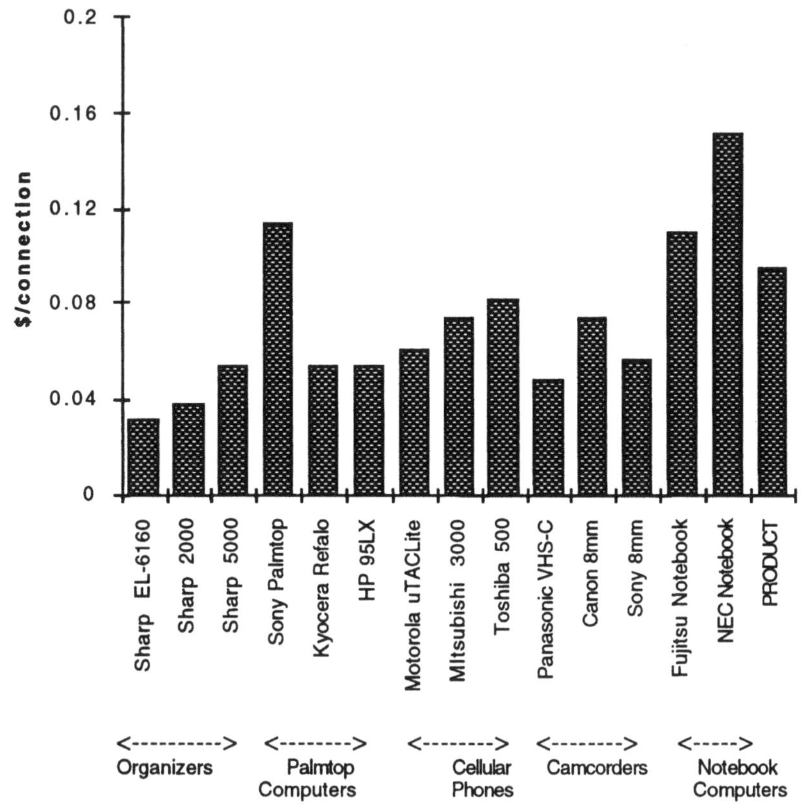

Fig. 4. Estimated electronics cost per connection. (Cost per connection for various portable systems. The product X under study and the camcorders have similar average pin counts.)

ing the area of various board technologies in the existing design by the wiring density for that particular technology we can estimate the total wiring length required to rout the design. In Table 4 we have indicated this total wiring length and have charted the required area of various board technologies needed to provide this amount of wiring length.

In Table 5 we have charted the corresponding substrate cost for each board technology to provide the wiring length needed for product X design. Obviously,

Table 3. Estimated wiring density of various board technologies [1].

Line Width (mils)	Space (mils)	Via Pad Dia. (mils)	Conductors Between	Wiring Density (in./sq.in./lyr)	Inches/sq. inch of wiring vs. number of layers (reserving one ground and one power layer)			
					4	6	8	10
10	10	30	1	33	67	133	200	267
7	10	30	1	35	70	140	211	281
7	10	30	2	41	81	162	243	324
6	6	18	2	63	125	250	375	500
5	5	18	2	70	140	279	419	558
3	5	18	2	77	154	308	462	615

Table 4. Estimated board area to route the product X with various PCB technologies.

Design: Product X PCBs				Board Area Required (sq. in.) vs. No. of Layers			
Inches of Wiring Required			2795				
Line Width (mils)	Space (mils)	Via Pad Dia. (mils)	Conductors Between	4	6	8	10
10	10	30	1	41.93	20.96	13.98	10.48
7	10	30	1	39.83	19.91	13.28	9.96
7	10	30	2	34.47	17.24	11.49	8.62
6	6	18	2	22.36	11.18	7.45	5.59
5	5	18	2	20.03	10.02	6.68	5.01
3	5	18	2	18.17	9.08	6.06	4.54

Table 5. Estimated bare board cost required for the product X with various PCB technologies.

Design: Product X PCBs				Board Cost (from MCC PCB cost model) for Number of Layers			
Estimated Board Cost for Different Technologies							
Line Width (mils)	Space (mils)	Via Pad Dia. (mils)	Conductors Between	4	6	8	10
10	10	30	1	$17.36	$11.10	$9.49	$8.61
7	10	30	1	$17.34	$11.10	$9.01	$8.27
7	10	30	2	$14.80	$11.12	$8.23	$7.70
6	6	18	2	$13.05	$8.61	$7.35	$6.85
5	5	18	2	$11.42	$8.41	$7.12	$6.68
3	5	18	2	$10.73	$7.90	$6.94	$6.43

in order to minimize cost and size, we would select a 10 layer board with 3 mil lines and 5 mil spaces since this yields a substrate size of 4.54 square inches and a cost of $6.43. Unfortunately, the footprint of these various components will not permit this choice, so we need to take a closer look at the component packaging strategy.

In Figure 5 we have summarized the situation with the existing design by showing the observed footprints for various components. We have also estimated the coverage ratio by dividing the total component footprint area by the total board surface area (2 × total substrate area). Estimated PCB weight and component weights are shown as well as the calculated connection density.

ACTUALS				(g)
(sq. inches)	IC pkgs	5.5	Board weight	68
	Discretes	1.26	component weight	19
	Connectors	1.14	Product weight	527
	Misc.	5.8		
Area required for components		13.7		
Board surface area		45.2	Corresponding connection density (conn./sq. inch)	47
Coverage ratio		30%		

Fig. 5. Estimated board metrics for existing product X.

In Figure 6 we have estimated the impact of several changes in the system packaging technology. We have changed the system IC packages from mostly 50 mil pitch lead pitch to 25 and 20 mil pitches. This reduces the footprint of the ICs by over 50% (assuming ICs are available in the fine pitch packages). We have changed the discretes from 1206 technology to 0603 technology. This reduces the discrete footprint by 75% to only 0.32 square inches (Figures 5 and 6). We have replaced the connectors with fine pitch SMT connectors and have reduced the total connector footprint to 0.6 square inches. We have left the miscellaneous parts (mostly large power supply components) unchanged from the existing design. Thus, the overall component footprint has been reduced to 9.22 square inches. If the existing coverage ratio of 30% is maintained, this implies that 30.4 square inches of board area are required to accommodate the component footprints. Therefore, 15.2 square inches of board are required to provide 30.4 square inches of mounting area-by mounting components on both sides of the PCBs.

In Table 6 we have shaded out the technology selections that do not provide the required board area to accommodate the component footprint. In Table 7 we have similarly shaded out the corresponding board costs that do not meet the minimum area requirements. If the designer wants to minimize cost, then the 3 mil line and 5 mil space PCB with 4 layers is the best choice at $10.73. If minimal area is desired then the best choice is 7 mil line and 10 mil space (with two conductors between) PCB with 6 layers at 17.24 sq. inch.

Shrink Components; Coverage constant				(gms)	
(sq. inches)	IC pkgs	2.5	Board weight	46	4.22%
	Discretes	0.32	component weight	14	0.95%
	Connectors	0.6			
	Misc.	5.8			
Area required for components		9.22	Corresponding connection density (conn./sq. inch)	70	
New board area (with corresponding 30% coverage area)		30.4			

Fig. 6. Estimated board metrics for product X with fine pitch components.

In Figure 7 we have made the same assumptions with regard to reducing the component footprints as in the previous example. We have, however, increased the coverage ratio to 70% corresponding to a connection density of 162 connections per square inch. Under these assumptions, 6.6 sq. inch of substrate are required to provide 13.2 sq. inch of mounting area.

As before, Tables 8 and 9 show the substrate area and cost options (values in shaded boxes) that meet these conditions. In this case, both area and cost are minimized (6.68 sq. inch and $7.12 respectively) with the 8 layer 5 mil line and space solution.

3.5. Comments on Plastic Housing

The total weight of the product X system is 527g, 295g of which are attributed to the plastic housing. Substantial savings in both housing weight as well as thickness can be achieved if other resin materials like CFRP (carbon fiber reinforced plastic) material is used instead of conventionally used ABS (acrylonitrile butadiene styrene). For the same housing area, wall thickness reduction of 50% and a weight reduction of 40% would be achieved if CFRP is used. Using such a material, the system could realize a weight savings of 40% × 295g or 118g. This amounts to an overall system weight reduction of 22%. Such materials are already being used in notebook and palmtop computers to reduce overall product weight without compromising the mechanical strength/rigidity of the product.

The use of smaller boards with higher density electronics packaging would allow the electrical and mechanical parts to be packed within a much smaller volume. This smaller volume would translate into a smaller housing and more total weight reduction.

Table 6. Estimated board area to route the product X with different PCB metrics and using finer pitch components, keeping board coverage constant.

Design: Product X PCBs							
Inches of Wiring Required			2795	Board Area Required (sq. in.) vs. No. of Layers			
Line Width (mils)	Space (mils)	Via Pad Dia. (mils)	Conductors Between	4	6	8	10
10	10	30	1	41.93	20.96	13.98	10.48
7	10	30	1	39.83	19.91	13.28	9.96
7	10	30	2	34.47	17.24	11.49	8.62
6	6	18	2	22.36	11.18	7.45	5.59
5	5	18	2	20.03	10.02	6.68	5.01
3	5	18	2	18.17	9.08	6.06	4.54

Table 7. Estimated board cost required for the product X with various PCB technologies and finer pitch components, keeping board coverage constant.

Design: Product X PCBs							
Estimated Board Cost for Different Technologies				Board Cost (from MCC PCB cost model) for Number of Layers			
Line Width (mils)	Space (mils)	Via Pad Dia. (mils)	Conductors Between	4	6	8	10
10	10	30	1	$17.36	$11.10	$9.49	$8.61
7	10	30	1	$17.34	$11.10	$9.01	$8.27
7	10	30	2	$14.80	$11.12	$8.23	$7.70
6	6	18	2	$13.05	$8.61	$7.35	$6.85
5	5	18	2	$11.42	$8.41	$7.12	$6.68
3	5	18	2	$10.73	$7.90	$6.94	$6.43

Increase coverage; shrink components	
Area required for components	9.2
Coverage ratio	70%
Required board surface area	13.2

	(gms)	
Board weight	20	9.14%
Component weight	14	0.95%
Corresponding conn. density (conn./sq.inch)	162	

Fig. 7. Estimated board metrics for product X with fine pitch and increased coverage ratio.

3.6. Recommendations

Based on the above analysis, we make the following recommendations for product X:

- Replace 1206 discrete components with 0603 sizes
- Replace 50 mil pitch packages with 20 and 25 mil pitch packages
- Replace through-hole connectors with SMT and SMT Flat Profile connectors
- Use miniaturized SMT power components where possible
- Surface mount IC components on both sides of the PCB
- Use 8 layer 5 mil line/space, 18 mil via PCB technology
- Utilize new plastic alloy formulations for housing construction

3.7. Resulting Improvements

- Board area will be reduced to under 14 square inches
- Single PCB will be required instead of 3 PCBs
- Board cost will be reduced by over 50%
- Product weight reduction will be 10% due to electronics improvements
- Product weight reduction will be 28% due to plastic alloy utilization

Table 8. Estimated board area to route the product with different PCB metrics and using finer pitch components, while increasing board coverage.

Design: Product X PCBs							
Inches of Wiring Required			2795	Board Area Required (sq. in.) vs. No. of Layers			
Line Width (mils)	Space (mils)	Via Pad Dia. (mils)	Conductors Between	4	6	8	10
10	10	30	1	41.93	20.96	13.98	10.48
7	10	30	1	39.83	19.91	13.28	9.96
7	10	30	2	34.47	17.24	11.49	8.62
6	6	18	2	22.36	11.18	7.45	5.59
5	5	18	2	20.03	10.02	6.68	5.01
3	5	18	2	18.17	9.08	6.06	4.54

Table 9. Estimated board cost required for the product X with different PCB metrics and using finer pitch components, while increasing board coverage.

Design: Product X PCBs							
Estimated Board Cost for Different Technologies				Board Cost (from MCC PCB cost model) for Number of Layers			
Line Width (mils)	Space (mils)	Via Pad Dia. (mils)	Conductors Between	4	6	8	10
10	10	30	1	$17.36	$11.10	$9.49	$8.61
7	10	30	1	$17.34	$11.10	$9.01	$8.27
7	10	30	2	$14.80	$11.12	$8.23	$7.70
6	6	18	2	$13.05	$8.61	$7.35	$6.85
5	5	18	2	$11.42	$8.41	$7.12	$6.68
3	5	18	2	$10.73	$7.90	$6.94	$6.43

Note that these results do not take into account additional benefits such as the reduction in the number of connectors needed through the consolidation into a single PCB or the compounded plastic housing weight reduction possible by reducing the product weight.

References

1. C. Capillo, *Surface Mount Technology*, McGraw-Hill Publishing Company, 1990.

Shekar Rao is an MCC employee working in the Packaging for Portable Electronics project. His primary responsibility in the Portables project is the reverse engineering analysis of different consumer electronic products as well as documentation and liaison with the different participant companies. He is also involved with the other group activities like the general technology reports, cost modeling, data base and trade-off analysis tool.

Shekar Rao has an M.S. and Ph.D. in Electrical Engineering from the Oregon Graduate Institute and a B.E. in Electrical Engineering from Mangalore University, India.

Bert Haskell is the project leader for the Portable Electronics Packaging Project at MCC. This project provides strategic information and technology analysis about leading portable electronic assemblies and their sources of manufacture. In addition, he has also been involved in High Density Connector, Quick Turnaround Interconnect, Optoelectronics Packaging, and Amalgam Flip Chip projects. Each of these projects were focused on Multichip Module applications.

Before joining MCC, he worked at Eastman Kodak Company, in thick-film hybrid and surface mount electronic assemblies. He has a B.S. degree in Mechanical Engineering from the University of Wisconsin and an M.S. in Mechanical Engineering from the University of Rochester in 1989.

Ian Yee is an Austin based consultant to the electronics packaging industry. Prior to that he was a Member of the Technical Staff in MCC's High Value Electronics Division (1985–93). He joined MCC from Veeco Instruments (1983–1985) where he held the position of product engineer for their ion beam etch systems. At MCC, Mr. Yee has developed TAB and MCM processes in the areas of dry etching, electroplating and wet chemical processing, thin-film deposition, photolithography, and has performed extensive cost modeling on various products and processes.

Mr. Yee received his B.S.E. in Chemical Engineering with honors from Princeton University, New Jersey in 1983. Mr. Yee is currently pursuing an M.B.A. at St. Edwards University, Texas.

JOURNAL OF ELECTRONIC TESTING: Theory and Applications, 5, 229-238 (1994)

Cost Based Surface Mount PCB Design Evaluation

M. ALEXANDER, K. SRIHARI AND C.R. EMERSON
Department of Mechanical and Industrial Engineering, Thomas J. Watson School of Engineering and Applied Science, State University of New York, Binghamton, New York 13902-6000

Abstract. In today's manufacturing field there is continuous demand for product flexibility and shorter product throughput time. Production planning and cost estimating activities need to be speeded up. Consequently, the traditional gap between the designer who designs new products and the estimator who evaluates the financial consequences of the design is fading. Cost optimization is being integrated in the design process, and the designer needs to take more responsibility for the cost of the product at the design stage itself. There is a lack of tools to assist the designer in analyzing a Printed Circuit Board's (PCBs) design from a manufacturing perspective, especially from an economic aspect. This is true in Surface Mount Technology (SMT) production where the integration of the design and manufacturing functions is often hampered by the lack of sufficient, in-depth manufacturing knowledge and associated cost information. This research developed a cost estimation tool for the designer in the surface mount PCB assembly domain by integrating Computer Aided Design (CAD), Computer Aided Process Planning (CAPP), and cost estimation techniques using a knowledge based framework. Cost estimation can be done at two design stages. First, a rough cost estimate can be obtained from knowing the component mix, the type of boad material, and the size of board. Then, after the detailed design, a more accurate assembly cost can be obtained by generating a macro-process plan from the given CAD data. The cost module considers tangible factors including the material costs, the equipment costs, process costs, and the labor costs. The model used also considers the intangible cost in a preliminary evaluation of design costs. The cost advisor and the Design For Manufacture (DFM) environment have been developed using C++ and object oriented programming constructs under the MS Windows operating system.

Keywords: Computer aided design, computer aided process planning, concurrent engineering, design for manufacture, knowledge based framework.

1. Introduction

The electronics industry is a critical segment of the industrial sector of the United States. Electronic components are important elements in a multitude of products in a variety of applications including telecommunications, computers, automobiles, entertainment, and the national defense. With annual sales of over $280 billion dollars, the electronic industry represents fifteen percent of the gross national product of the United States [1].

In the Printed Circuit Board (PCB) assembly domain, manufacturers must deal with rapid technological changes of both products and processes, multi-plant organization, high cost of materials, and a continuing trend of miniaturization [2]. Lengthy and variable vendor lead times make it difficult to supply customer demands with low levels of inventory within acceptable cycle times. A predominant percentage (sixty to eighty percent) of a product's manufacturing cost is determined during its design [3]. Studies indicate that the product assembly cost often accounts for over 40% of manufacturing cost [3]. Consequently, there is a compelling need in today's advanced manufacturing environment to influence a product's design and reduce its manufacturing cost early in its design cycle. This research develops a cost advisor to assist the PCB designer in predicting the cost of the product during the early design phase enabling the designer to make appropriate decisions that will reduce the cost of the PCB assembly.

2. Concepts Used

2.1. Printed Circuit Board Assembly and Surface Mount Technology (SMT)

Advanced manufacturing systems, like the electronics packaging domain, are characterized by a growing range of variants, a reduction in product time and a growing complexity coupled with the demands for shorter delivery times [4]. There are three levels in PCB assembly namely the component level, the board level, and the system level.

The use of SMT (when compared to through-hole technology) is increasing because of benefits such as improved functionality and quality, significant savings in weight, reduction in electrical noise, standardization of assembly machines, standardization of component shapes, and packaging and savings in real estate on the PCB. Surface mount PCB assembly is expected to reach 55% by the early 1990s and to be over 70% by late 1990s [5]. They can be classified into three different type of processes based on the type of components used and the board design namely Type I, Type II, and Type III PCBs [6]. A Type I SMT assembly contains only Surface Mount Components (SMC) either on one side or both sides of the PCB. A Type III SMT assembly contains only passive components (resistors, capacitors, and transistors) glued to the bottom side of the board, while a Type II is a combination of Type I and Type III PCBs. Each PCB follows a relatively fixed path through the manufacturing system. A typical assembly system consists of a solder paste dispenser, one or more automatic component placement stations, a solder reflow station, a cleaning station, and a test, diagnosis and rework station.

2.2. Design in PCB Assembly

The design of a PCB assembly can be classified into two parts: mechanical design and electrical design. The former deals with the board space available for the circuit that determines the number of layers and component densities of the board. The designer must also keep in mind issues such as solderability, cleanability, testability, and heat dissipation facility. The latter takes into account factors such as length and width of the conductors, transmission delays, etc. The designers should also take into consideration constraints such as operational environment in which the circuit board is likely to be used.

2.3. Knowledge Based Systems

Steps involved in the development of an expert system would be problem definition, knowledge acquisition (representation and coordination), inference engine development, implementation, testing and validation, and the development of an update mechanism. Design For Manufacture (DFM) is an attractive area for research because it could be used to embody the knowledge of experienced manufacturing expert in order to design for manufacturability. Most Compuer Aided Design (CAD) systems tend to concentrate on object modeling, analysis, and drafting [7]. Expert (or knowledge based) systems can be used to provide facility specific process information to the PCB assembly designer. A knowledge engineer can capture the manufacturing knowledge from an "expert" manufacturer, and subsequently store it in the form of rules and facts within a knowledge based system. A design can be checked with respect to the given manufacturing facility, and the designer can be informed about potential manufacturing problems and alternatives. This will reduce the redesign of the PCB after it is sent to the manufacturing floor. A manufacturing expert system can be developed to generate the process parameters for every stage in the assembly of the PCB.

3. Problem Statement and Research Objective

The electronic industry enjoys the advantage of modern manufacturing technologies. At the same time, it is subjected, as are other industries, to the limitations and problems that the state-of-the-art causes. Manufacturers in the electronics industry face aggressiveness global competition with respect to cost and product technology. Electronic products typically have a short life span, exacerbating the problems related to bringing new products into production as well as those involved in concurrently manufacturing a number of products, each in a unique stage of its life cycle. The never ending quest for miniaturization drives both manufacturing and assembly technologies. In addition, multiplant organization and multiechelon product structure common to the industry intensifies management and material flow coordination problems.

PCB assembly operations face unique problems associated with material flow management, equipment operation, and product testing. For example, lengthy and variable vendor lead times make it difficult to supply cutomer demands with low levels of inventory and with acceptable cycle time. The sheer number of high cost components necessary to assemble a batch of PCBs creates exceptional inventory management and material flow problems. Apart from the technology of the equipment itself, the process engineer needs to deal with the problems associated with fine pitch components. The management of equipment—for example, in devising effective setup management capabilities—presents unique challenges not found in other industries. Also testing state-of-the-art products entails industry-specific strategies, technologies, and material flow considerations.

While Computer Integrated Manufacturing (CIM) has helped bridge the gap between design and manufacturing, it has not tightened all the checks and balances within the product design cycle itself. Since 70% of a product's cost is set during the design cycle [3], cost-cutting measures must come into play on the front end of the product life cycle. Consequently, there must be simultaneous design of a product and the process required to produce it. This involves the participation in the design engineering stage from downstream functions such as manufacturing, third party suppliers, quality control, and customer support.

The main obstacle to DFM is that design engineers are not manufacturing experts. Most products' designs use components made from diverse manufacturing processes. It is humanly impossible for a design engineer to be well-versed in all areas. Many companies tried to solve this problem by creating libraries of standards. These standards cover everything from allowable tolerances, machine capacity, accuracy, repeatability, required coating, plating, and finishes. The problem with these standard libraries is that there is more information available than can be digested and used effectively.

Cost is the ultimate measure of manufacturability. An expert system combining manufacturability and costing information is needed at several levels within a company. Sales needs cost information to establish pricing. Marketing needs cost information in the planning of new products. Purchase department needs costing information for vendor negotiations. Cost information is also needed to do analysis as part of a make or buy decision, or to determine model, type and size.

This research develops a cost advisor for a DFM environment. It evaluates the design on the basis of cost of manufacturing (assembly) the product (PCB design). In order to arrive at the assembly cost a macroprocess plan will be developed to select the machines to assemble the given PCB depending on the size and type of components, size and type of board, the solder paste used, etc. The input to this system is a CAD drawing of the PCB. The output of the system is the macro-assembly sequence and the estimated manufacturing (assembly) cost.

4. Research Methodology

It has long been known, though rarely put into practice, that design and manufacturing are interrelated activities and that efficient, cost-effective production can be achieved when product design considers the manufacturing equipment to be used, and vice versa [8]. However, implementing this philosophy is not simple. There is also a lack of information that is being provided to the designer on the manufacturing process in a detailed and systematic way. Selecting the most appropriate manufacturing process in terms of technological feasibility and cost for a component design is one of the most important decision-making tasks. Failure to get it right normally results in components that are of variable quality and/or expensive to make. The integration of knowledge based expert systems and CAD systems is one such method of providing the designer with a rapid means of investigating alternative design solutions as it helps in the automated evaluation by extracting the necessary information directly from the CAD representation.

4.1. Design for Manufacturing System

A DFM expert system environment has been designed and developed in related research. This DFM system provides the PCB designer with a feel for the manufacturing related problems. This will help in reducing the redesign of the board after the design is sent to the manufacturing floor. The DFM system's architecture is shown in Figure 1. The design is reviewed from two perspectives: manufacturing problems due to thermal considerations and manufacturing problems due to spacing or location considerations. With respect to the thermal considerations, the PCB design is checked from a solderability as well as reflow perspective. As far as the spacing considerations are concerned, the components are checked if they are cleanable within the limitations of the cleaning equipment chosen. The interaction between the designer and the DFM system is through a blackboard. Information from the PCB drawing (CAD) is read into the system in a .dxf file

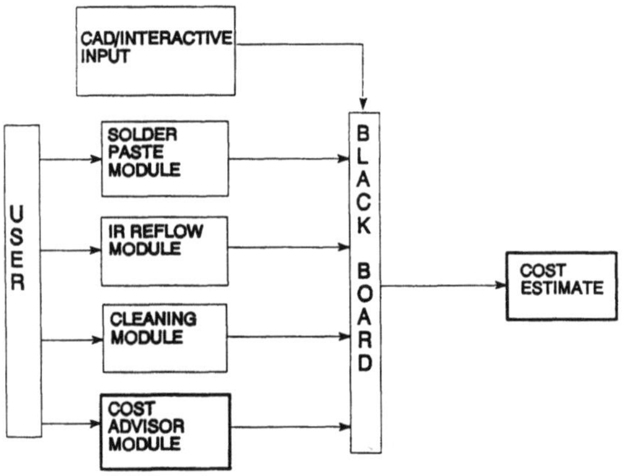

Fig. 1. Architecture of DFM system.

format (data exchange file). Information from the databases is used along with rules in the rule base by the system to deduce its evaluations and recommendations.

Decisions made during the engineering design phase substantially impacts the manufacturing processes used. These decisions may determine the degree to which the production process can be automated; they could impact the type and complexity of equipment involved in production, automation and integration [9]. DFM is defined as analyzing the product and process features to optimize production of the product at minimum cost, maximum quality, maximum functionality, and maximum customer satisfaction. DFM can also lead to shortened product design time by reducing the number of design changes. By automating the DFM evaluation procedure, feasibility and ease of manufacturing the product can be evaluated simultaneously and it saves the designer's valuable time.

4.2. *The Role of Cost in DFM*

The cost advisor is a part of the DFM system. It is used during the conceptual design stage to predict the cost of assembling a PCB. It helps determine a product's profitability. While a DFM advisor could provide process related suggestions, the cost of manufacture is a critical factor in accepting (or rejecting) a design. Cost is the ultimate measure of manufacturability. An expert system combining manufacturability and costing information can be used at several levels within a company. Sales can use cost information to establish pricing. Marketing can use cost information in the planning of new products. Expert systems used in purchasing provide costing information for vendor negotiations. Cost information can also be used to do analysis as part of a make or buy decision, or to determine model type and size.

The cost model required to track cost in a concurrent engineering facilities will be different from the usual models for estimating product costs. Cost elements, such as overhead, must be held to a minimum so that as many costs as possible can be assigned to products as direct costs. Also, cost changes in machining, quality control, and inspection must be tracked. Implementation of this cost model will allow the design engineer to observe the effect on cost of manufacturing the PCB by altering the elements of design. The cost model also helps to determine the throughput rate of the system. In addition the designer can identify the most costly elements of the manufacturing process. This will allow the designer to concentrate in the areas which have the most potential for cost reduction. Identifying these areas will assist the design engineer in making cost-benefit analyses of design changes. The cost per component could be used for comparing two packaging alternatives. Cost per module/board is the overall cost of the module/board and is used as a baseline for comparison of two alternate PCB designs.

5. System Framework

A knowledge based cost estimator framework for PCB assembly is described below. The system combines traditional cost estimation techniques and the advanced manufacturing cost estimation techniques along with expert system capabilities resulting in a powerful and easy to use system. The system format is a Decision Support System (DSS) that guides the designer to arrive at the best PCB design which in turn results in the lowest assembly cost of that PCB. The system offers a quick and accurate cost estimate of assembling the PCB. The architecture of the cost advisor is shown in Figure 2. The cost advisor consists of the central processing module, conflict resolution module, product module, utility module, machine database, material database, and the labor database.

The Central Processing Module is responsible for all the necessary control functions and calculations. The rules used by the system are stored here. It controls and keeps track of the messages passed to and from other modules such as product module. The Product Module is the input module to the system. The CAD data must be provided to the cost advisor in a .dxf format. It must contain information about the PCB, the

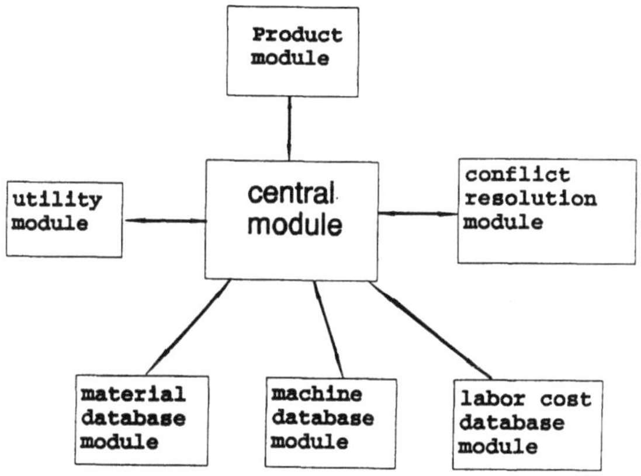

Fig. 2. Architecture of cost module.

components, and the other inputs needed to assemble the PCB. The Conflict Resolution Module decides which machine is to be selected when there is a conflict between two or more machines that can perform the same operation. This is done by selecting the machine which has a lower operating cost, less scrap rate, better quality, more reliability, etc. The Utility Module consists of a set of utility programs. The main function of this set of programs is to define and maintain user supplied databases.

Database Construction is an important part in this system. The usefulness of this system depends on the extensiveness and reliability of the database. Therefore it is necessary to make it extremely user friendly so that the user can easily modify, search, add or delete data, create new database, or add new fields to existing database. This was done by using CodeBase 5.0 library functions for database management. The data is stored in dBASE IV format. The database contains information about components, solder paste, cleaning agents, board dimensions, machine capabilities such as base rate, cost, accuracy, repeatability, setup time, etc.

The three steps necessary for the system to arrive at the manufacturing/assembly cost of a given PCB design are the extraction of assembly features, proposing an assembly sequence, and arriving at the manufacturing/assembly cost of the given PCB design. The interaction between the designer and the cost advisor modules is through a blackboard (Figure 1). Information from the PCB drawing is read into the system in a .dxf format. The geometric features (component geometry and board geometry) are read from this file. Then, using the information from the databases (material, machine and labor), the rules from the rule base (central processing module) are applied to generate the assembly sequence. If there is a conflict between two or more machines that can do the job then, the conflict resolution module determines which is the best machine based on nonmonetary factors such as quality, scrap rate, and reliability.

6. Cost Terminology

This cost advisor is a hybrid of computational and rule based methods. Direct calculation is used wherever possible and heuristic rules are used to complete the model. The assembly cost is calculated for each process step. The total cost of a module/product consists of material cost, assembly cost, inventory cost, test and rework cost, scrap cost [10].

The assembly cost can be split into operating cost, inventory cost. The operating cost is facility dependent and board dependent. Facility dependent factors include the labor rate, the overhead cost, and the machine cost. The board dependent factors include the assembly cost, the material cost, the scrap cost, and the cost of repair and test. The total cost model is as shown below:

$$C_{pr} = CM + CI + CA + CTR + CS$$

where

C_{pr} is the cost of manufacturing one PCB.
CI is the sum of all inventory terms.
CA is the sum of all assembly terms.
CTR is the sum of all test and rework terms.
CM is the sum of all the raw materials used.
CS is the scrap cost (which cannot be reworked).

6.1. *Inventory Cost Terms (CI)*

This model will consider only those inventory costs associated with maintaining the supply of surface mount components used in manufacturing the PCBs. Work-in-progress inventory costs are not included since they are assumed to be small. In general, the inventory costs associated with any manufacturing system have four main elements ordering cost, inventory-carrying cost, out-of-stock cost, capacity-associated cost [11]. For simplicity the cost model used in this research only includes ordering cost (TOC) and inventory-carrying cost (TCC).

$$CI = TIC/P_{yr}$$

where

CI is the inventory cost.

TIC is total inventory cost, which is composed of TCC and TOC.

P_{yr} is the total number of PCB per year.

To implement this model the designer may use these figures directly, if known, or estimate them from the Economic Ordering Quantity (EOQ) model (described below).

$$EOQ = \sqrt{\frac{2ASC}{I}}$$

where

A is the annual usage in dollars.

SC is the setup or ordering costs in dollars (fixed cost to place one order).

I is the inventory carrying cost coefficient, a decimal fraction of the average inventory value. It includes elements for obsolescence, deterioration, taxes, insurance, storage, retrieval, and capital costs.

The total carrying cost (or TCC) is given by:

$$TCC = I(R + EOQ/2)$$

and Total Ordering Cost (or *TCC*) is given by:

$$TOC = SC(A/EOQ)$$

where *A/EOQ* is the number of orders placed in one year, *R* is the dollar value of the inventory reserve. *I* is the inventory carrying cost coefficient, and *SC* is the setup or ordering cost.

6.2. *Assembly Cost Terms (CA)*

The total assembly cost per completed assembly consists of cost of setup and assembly at each station.

$$CA = \sum_{i=1}^{S} CA_i$$

Where *s* is the number of stations in the assembly system and CA_i is the setup and assembly cost incurred at assembly station *i*. The general form for assembly cost is [11].

$$CA_i = t_{pri}\left(W_{tj} + \frac{M_i}{2000*N*S}\right)$$

where

CA is the assembly cost per completed assembly incurred at station *i*.

t_{pri} is the average production time of acceptable assemblies at station *i*.

W_{tj} is the wage rate of operator(s) assigned to machine this should include direct labor rate, indirect labor rate and overhead rate.

M_i is the cost of machines.

N is the payback period for the equipment. This ignores the time value of money, tax effects, and the effect of production (cash flow) occurring after the payback period.

S is the number of shifts worked per day. If it is assumed that a year consists of 50 weeks and there are 40 hours per week, then 2000**N***S* determines the hours per shift year.

6.3. *Sencil Printing Station*

Since the assembly system operates on small lot size, setup time is significant. The average time a PCB spends at this station, then, depends on the setup time, the cycle time and the downtime. Therfore [11]:

$$t_{pri} = t_{su}/L + t_s + x_i T_i$$

where

t_{pri} is the average production time of acceptable assemblies at station *i*.

t_{su} is the screen printing machine setup time.

L is the lot size.

x_i is the fraction of unsuccessful screening operation.

T_i is the time required to correct one unsuccessful operation.

t_s is the average time for screen printing operation.

6.4. *High-Speed Pick and Place Chip Placement Machine (Chip Shooter)*

Each machine can pick from about one hundred different surface mount component types. Therefore:

$$t_{pri} = n_i(t_p + x_i T_i)$$

$$M = C_i + n_{tj}C_{tj} + C_{cari}$$

where

M is the placement machine cost.

C_i is the cost of the machine, transfer device, buffer device and controller.

n_{tj} is the total number of different part magazines.

C_{ti} is the cost of one part feeder.

C_{cari} is the cost of the part carousel and controller.

n_i is the number of parts to be assembled.

t_{pri} is the average production time of acceptable assemblies at station i.
t_p is the average placement time for one successful placement.
x_i is the fraction of unsuccessful placement operations.
T_i is the time required to correct one unsuccessful operation.

If vision assisted placement is required and twin head moveable placement heads chip-placement machine is used then:

$$t_{pri} = n_i(t_p + x_i T_i)$$

where

n_i is half the total number of parts to be assembled.
t_{pri} is the average production time of acceptable assemblies at station i.
t_p is the average placement time for one successful placement.
x_i is the fraction of unsuccessful placement operations.
T_i is the time required to correct one unsuccessful operation.

6.5. *Infrared Oven and Cleaning Station*

These stations are integral to the assembly process. While no assembly takes place here, every station contributes to the overall assembly cost.

t_{pri} = average processing time of acceptable assemblies at station i.

6.6. Test and Rework Cost Terms

In this model, it is assumed that all PCBs undergo a burn-in and test cycle. All faults detected here are repaired. Detected faults include an incorrect SMT component that is successfully assembled, an incorrectly assembled component in which the electrical connection has not been made during the soldering process, etc. The test cost (C_{test}) depends on the time the test takes and on the cost of the test equipment. Rework cost is determined by the fault rate for all components, the number of components in the PCB and the cost of reworking one faulty PCB. In this model, the rework cost is the same for all faults and virtually all faults are corrected successfully during the rework step. The rework cost is function of the fault rate. For a given PCB design, the probability of developing a fault during the manufacturing process increases with the number of components on each PCB.

If First Pass Yield (FPY) is the fraction of PCBs which pass the first circuit test, then [11]:

$$FPY = (1 - FR)^n$$

where

FR is the fault rate.
n is the number of components on each PCB.

The proportion of PCBs that fail the first circuit test is $1 - FPY$. Each reworked PCB must be tested again, so the rework cost (CR) is given by

$$CR_k = (CPF + C_{test})(1 - FPY_k)*n$$

The total cost of rework is given by:

$$CR = \sum_{k=1}^{l} (CR_k)$$

where

$$FPY_k = (1 - FR_k)^{nk}$$

n_k is the number of components k in the PCB.
l is the total number of component types in the PCB.
CR_k is the rework cost per PCB for kth component type.
CPF is the average cost of repairing one failed PCB (in dollars).
C_{test} is the cost of running the retest on the rework PCB (assumed to be the same as the original test).

The total test and rework cost CTR is given by:

$$CTR = C_{test} + CR$$

where

C_{test} is the cost of running the retest on the rework PCB (assumed to be the same as the original test).
CR is the cost of rework.

6.7. *Availability Correction Factor*

Availability compensates for the time that a machine is unavailable due to maintenance requirements. This includes time required to perform maintenance and the lost time waiting for spare parts is considered. Equipment availability is modeled as:

$$A = 100\% - \%T_{pm} - \%T_{sp} - \%T_{dt}$$

where

A is the machine availability fraction.
T_{pm} is the preventative maintenance time %.
T_{sp} is the spare parts caused idle time %.
T_{dt} is the equipment caused downtime %.

6.8. *Utility Correction Factor*

Utility is determined with consideration of the setup time, the operator unavailability, and capacity compensation. The utility function is modeled as:

$$U = OA \times CC \times (100\% - [N_s \times T_s/T_r])$$

where

OA is the operator availability (10% derating factor if manual board feeding or 100% if automatic).
CC is the capacity compensation fraction.
U is the machine utilization fraction.
N_s is the number of setups per day.
T_s is the time per setup.
T_r is the production hours per day.

6.9. *Capacity Compensation (CC):*

Is the fraction of time that work is available for the machine. The CC term is given by:

$$CC = 100\% - D_f/C_a$$

where

D_f is the forecast demand parts per hour.
C_a is the available capacity PPH.

6.10. *Efficiency Correction Factors*

A correction for part placement rate is required because of inconsistent production claims made by manufacturers. The claimed rates are based on unrealistic assumptions that placements have minimum head travel, placement is confined to a small placement area, and the position of the part feeder is assumed to be near the placement head. Four additional sources of efficiency reduction exist. They are the loss due to part complexity, loss due to inefficient board design, loss due to board transfer, and loss due to suboptimal machine setups. The efficiency compensation is modeled as:

$$E = PC \times BC \times T_t$$

where

E is the machine efficiency fraction.
PC is the part complexity factor.
BC is the board complexity factor.
T_t is the transfer time.

Efficiency loss due to part complexity:

$$PC = \frac{(\%S \times R_b + (1 - \%S) \times R_c)}{R_b}$$

where

PC is the part complexity.
S is the fraction of simple parts used (passive components e.g., resistors, capacitors, etc.)
R_b is the base rate.
R_c is the complex part placement rate (e.g., PLCC, ICs).

Efficiency loss due to board complexity: This is a function of the board design. Board Complexity factor compensates for two significant reductions in efficiency from board design. One, the board is often larger than those used on baseline evaluations and second, boards have components at locations far away from the feeder.

Efficiency loss due to board transfer: The overall transfer time range for surface mount PCB assembly equipments is 4 to 15 sec [12]. This will be a significant factor in the case of small bath size. Therefore, the corrected throughput (t) is [12]:

$$t = R_t \times Avail \times Util \times Effi$$

7. Example

The PCB design considered (pcb.dxf) during a sample session is shown in Figure 3. The facility represented in the software system (chosen during system configuration) consists of a stencil printer, a high speed placement machine, a flexible high accuracy placement machine, a reflow oven, and a cleaning machine. The final application of the PCB selected was industrial and the solder paste selected was paste 3 (with a Sn63Pb37 composition). The quality level required was "excellent." The output includes the macro process plan for the given board, the list of machines selected to assemble the board along with the assembly cost, the material cost, the cost of test and rework, and the inventory cost (Figure 4).

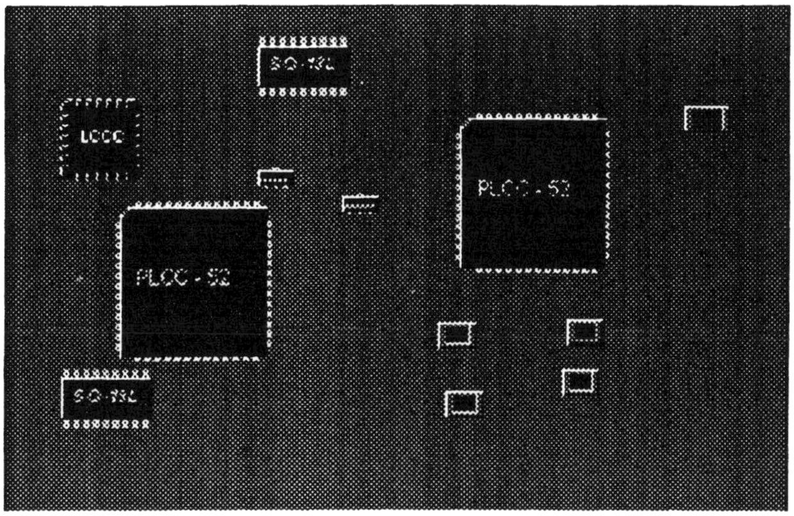

Fig. 3. PCB.DXF.

MACRO PROCESS PLAN

quit

THE PROCESS PLAN FOR THE GIVEN PCB DESIGN

MACHINES SELECTED	ASSEMBLY PROCESS	ASSEMBLY COST
UNIVERSAL_4232	PRINT THE SOLDER PASTE	1.75
SONY_SS	PLACE COMPONENTS	0.44
IAR_350c	REFLOW SOLDER THE BOARD	3.06

THE TOTAL MATERIAL COST PER PCB =	202.00
THE TOTAL INVENTORY COST PER PCB =	2.01
THE TOTAL TEST AND REWORK COST PER PCB =	5.66
THE TOTAL ASSEMBLY COST PER PCB =	5.25

Fig. 4. Output screen of cost advisor.

8. Conclusion

The product's design significantly impacts the finished product's cost. Therefore it is of utmost importance to provide the designer with decision support tools that will help in lowering the cost of product. The advancement in SMT components, multilayer boards, increased board density, enhanced production rates, shorter product life cycle, and higher yields mandate the use of design for manufacturing principles. This article describes a cost estimator tool that would present the manufacturing/assembly cost to the designer during the design phase. The cost analysis can be best applied as an aid for the designers, by providing rapid comparisons at an early design stage. The system described above uses a knowledge based framework. Object oriented programming techniques are used to develop the system. The main advantage of using this approach is that it allows for a step wise, modular development of the system.

Acknowledgments

This research was sponsored by a consortium funded by the National Science Foundation (grant number EEC-9108850), the New York State Science and Technology Foundation, Digital Equipment Corporation, Universal Instruments Corporation, Martin Marietta Control Systems, and IBM-Owego.

References

1. W.E. Wilhelm and J. Fowler, "Research Directions in Electronic Manufacturing," *IEE Transactions*, vol. 24, pp. 6–17, 1992.
2. W.E Wilhelm, "Scheduling and Logistics on Electronics Manufacturing and Assembly: An Introduction," *IIE Transactions*, vol. 24, pp. 2–5, 1992.
3. R.K. Li and C. Hwang, "A Framework for Automatic DFA System Development," *Computers in Engineering*, vol. 22, pp. 403–413, 1992.
4. W. Eversheim and M. Baumann, "Assembly-Oriented Design Process," *Computers in Industry*, vol. 17, pp. 287–300, 1991.
5. C. Lea, *A Scientific Guide to Surface Mount Technology*, Electrochemical Publications, Ayr, Scotland.
6. R.P. Prasad, *Surface Mount Technology—Principles and Practice*, Van Nostrand Reinhold, New York, 1988.
7. H.S. Luong, "Applications of Expert Systems in Manufacturing: A Case STudy," *Computers in Industry*, vol. 18, pp. 193–198, 1992.
8. A.J. Allen and K.G. Swift, "Manufacturing Process Selection and Costing," *Proceedings Institution of Mechanical Engineers*, vol. 204, pp. 143–148, 1990.
9. L. Galbraith, "The Role of Design for Manufacturing in System Integration," *Computers and Industrial Engineering*, vol. 21, 1–4, pp. 11–15, 1991.
10. C. Bloch and R. Ranganathan, "Process-Based Cost Modeling," *IEEE Transactions on Components, Hybrids, and Manufacturing Technology*, vol. 15, pp. 288–294, 1992.
11. M.R. Driel and J.S. Klegka, "Cost and Performance Analysis of Printed Circuit Board Manufacture Using Surface Mount devices," *International Journal of Advanced Manufacturing Technology*, vol. 4, pp. 72–94, 1989.
12. D.D. Price and C.R. Emerson, "3 Steps to 'Improved Performance' Equipment Selection," *Surface Mount Technology*, pp. 23–28, July 1992.

Mathew Alexander was a graduate research assistant in the Department of Mechanical and Industrial Engineering at the State University of New York, Binghamton, New York. He completed his M.S. in the Fall of 1993. Prior to that, Mr. Alexander completed his B.S. in Mechanical Engineering from the University of Bangalore, India in 1991. His research interests are in the application of knowledge based systems to manufacturing problem solving in the electronics packaging domain. Mr. Alexander is currently employed as a software engineer.

K. Srihari is an Assistant Professor in the Department of Mechanical and Industrial Engineering at the State University of New York, Binghamton, New York. He completed his M.S. (1985) and Ph.D. (1988) in Industrial Engineering and Operations Research from VPI & SU, Blacksburg, Virginia. His reearch interests are in electronics manufacturing.

C. Robert Emerson is a Professor and the Director of Industrial Engineering in the Department of Mechanical and Industrial Engineering, State University of New York, Binghamton, New York. An alumnus of Purdue University, Dr. Emerson's research interests are in quality and reliability issues in manufacturing systems engineering, especially in the electronics pacakging domain.

JOURNAL OF ELECTRONIC TESTING: Theory and Applications, 5, 239-252 (1994)

Sensitivity Analysis in Economics Based Test Strategy Planning

J.H. DICK AND E. TRISCHLER
Siemens-Nixdorf-Informationssysteme AG, P.O. Box 830951, 8000 Munich 83, Germany

C. DISLIS AND A.P. AMBLER
Brunel University, Uxbridge, Middlesex, UB8 3PH, United Kingdom

Abstract. This article will present methods to analyze the sensitivity of test costs to the inaccuracy of the individual costing parameters. The results show that a few parameters—e.g., the gate count—need very detailed estimates, whereas the accuracy of many other parameters is insignificant in 99% of all cases. The techniques presented allow an in-depth evaluation of what is perceived as the main drawback in the use of economic modeling methods, namely, the element of risk associated with inaccuracies in the input data.

Keywords: Costing parameters, Monte Carlo method, scan path, sensitivity analysis.

1. Introduction

This article addresses one of the main drawbacks associated with the use of cost (economics) models in general and specifically for making test strategy decisions, namely, the risk of arriving at misleading decisions as a result of using inaccurate input parameters. This work was based on ECOTEST [1], [2], a test strategy planning system for VLSI design, developed by Siemens-Nixdorf and Brunel University under the European commission's ESPRIT EVEREST program. The system provides both manual and automatic methods to determine the most economic design for test strategy for the target design. This is achieved by using a financially based test economics model for the objective evaluation and comparison of alternative strategies. Other systems also use cost evaluators to optimize the test strategy [3], but their cost measures are not financially based. The test economics model provides a set of equations, which estimate design, manufacture, and test costs [2]. The cost estimation is based upon parameters that include the gate count, the die size, or the production volume of the VLSI. For practical reasons, the estimation of the parametric data will be subject to inaccuracy in most cases. This is due to the trade-off between the accuracy of the data and the effort or cost of the data gathering task.

The objective of this article is to study the impact of inaccurate costing parameters on the resulting cost. This is a complex problem, due to the complexity of the cost model. Therefore we have developed efficient methods to study this impact. In Section 2 the description of the problem is presented, the need and the gain of this work is discussed, and the different applications of the sensitivity analysis are introduced. Section 3 describes the method to analyze the variation of all parameters at the same time. Section 4 describes the method used to handle the correlation of input parameters. Sections 5 to 8 present three applications of sensitivity analysis and related results, and a summary of the article and conclusions are given in Section 9.

2. Description of the Problem

In ECOTEST, test strategy planning is based upon economic evaluations, calculated by a cost model. The cost model can be seen as a decision model, where the decision is driven by minimizing the resulting cost value. The cost model uses a number of primary parameters supplied by the user as the basis for calculations. This data acquisition process can present significant problems. The most likely problems are summarized as follows:

- Data gathering can be expensive.
- Due to being estimates, the input data can be very inaccurate.
- Some of the input data may not be defined yet; several alternatives are possible.

The problems mentioned above are addressed in several publications [4]–[6]. Dinkelbach states, that these problems are the major reason, why the practical usage of cost modeling techniques is still very limited [7]. Extending cost modeling techniques by integrating sensitivity analysis techniques is therefore of major importance. The objective of a sensitivity analysis is to study the effect of parameter changes on the decision to be made.

In previous work [1], [4], [6]–[8] sensitivity analysis was always performed by varying one parameter and keeping all other parameters constant. This implementation does not show the real impact of inaccuracy of input data. The reason for this is that the sensitivity for the parameter under consideration is calculated statically, i.e., with a specific value for all other parameters. However, the assumption that only one parameters varies at any one time is not realistic, and a different evaluation method is needed, which takes into account combinations of input parameters in order to evaluate the sensitivity of the output. This article will describe a novel method of sensitivity analysis, which considers the variation of all parameters during the sensitivity analysis for one parameter. This type of sensitivity analysis is called *dynamic sensitivity analysis*, because the sensitivity calculation is performed by dynamically varying all other parameters.

The basic algorithms and methods used for the dynamic sensitivity analysis will be developed in the next section. Three different sensitivity analysis applications were performed, all based on the dynamic sensitivity analysis:

- A *general sensitivity analysis* to classify each parameter of the cost model concerning its sensitivity to the cost in general, i.e., independent of a specific case.
- An *iterative sensitivity analysis* to perform a detailed examination of the input data of the cost model iteratively. This is designed to find a set of parameters whose accuracy is critical, i.e., will always influence the accuracy of the total cost.
- The *total variation sensitivity analysis* to study the probability density function of the total cost, which is based upon the inaccuracy of the estimated input values, where the inaccuracy is handled as a variate with a known distribution function.

3. Monte Carlo Methods for Sensitivity Analysis

Dynamic sensitivity analysis attempts to handle the inaccuracy of the input data, which are likely to be estimates. This inaccuracy can be seen as a variate with a defined distribution. The type of the distribution is an equal distribution, if the estimate defines a range, or a limited normal distribution, if the estimate is given by a mean value and a deviation from that mean value. The limited normal distribution is used to account for the fact that certain parameters, such as gate count, have defined limits (e.g., gate count cannot be smaller than zero).

The problems to be solved with the sensitivity analysis are the following:

1. Determine the mean value of the final cost and its variance by solving the integral of the cost model in order to answer the questions:
   "How sensitive is the resulting cost concerning the inaccuracy of the input values?"
   "What is the sensitivity behavior of the resulting cost considering the variation of one parameter with inaccuracy of the other parameters?"
2. Solve the differential of the cost model in order to answer the following question:
   "What is the maximum sensitivity of the resulting cost concerning the variation of one parameter with constraints for the other parameters?"
   This problem is an optimization problem, which can be solved by an extrema analysis.

Problem 1 can be solved analytically, but this would require the solution of multiple integrals over a very complex, discontinuous function, i.e., the cost model. Problem 2 cannot be solved analytically without modifications, which are related to discontinuities in the cost model, because they cannot be differentiated. Although the differentiation of the cost model could be performed analytically within the continuous regions, this approach is not practical for the following reasons:

First, the analytical method implies the solution of very complex integrals and differentials, which are strongly tied to the related function. However, the function changes from test strategy to test strategy, because parts of the function are related to the test methods form the test strategy. This means that the function to differentiate or to integrate is different for each test strategy, and therefore the integral or differential to be solved analytically varies. The multitude of test strategies, which is based upon the multitude of test methods would require the solution of many complex integrals and differentials, an extremely complex and error prone process. Second, this approach is very inflexible, because for each modification of the cost model or for each new test method to be considered, the whole manual integration and differentiation work has to be redone.

For these reasons, the Monte Carlo method was chosen for solving the integration problem and the optimization problem:

- The method is independent from the function to analyze, and therefore it is very flexible concerning its application to changing functions.
- The nature of the input parameter values are variates. Monte Carlo simulation is also based on variates [10], and therefore the method fits very well to the nature of the underlying problem.
- The method can be used for both the integration and the optimization problem [10], [11].

In the following, the Monte Carlo method as used here will be introduced. In the subsequent sections the methods will be adoped to the related problems to be solved.

"Problems handled by Monte Carlo are of two types called probabilistic or deterministic according to whether or not they are directly concerned with the behavior and outcome of random processes." [11] The type of Monte Carlo used in this thesis is deterministic, because the process, which is simulated by the cost model, is deterministic. The behavior of the cost model is not random, and the behavior of the input data is also known. Also, in theory, the problem can be solved deterministically, as shown above. Hammersley [11] defines deterministic Monte Carlo as a numerical solution of a deterministic problem. Deterministic Monte Carlo simulation is also called *Sophisticated Monte Carlo.*

The essential feature of Monte Carlo simulations is that a random variable is replaced by a corresponding set of actual values, having the statistical properties of the random variable [11]. This random variable can be part of the process under consideration, or it can be modeled from a deterministic variable. The actual set of values, so-called the random numbers, are used to analyze the process. This procedure is called the Monte Carlo simulation. Consider the following example:

A process is defined by the cost model and some input data, which are normal distributed. The question to answer is: What is the probability that the resulting cost exceed a certain limit? This question can be answered through Monte Carlo simulation as follows:

- Generate random numbers for the normal distributed input parameters having the statistical properties of the given normal distribution.
- Calculate the related resulting cost value.
- Repeat the above procedures many times.
- Measure the probability of exceeding the limit by calculating the relation of resulting cost values exceeding the limit to the number of calculations.

Instead of solving the complex multiple integral of the cost model over the distribution function of the input parameters, we simply measure the result by performing Monte Carlo simulation. Summing up, the Monte Carlo method is based upon the generation of random numbers relating to given distribution properties.

4. Correlation of the Input Parameters

When the input parameters of the cost model are varied randomly, we have to take into account, that some of the paramters are correlated to each other. For example, the number of cells of the design is correlated to the gate count of the design. This means, that if the gate count is high, there is a high probability, that also the cell count has a high value. If the correlation is not taken into account for the generation of the random numbers, we can get unrealistic or even invalid combinations of input parameters. For example, if the gate count is uniformly distributed between 1,000 and 10,000, and the cell count is uniformly distributed between 300 and 3,000, there is a probability of 0.22, that the cell count will be higher than the gate count, if these two values are independent. As we know, this cannot happen in reality, because in cell based designs a cell contains at least one gate equivalent. In order to get realistic results for the Monte Carlo simulation, we must take into account the correlation of parameters for the computation of the random numbers. If we see the input parameters as a vector x, and we can produce random numbers x_0 that follow a normal distribution with a mean of 0 and standard deviation of 1, we can generate a random number vector as follows:

$$x = C^* \cdot x_0 + \mu \qquad (1)$$

where C^* is given by

$$C^* \cdot C^{*T} = C \qquad (2)$$

where C represent the covariance matrix, and μ is the desired mean value. A proof is given in [12]. Because C is a positive definite matrix, C^* can be derived by using the Cholesky decomposition method see [12].

This method is used to calculate correlated multivariates, and it allows a correlation between all variates. The correlation is defined by the covariance matrix. In the case of the cost model, we have the following correlations:

gate count <=> cell count
gate count <=> number of flip-flops
number of flip-flops <=> sequential depth

This leads to the following covariance matrix for these three parameters:

gates:	μ_{11}	μ_{21}	μ_{31}	0
cells:	μ_{12}	μ_{22}	0	0
flip-flops:	μ_{13}	0	μ_{33}	μ_{43}
sequential depth:	0	0	μ_{34}	μ_{44}

where μ_{ii} is the variance σ_i^2 of the single parameter i, and μ_{ij} is the covariance of parameter i and parameter j. This is a special case of the covariance matrix, where the parameters are correlated in pairs. In this case the method can be simplified as follows:

$$X_{Gate} = \mu_{Gate} + X_{0Gate} \cdot \sigma_{Gate} \tag{3}$$

$$X_{Cell} = \mu_{Cell} + (X_{0Gate} \cdot \rho_{Cell} + X_{0Cell} \cdot \sqrt{1 - \rho_{Cell}^2}) \cdot \sigma_{Cell} \tag{4}$$

$$X_{Flip\text{-}Flop} = \mu_{Flip\text{-}Flop} + (X_{0Cell} \cdot \rho_{Flip\text{-}Flop} + X_{0Flip\text{-}Flop} \cdot \sqrt{1 - \rho_{Flip\text{-}Flop}^2}) \cdot \sigma_{Flip\text{-}Flop} \tag{5}$$

$$X_{SeqDepth} = \mu_{SeqDepth} + (X_{0Flip\text{-}Flop} \cdot \rho_{SeqDepth} + X_{0SeqDepth} \cdot \sqrt{1 - \rho_{SeqDepth}^2}) \cdot \sigma_{SeqDepth} \tag{6}$$

This can be proven by using (1) and (2) for each pair of the correlated parameters.

In order to be able to generate the correlated random numbers, we must know the distribution functions X^* per parameter, the mean value μ per parameter, the root to the variance σ per parameter, and the correlation factor ρ per correlation. X^*, μ and σ are variables. The correlation factor was derived using data from existing designs. On the basis of the calculated correlation factors, the following values were used for the sensitivity analysis:

p_{cell}: 0.78
$p_{flip\text{-}flop}$: 0.81
$p_{seqdepth}$: 0.23

5. General Sensitivity Analysis

A general study of the cost model is performed in order to classify all parameters of the cost model concerning their sensitivity impact on the total cost value. This analysis gives a general idea, which parameters must be estimated very accurately, even if the cost for this estimation is high, and for which parameters a rough estimate may fulfill the accuracy requirements concerning the resulting cost value. An outcome of this study may even be that some of the parameters can be neglected for the cost evaluation. This fact would allow a simplification of the cost model by cutting out the effect of these parameters. The refined cost model would provide the same results with lower costs in data acquisition and test strategy planning.

The sensitivity classification of the parameters will be performed by estimating for each parameter:

1. The mean value and the variance of the sensitivity in a constrained space $D \in R$ of the input parameters.
2. The maximum sensitivity of the total cost in a constrained space $D \in R$ of the input parameters.

In our approach all three characteristics will be estimated by performing a Monte Carlo simulation. Due to the fact that parts of the cost model are test method dependent, i.e., the cost model is different for different test strategies, the sensitivity analysis was performed for three different representative test strategies. These are *no DFT*, *scan path* and *circular self-test path*.

The integrals to calculate the mean value and the variance will be estimated by a Monte Carlo simulation as follows:

$$\mu = \frac{1}{n} \cdot \sum_{i=1}^{n} C'(x_i);$$

$$\sigma^2 = \frac{1}{n-1} \cdot \sum_{i=1}^{n} (C'(x_i) - \mu)^2$$

where μ is the mean value, σ^2 is the variance, $C'(x)$ is the sensitivity, n is the number of simulations and x_i is the vector of random numbers for the input parameters of the cost model. The random numbers per parameter are uniformly distributed. The range of the distribution covers all typical values. They are defined in Table 1.

The sensitivity $S_j(x_i)$ must be calculated for each input parameter. The sensitivity value is defined here as the relative difference in the total cost of a relative difference of the analyzed input parameter.

$$s_j = \frac{C'(x_j \cdot s) - C'(x_j)}{C'(x_j)}$$

where s is the sensitivity factor, which defines the increase of the parameter x_i as a percentage of x_j.

Table 1. Distribution characteristics of cost model parameters.

Parameter Name	Abbreviation	Lower Limit	Upper Limit
Number of cells	cells	3	64000
Complexity exponent	cexp	0.8	1.0
Number of gates	cgate	1,000	100,000
Labor cost rate	costrate	$500	$5000
Performance complexity	cperf	1	3
CPU time	cputime	0	1,000 h
Design center cost rate	descentrate	$10,000	$40,000
Computer equipment rate	equrate	$25	$1000
Designer's experience	exper	0	100
Required fault coverage	fcreq	70%	100%
Average number of faults per gate	fpg	2	5
Constant factor concerning designer's productivity	kdes	1	2
Constant factor concerning total productivity	kp	70,000,000	90,000,000
Manual test generation time per fault	mtgtime	0.05 h	1 h
Design originality	or	0	1
Productivity of the CAD system	pcad	1	5
Percentage of design time an external design center is used	percuse	0%	100%
Number test pattern for which test application cost increases	pms	64,000	640,000
Test application cost per step	pps	0	$25
Production unit cost per gate	puc	$0.5	$2
Sequential depth	seqdepth	0	103
Production volume	vol	1,000	1,000,000

Monte Carlo simulations are estimates that converge to the exact value of the integral. The number of simulations needed to achieve a satisfying accuracy of the Monte Carlo estimate can be derived from the standard error. We have implemented a method to calculate the standard error and to calculate the confidence interval for the estimations. The authors have found that 10,000 simulations are sufficient for this analysis.

In Figure 1 for each parameter j the sensitivity S_j as defined above is printed. The sensitivity factor was set to 1.1, which relates to a 10% variation of the parameter. The vertical line marks the range, in which the sensitivity will be in 99% of all cases. The small horizontal bar marks the mean value of the sensitivity. The figure presents the results for the test method *no DFT*. Figures 2 and 3 provide results for scan-path and circular self-test path respectively.

We distinguish between the three cases no DFT, scan path, and self-test. For each case the parameters are classified in Table 2.

See Table 1 for the meaning of the parameter abbreviations. From this classification and Figures 1 to 3, the following conclusions can be drawn:

- The parameters, number of gates, sequential depth, production unit cost per gate and production volume, are the most important parameters for the sensitivity and accuracy of the cost estimation.
- The parameters, constant factor concerning designer's productivity, designer's experience, number of cells, productivity of the CAD system, constant factor concerning productivity, design center cost rate, computer equipment rate, percentage of design time an external design center is used, CPU time, design originality, performance complexity and manual test generation time per fault, do not affect the sensitivity of the total cost very much in most cases. Inaccurate data for these values may fulfill the accuracy requirements of the total cost.

The sensitivity to the total cost for some parameters remains negligible in at least 99% of all cases. This leads to the question, whether the total cost are insensitive to these parameters at all. To answer this question, the maximum sensitivity for the parameters must be determined. If the maximum sensitivity is negligible, the cost model can be refined by taking out the insensitive parameters. The determination of the maximum sensitivity is an optimization problem. As we are using the Monte Carlo method, we had to implement a method to reduce the sample size of the Monte Carlo simulation, because crude Monte Carlo simulation [11] would lead to an extremely large sample size. We have implemented a technique called *importance sampling*. The algorithm developed is described in the next section and enables the finding of the maximum sensitivity conditions with only 1,000 simulations.

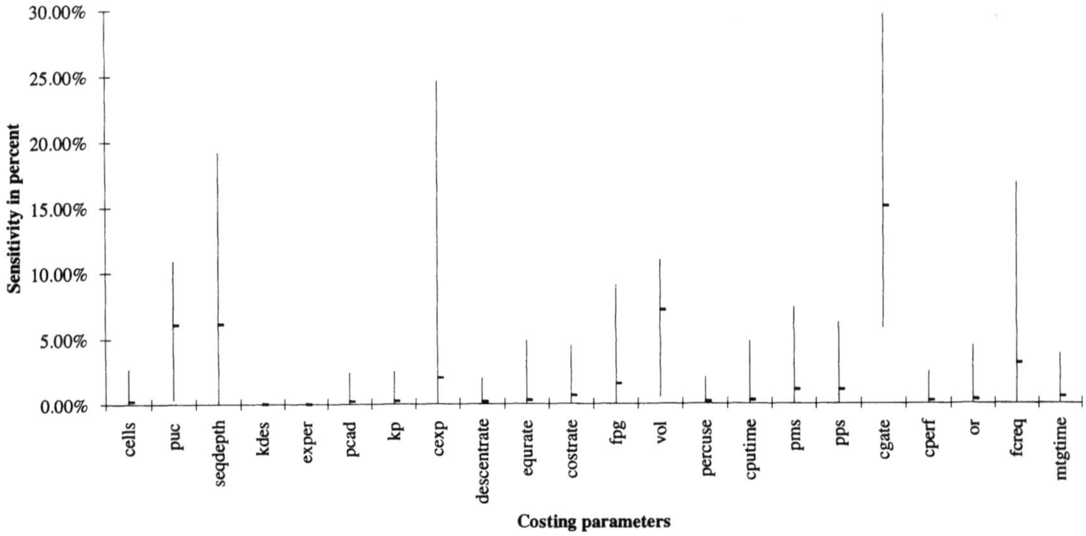

Fig. 1. Sensitivity of total cost with a sensitivity factor of 1.1 (10% variation), no DFT.

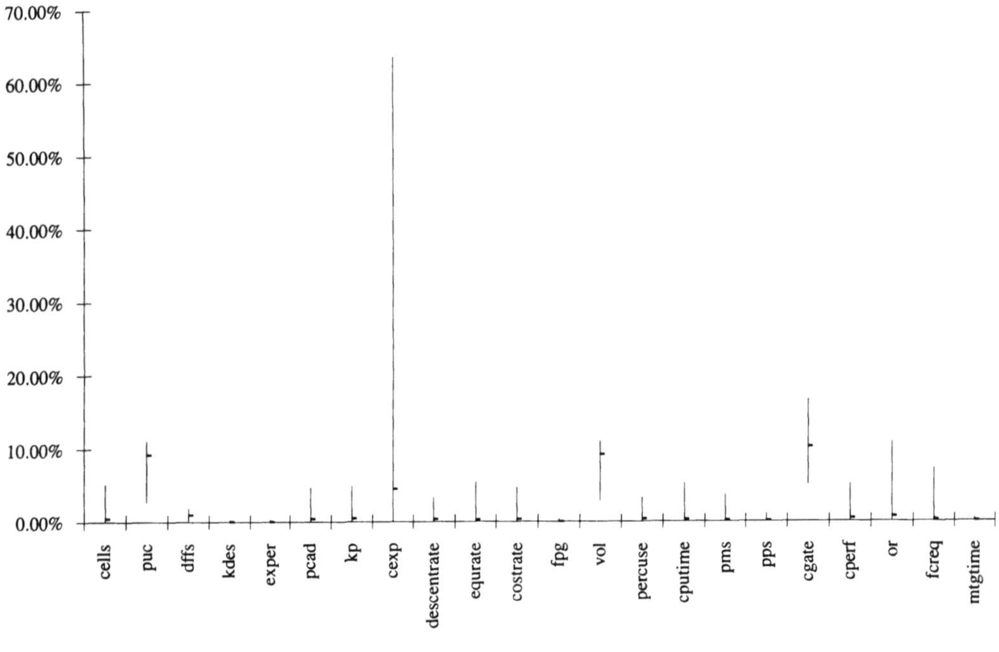

Fig. 2. Sensitivity of total cost with a sensitivity factor of 1.1 (10% variation), scan path.

6. Estimation of Maximum Sensitivity

As already mentioned in the previous section, the sensitivity to the total cost for some parameters remains negligible in at least 99% of all cases. This leads to the question whether the total cost is insensitive to these parameters at all. To answer this question, the maximum sensitivity for the parameters must be determined. If the maximum sensitivity is negligible, the cost model can be refined by taking out the insensitive parameters.

This section will describe the methods developed in order to determine the maximum sensitivity and will present the results.

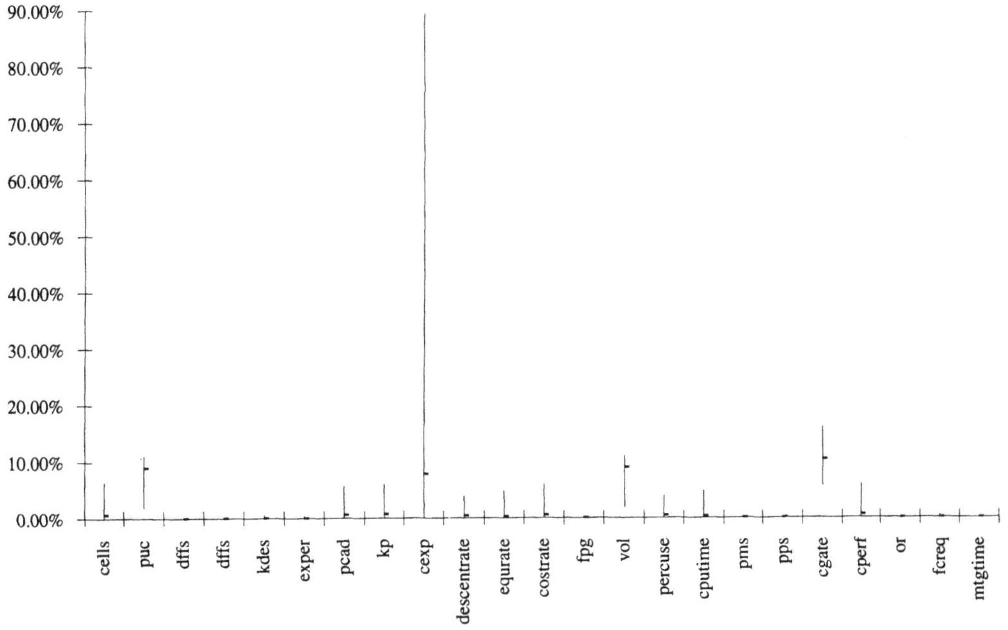

Fig. 3. Sensitivity of total cost with a sensitivity factor of 1.1 (10% variation), self-test.

Table 2. Parameter classification according to sensitivity.

Class	No DFT	Scan Path	Self-Test
High Mean, High Maximum	cgate, seqdepth	cgate	cexp
High Mean, Av. Maximum	puc, vol	puc, vol	cgate, vol, puc
Av. Mean, High Maximum	cexp, fcreq, fpg, costrate, pms, pps	cexp	
Low Mean, Av. Maximum	cells, pcad, kp, descentrate, equrate, percuse, cputime, or, cperf, mtgtime	cells, or, fcreq, pms, cperf, cputime, pcad, kp, equrate, costrate, descentrate, percuse,	cells, pcad, kp, descentrate, equrate, costrate, percue, cputime, cperf
Low Mean, Low Maximum	kdes, exper	dffs, kdes, exper, fpg, pps, mtgtime	dffs, kdes, exper, fpg, pms, pps, or, mtgtime

6.1. *The Algorithm*

The following definition of the maximum sensitivity will be used:

- The total cost difference for parameter i is defined to

$$CD_i(x) = C(x, x_{i0}) - C(x, x_{i1}) \text{ with } C(x, x_{i0}) \geq C(x, x_{i1}) \quad (7)$$

- The relative difference for parameter i is defined to

$$RD_i(x) = \frac{C(x, x_{i0}) - C(x, x_{i1})}{C(x, x_{i0})} \quad (8)$$

- The maximum total cost difference for parameter i is defined to

$$CD_{max_i}(x, x_i) = max(CD_i(x, x_i))\forall (x_{i0}, x_{i1} \in D \in R) \quad (9)$$

where D is the constrained search space of x_i. This means, that CD_{max} is such that, if all parameter values beside x_i are kept constant, x_{i0} is the maximum and x_{i1} is the minimum of the function $CD_i(x, x_i)$.

- The relative maximum difference for parameter i RD_{max} is defined to the relative difference of the maximum total cost difference.
- The maximum sensitivity for parameter i is defined to

$$S_{max_i}(x) = max(RD_{max_i}(x))\forall (x \neq x_i \in D \in R) \quad (10)$$

This means that the maximum sensitivity is determined by that combination of input parameters which leads to the relative maximum difference in the total cost which can be caused by the variation of parameter *i*.

This definition of the maximum sensitivity will be used for deciding whether a parameter can be taken out of the cost model.

In order to determine the maximum sensitivity, we need to determine the relative maximum difference. If the function $CD(x_i)$ is monotone, the maximum and the minimum of CD is given by the upper and the lower limit of the constrained space of the parameter. Because this is not always the case, the following approximation method was selected to determine the maximum and minimum of CD:

```
calculate_CDmax()
 x_min is the lower bound of range
 x_max is the upper bound of range
 minind = maxind = 1
 while ((recursion depth less than 5) and (minind and
   maxind are not 0 or 10))
   C_min = 0
   C_max = maximum float number
   minind = 0
   maxind = 10
   for i from 0 to 10
     Cx = C(x_min + i · (x_max − x_min)/10
     if (Cx > C_max)
       C_max =Cx
       maxind = i
     else if (Cx < C_min) then
       Cmin = Cx
     minind = i
     endif
   endfor
   if (minind and maxind are not 0 or 10) then
     x_min = x_min + max (0, min (minind, maxind)
       − 1) · (x_max − x_min)/10
     x_max = x_min + min (10, max (minind, maxind)
       + 1) · (x_max − x_min)/10
     increment recursion depth
   endif
 endwhile
RDmax = (C_max − C_min)/C_max
```

This algorithm will give a good approximation of the relative maximum difference, if the sensitivity of the function does not vary too much, i.e., there are no "sharp peaks" in the function. This assumption can be made for the cost model.

The maximum sensitivity can be determined by a Monte Carlo simulation as follows:

$$S_{max} = max \bigcup_{j=1}^{n} CD_{max}(x_j) \qquad (11)$$

The values of x are sampled from uniformly distributed variates as defined in Table 1. This method requires a large amount of simulations to achieve sufficient results, because we want to determine a specific element out of an infinite set of parameter combinations. The following example should illustrate this:

The number of parameters is 23. If we divide each parameter range into four subranges, and if we want to combine all subranges for all parameters, the number of combinations is 4^{23}. There is no chance to run so many simulations.

Therefore a technique was developed to reduce the sample by keeping the accuracy of the result. This technique will be described in the following section.

6.2. *Reduction of the Sample Size*

There are several publications that are related to Monte Carlo optimization ([3], [10], [14]). All of them address the problem of reducing the sample size by using a technique called *importance sampling*. This technique exploits the random samples of the past for generating the random samples in the future. In [13] and [14] the importance sampling techniques closely linked to the problem to solve, which is different from the problem to be solved here. In [10], the techniques presented are more general. The authors have therefore implemented a different algorithm. In this algorithm the Monte Carlo simualtion is divided into two phases:

1. In the first phase Monte Carlo simulation is performed with the input parameters uniformly distributed. In this phase we make an estimate of the values of the input parameters, which will lead to a high sensitivity. We call these values *maximum sensitivity values*.
2. In the second phase Monte Carlo simulation is performed with a distribution function for the input parameters, which is adopted to the maximum sensitivity values. In this phase the sensitivity values are updated with each simulation. This means, that in this second phase the parameter values are sampled in the region where we expect the actual maximum sensitivity value.

The algorithm is implemented as follows:

calculate_max_sens()
for L times
perform Monte Carlo simulation with uniformly distributed variates as input parameters and determine RD_{max}
endfor
select M samples from all previous simulations which cause the M maximum RD_{max}
for N times
perform Monte Carlo simulation with importance samples and determine RD_{max}.
update the M maximum samples with the sample which just was calculated.
endfor

The important samples are based upon the input parameter values of the *M* maximum samples, and they are calculated as follows:

importance_sampling()
determine the distribution of the parameter values which are related to the M maximum samples.
constrain this distribution by the lower and the upper limit of the parameter range
generate a variate from this distribution.

The generation of a variate based on the determined distribution function was implemented by applying a table method that is similar to the Marsaglia table method [11]. Instead of using the table for a normal distribution, we take the table of the values for the *M* maximum samples.

The object in this method is to concentrate the distribution of the sample points in the parts of the parameter range that are of most importance instead of spreading them out evenly. Nevertheless, the *L* simulations with a uniform distribution function are needed to make a good estimate about which parts are important. If a certain parameter has influence on the maximum sensitivity, the related importance sampling distribution will converge to a certain value, which we expect to be the value for the maximum sensitivity. The speed of the convergence and therefore the sample size and the risk of running into a local maximum is driven by the values *L* and *M*.. A high value for *M* will cause slow convergence but a low risk of running into a local maximum and vice versa. A higher value for *M* will lower the risk of running into a local maximum but will increase the sample size and therefore computer runtimes. The authors performed some experiments with different values for *L* and *M* to find a good trade-off. In the next section, the values for *L*, *M* and *N* will be presented and the maximum sensitivity values for the parameters which are expected to be insensitive will be presented and discussed.

6.3. Results

It was found that $M = 10$ and $L = 300$ was a good trade-off between fast convergence and a low risk of running into a local maximum. The total number of simulations $L + N$ was set to 1,000.

The maximum sensitivity was determined for the following parameters:

performance complexity, design center, cost rate, designer's experience, average number of faults per gate, constant factor concerning designer's productivity, constant factor concerning total productivity, manual test generation time per fault, number of flip-flops, productivity of the CAD system, percentage of design time an external design center is used.

The analysis was made for the test strategies no DFT, scan path and self-test. The resulting maximum sensitivity in percentage of the related maximum cost is shown in Figure 4. This graph shows that the maximum sensitivity exceeds 20% for all parameters, which is by far too high to neglect. This means, that for *all* parameters situations may occur, where a variation of the parameter within the range defined in Table 1 can lead to more than a 20% variation of the total cost. A simplification of the cost model in general by removing parameters is therefore not possible.

7. Iterative Sensitivity Analysis

This second application of sensitivity analysis leads to a test strategy planning procedure in iterative steps as follows:

1. Make a rough and inexpensive estimate of the input data.
2. Perform test strategy planning.
3. Perform a sensitivity analysis in order to find out, which parameters need to be defined more accurately. These are those parameters, which have a high sensitivity.
4. If accuracy of the result is OK, then the test strategy planning procedure is finished, else make more accurate estimates for the most sensitive parameters and go to step 2.

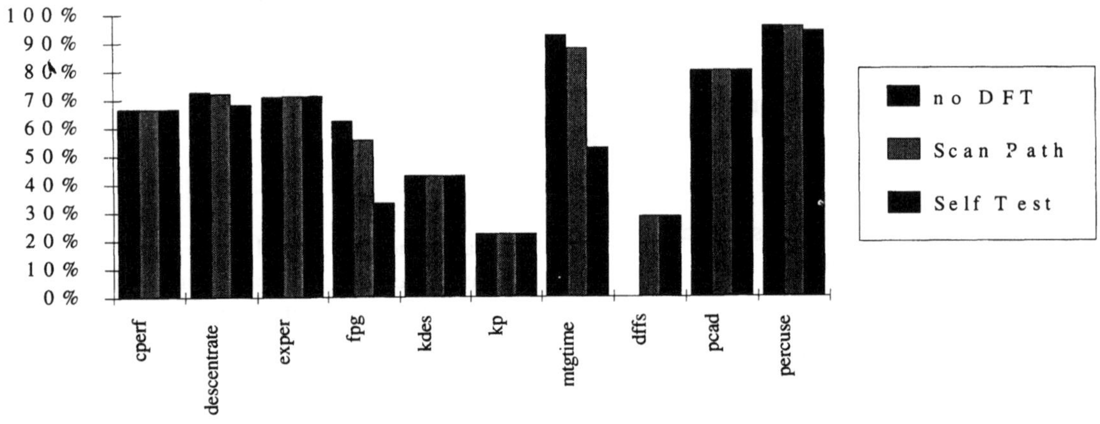

Fig. 4. Maximum sensitivity values.

In the iterative sensitivity analysis, the same methods are used as for the general sensitivity analysis. The main differences are the iterative procedure and the different distribution of the parameter values. For the general sensitivity analysis, we used a uniform distribution for all parameters in the range of all possible values. For the iterative sensitivity analysis, we have inaccurate estimates for the parameters, which can be either a uniform distribution or a normal distribution.

8. Total Variation Sensitivity Analysis

The third application considers the fact that there is a certain cost limit for data acquisition, which makes estimated input data remain inaccurate. This means that the decision criteria, i.e., the resulting cost value, is subject to inaccuracy. This inaccuracy is different from test strategy to test strategy. The following eample will illustrate this fact.

Comparing the test method scan path in combination with combinational ATPG to *no* scan path in combination with sequential ATPG, the related test generation cost as part of the resulting cost are much more uncertain for no scan path. The reason for that is the inaccurate parameter, sequential complexity, which affects sequential ATPG cost but not combinational ATPG cost. This fact may lead to a situation where the resulting costs for the sequential ATPG approach are lower as those for the combinational ATPG assuming that the estimated value for the sequential complexity is exact, but that there is a high probability, that the costs will be much higher for the sequential ATPG approach, considering the high probability that the sequential complexity will be higher than estimated.

In order to take the inaccuracy of the cost estimation into account, the total cost value will be replaced by the distribution function of the total cost value as the optimization criteria.

8.1. The Algorithm

The distribution function $F(x)$ will be used to determine with a given cost probability p_c the upper cost limit CL:

$$CL = F^{-1}(p_c) \tag{12}$$

The cost probability has to be defined by the user of the test strategy planner. This cost probability is defined as the probability, that the total costs will be less than the related cost limit, due to inaccuracy of the input data. The value for the cost probability represents the *risk of the cost estimation* and by that the risk of the decision. In other words, the risk of the decision is the risk that the chosen test strategy is not the optimum.

The calculation of the cost limit is implemented by performing Monte Carlo simulation. if the n cost values resulting from n simulations are stored in a sorted list $c[n]$ in rising order, the cost limit can be estimated by a linear approximation as follows:

$$CL = c[int(n \cdot p_c)] + (c[int(n \cdot p_c) + 1] - c[int(n \cdot p_c)]) \cdot (n \cdot p_c - int(n \cdot p_c)) \quad \text{if } p_c < 1$$

$$CL = c[int(n \cdot p_c)] \quad \text{if } p_c = 1$$

where $int(np_c)$ is the integer of the product np_c. It represents the index in the list of cost values, which points to the cost value which is closest to the estimated cost limit.

8.2. *Determination of the Number of Simulations Needed*

The accuracy of the determination of the cost limit as described in the previous section depends on the number of simulations for two reasons:

- The Monte Carlo simulation is an unbiased estimator for the distribution of the total costs. This estimator is subject to inaccuracy as long as the number of simulations is finite.
- The linear approximation is subject to inaccuracy that depends on the curvature of the distribution function in the region where the linear approximation is made, and it depends on the number of simulations.

It is therefore necessary to study the impact of the number of simulations on the accuracy of the cost limit estimation in order to determine an appropriate number for the Monte Carlo simulations.

It was decided to study the convergence of the Monte Carlo simulation for analyzing the estimation error and the number of simulations needed to keep the error under a certan limit. Due to being an unbiased estimator, we can assume, that the Monte Carlo estimation will converge to exact value by increasing the number of simulations. If the difference between succeeding estimations during the Monte Carlo simulation become very small, we take the related value as the exact value. Based on this exact value *CL* the relative estimation error for an estimate CL_{est} is given by

$$\vartheta_{est} = \left| \frac{CL - CL_{est}}{CL} \right| \tag{13}$$

The convergence study was performed for the test strategies no DFT, scan path, and circular self-test path, and for the following cost probabilities: 99%, 90%, 75%, 60% and 50%.

The relative estimation errors are shown in Figures 5 to 7. The authors selected the cost limit for 10,000 simulations as *CL*. The graphs show that for cost probabilities of 90%and 99% the estimation error converges slower to zero than for cost probabilities of 50% or 60%. This is because of the larger gradient of the inverse distribution function $F^{-1}(p_c)$ for the high end and low end regions of the function, for which an argument difference affects the related cost limit more than for lower gradients. The relative error remains for all curves under 5% for simulation number greater than 200. Therefore the authors propose a simulation number of 200 to perform Monte Carlo simulations for total variation sensitivity analysis.

9. Summary

In this article the authors presented the techniques they have developed to perform sensitivity analysis for three different applications. All applications are based on Monte Carlo simulations. We have implemented the methods in the test strategy planner ECOTEST. The system is implemented in C++, and it runs on SUN SPARC and HP-Apollo workstations under UNIX.

We have shown that in 99% of all cases the estimation accuracy for one third of the costing parameters is insignificant. In most cases only five parameters, the

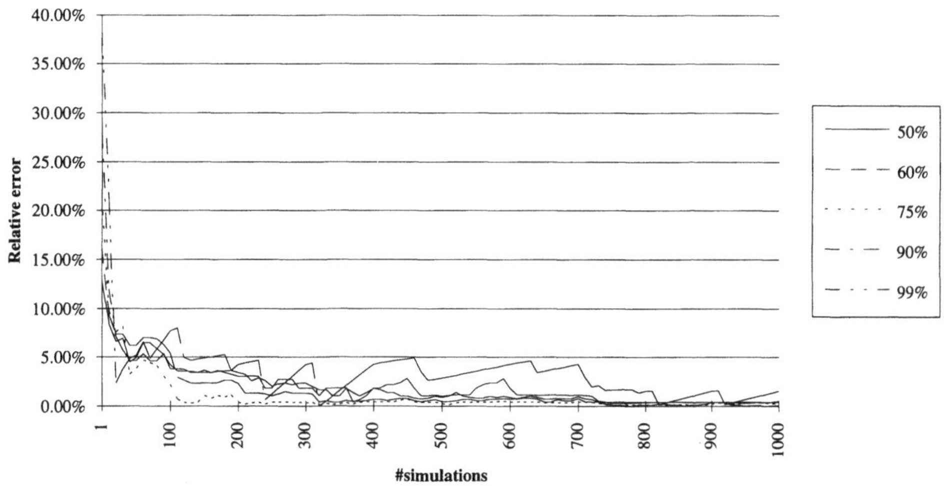

Fig. 5. Convergence of Monte Carlo simulation for no DFT.

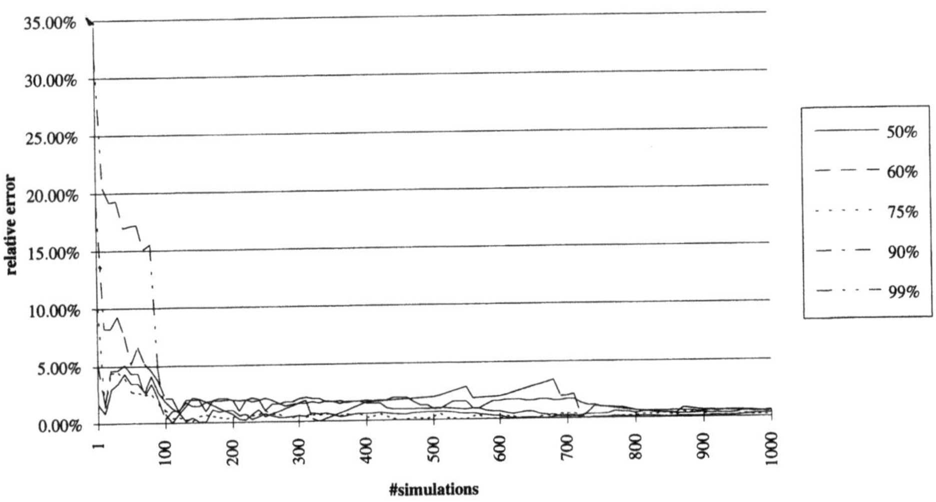

Fig. 6. Convergence of Monte Carlo simulation for scan path.

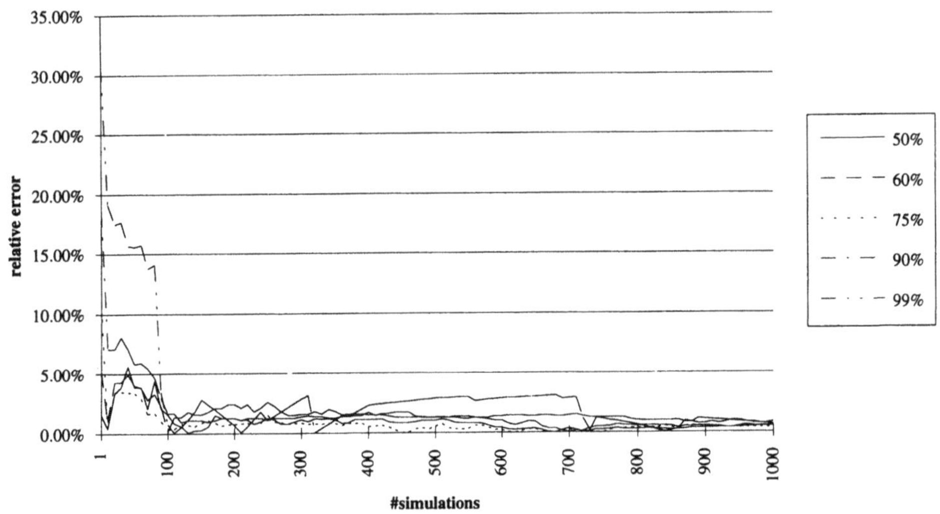

Fig. 7. Convergence of Monte Carlo simulation for self-test.

production unit cost, the sequential depth, the production volume, the gate count, and the required fault coverage, control the accuracy of the total cost. This knowledge enables a drastic reduction of the effort in data gathering for economics based test strategy planning.

The general sensitivity analysis showed that for each parameter configurations may occur, which lead to a maximum sensitivity of more than 20% of the total cost. But in most cases, the total cost is sensitive to only a few parameters of the cost model. These are the number of gates, the sequential depth, the production unit cost per gate, the production volume, and the complexity exponent. Based on this outcome, it is proposed that economics based test strategy planning is performed as follows:

- Provide values for the "important parameters" listed above.
- Perform iterative sensitivity analysis until the accuracy requirements are fulfilled.

This method can drastically reduce the effort spent in data gathering, because in most cases the accuracy requirements can be fulfilled by providing values only for a few parameters.

References

1. C. Dislis, I.D. Dear, J.R. Miles, S.C. Lau and A.P. Ambler, "Cost Analysis of Test Method Environments," IEEE International Test Conference, pp. 875–883, 1989.
2. I.D. Dear, C. Dislis, A.P. Ambler, J. Dick, "Economic Effects in Design and Test," *IEEE Design and Test of Computers*, vol. 8, no. 4, pp. 64–77, Dec. 1991.
3. M.S. Abadir, "TIGER: Testability Insertion Guidance Expert System," *Proceedings IEEE International Conference on Computer Aided Design*, pp. 562–565, 1989.
4. M.A. Myers, "An Analysis of the Cost and Quality Impact of LSI/VLSI Technology on PCB Test Strategies," IEEE International Test Conference, 1983, pp. 382–395.
5. S.A. Szygenda, "Profit, Liability, and Education: Influencing Factors on the Economics of Non-testing," *Proc. Economics of Design and Test for Electronic Circuits and Systems*, Ellis Horwood Limited, Chichester, 1992.
6. B. Davis, *The Economics of Automatic Testing*; McGraw-Hill, London, 1982.
7. W. Dinkelbach, *Sensitivitätsanalysen und parametrische Programmierung*; Springer-Verlag, Berlin, Heidelberg, New York, 1969.
8. C. Dislis, *A Financially Based Automated Advisor for Design for Test Strategy Generation*, Ph.D. thesis, Brunel University, United Kingdom, 1992.
9. P. Varma, A.P. Ambler, and K. Baker, "An Analysis of the Economics of Self-Test," *IEEE Proceedings of the International Test Conference*, 1984.
10. R.Y. Rubinstein, *Monte Carlo Optimization, Simulation and Sensitivity of Queueing Networks*, John Wiley & Sons, New York, 1986.
11. J.M. Hammersley and D.C. Handscomb, *Monte Carlo Methods*, Methuen & Co. Ltd., London, 1965.
12. J. Armaos, *Zur Optimierung elektrischer Schaltungen unter Berücksichtigung der Parametertoleranzen*. Ph.D. thesis, Institute for Computer Aided Circuit Design, Technical University of Munich, 1982.
13. K.J. Antreich, and R.K. Koblitz, "Design Centering by Yield Prediction," *IEEE Transactions on Circuits and Systems*, vol. CAS-29, pp. 88–95, Feb. 1982.
14. G. Kjellstrom and L. Taxen, "Stochastic Optimization in System Design," *IEEE Transactions on Circuits and Systems*, vol. cas-28, July 1991.

Jochen Dick is a staff member of Siemens-Nixdorf-Informationssysteme in Munich. His work has included research in automatic test pattern generation and the development of a DFT rule checker. He gained his PhD from Brunel University on test economics and test strategy planning and holds an MS in Electrical Engineering from the Technical University of Munich.

Erwin Trischler received the Dipl.-Ing. degree in electrical engineering from the University of Zagreb, Croatia and the Dr.-Ing. degree in electrical engineering from the Technical University of Munich, Germany.

He was the manager of Computer Aided Test Engineering (CATE) research and development at SMI AG, AP division, Munich, Germany and involved in the European Community Project EVEREST (ESPRIT 2, #2318) and JESSI AC6 project "Test Generation and Design for Testability Support." Presently he is manager in the area of telecommunication management network (TMM) operations systems development at SNI AG, Munich, Germany.

In the area of CAE his interests include concurrent engineering, design, and test automation for complex digital systems, especially ATPG, design for testability, built-in self-test, and economics based test strategy planning. Previously, he was also a senior researcher at the Siemens' Corporate Research and Technology Laboratories in Princeton, NJ.

Dr. Trischler received the 1988 best paper award of the IEEE Transactions on Computer-Aided Design of Integrated Circuits and Systems. He was also a member of the Technical Program Committee of the 2nd European Test Conference (1991), 1st Asian Test Symposium (1992) and the topic chairman at the European Design Automation Confrence (1993). He has published over 25 papers in the area of testing digital systems.

Chryssa Dislis is a Research Fellow in the Department of Electrical Engineering and Electronics at Brunel University and currently works on economics modeling in the design of MCMs. Her research interests include design for testability, CAD for VLSI devices, and the economics of test at chip and board levels. Dislis holds a PhD from Brunel University. She is a member of IEEE and IEE. She is a Chartered Engineer.

Tony Ambler holds the Racal-Redac Chair in Test Technology at Brunel University. He currently serves as a member of the editorial board of JETTA, is Programme Chair for the 1994 IEEE International Conference on Computer Design (ICCD) and will be General Chair for the 1995 European Design and Test Conference. He is the co-founder of the International Workshops on Economics of Design, Test and Manufacturing held in Austin, Texas. His research interests include design for testability, test economics, CAD for electronic systems and concurrent engineering. He is a member of the IEEE Computer Soeicty, ACM, and IEE. He is a Chartered Engineer.

JOURNAL OF ELECTRONIC TESTING: Theory and Applications, 5, 253-261 (1994)

Improving Quality: Yield Versus Test Coverage

STEVEN D. MILLMAN
Motorola, Inc., 2100 E. Elliot Rd., MD EL510, Tempe, AZ 85284

Abstract. Until now, it has been thought that the best way to improve quality was to improve yield, that is, to improve the manufacturing process so that fewer defective parts are manufactured. This philosophy has been applied to all areas of manufacturing, from widgets to whalers. One of the tenets of this philosophy is that quality should not be tested in, but should be built in via a well-controlled process that is continuously improving. However, when it comes to semiconductors, significant improvements for a process in order to improve yields can cost millions of dollars. In such a capital intensive industry, where the life of a process is only a few years, such investments are often not in the best economic interests of a semiconductor manufacturer. How, then, can the customers' requirements for quality be met? This article shows that for typical values of yield and test coverage, quality improves faster due to improvements in test coverage than it does for improvements in yield. As a result, testing can be used to achieve improvements in quality at lower costs, and therefore more economically, than improvements in the manufacturing process.

Keywords: Fault modeling, physical defects, quality, test economics, yield.

1. Introduction

Until now, it has been thought that the best way to improve the quality of products shipped to customers was to improve the yield of the manufacturing process. This view of improving yield rather than fault coverage is also important from an economic standpoint since higher fault coverage results in fewer parts that can be sold. Improving the test coverage simply detects a higher percentage of the faulty products before they reach the customer. Unfortunately, the parts identified as faulty typically cannot be repaired and are thrown away. Hence, money is spent manufacturing products that never see the customer. On the other hand, by increasing the manufacturing yield, a smaller percentage of the parts are defective and fewer parts are thrown away. As a result, the manufacturing cost per unit decreases allowing higher profits and/or a better competitive position.

However, customers are requiring higher quality and are not concerned with vendors' profit margins. As a result, manufacturers need to increase quality and, more importantly, the customers' *perceptions* of quality. This article shows that increasing test coverage can be a more economical means to increasing quality than increasing yield. This will be shown to be true even if test coverage is already high and even if increasing test coverage is more expensive than a similar increase in yield.

The next section describes the two models used in this article to relate test coverage and manufacturing yield to customer quality. Since test coverage is shown to be so critical to improving quality, the effect of incorrectly modeling defects is then presented. Next, it is shown that, for typical values of test coverage and manufacturing yield, quality improves faster due to changes in test coverage than it does for changes in yield. Finally, when to improve test coverage and when to improve manufacturing yield is discussed.

2. Models for Quality

The discussion on quality that follows is based on two models that relate test coverage and manufacturing yield to quality level. The first model was presented in [1], and is shown in Equation 1, where $T \equiv$ Test coverage, $Y \equiv$ manufacturing Yield, and $QL \equiv$ Quality Level.

While this is a simple and elegant model, it has been shown to be overly pessimistic [2], [3]. (The possible explanations for this pessimism are many and are beyond the scope of this article.) However, due to its simplicity, the following analyses are done first with this model and then with the more complex Agrawal model.

$$QL = Y^{(1-T)} \tag{1}$$

The second model to be used was first presented in [4], and is shown in Equation 2, where $T \equiv$ Test coverage, $Y \equiv$ manufacturing Yield, $n_0 \equiv$ the average number of faults per faulty part, and $QL \equiv$ Quality Level. The parameter n_0 takes into account the fact that defects tend to cluster rather than occur uniformly across the wafer. It was shown in [3] that the Agrawal model accurately predicted the quality level for an actual part with $n_0 = 3.5$. Naturally, the correct value of n_0 will vary from part to part and process to process. For the work presented below, the value $n_0 = 3.5$ is used. It will be seen that the conclusions do not change for different values of n_0.

$$QL = 1 - \frac{(1 - T)(1 - Y)e^{-(n_0-1)T}}{Y + (1 - T)(1 - Y)e^{-(n_0-1)T}} \tag{2}$$

Before proceeding, it is important to define exactly what is meant by test coverage, manufacturing yield, and quality level, especially with regard to the models for quality level presented above. *Test coverage*, as defined in Equation 3, is the percentage of modeled faults that are detected by a set of test vectors. As shown in the next section, which faults to model and which fault model to use can have a significant impact on the accuracy of the prediction of quality. For the work presented below, the tests considered are for functional, or Boolean, testing only. Parametric and other forms of testing that are not functional, such as IDDQ, are not considered. However, there is no reason to expect that failures detected by such tests will affect quality differently than failures detected via Boolean methods.

$$T = \frac{\text{Number of Modeled Faults Detected}}{\text{Total Number of Modeled Faults}} \tag{3}$$

Manufacturing yield, as defined in Equation 4, is the percentage of parts available for testing that were manufactured correctly. This is not the same as the true yield since parts that have failed previous tests have already been eliminated from consideration.

$$Y = \frac{\text{Number of Defect-Free Parts Tested}}{\text{Total Number of Parts Tested}} \tag{4}$$

Quality level, as defined in Equation 5, is the percentage of parts that pass the test and are defect-free. Note that this percentage does not include defective chips whose defects are not detectable by the set of test vectors included in the test. One of the most critical goals in business is to get the quality level as close to 100% as possible.

$$QL = \frac{\text{Number of Defect-Free Parts That Pass the Test}}{\text{Total Number of Parts That Pass the Test}} \tag{5}$$

It can be concluded from the above definitions that the work presented below focuses on a particular step of the manufacturing process, namely functional test. However, the analyses done below could be applied to any step of the manufacturing process where a choice between increasing yield or test coverage must be made.

3. Fault Models and Test Coverage

For either model of quality presented above, it is clear that test coverage has a significant impact. Unfortunately, test coverage numbers are typically derived from simulations of fault models that have little to do with the way that faulty chips behave. For example, while it has been shown that bridging faults are the most common type of failure in some technologies [5]–[7], the stuck-at fault model is still used more often than any of the various bridging fault models. Even worse is the situation where no fault model is used and a simple toggle test is applied to the chip.

Figure 1 shows the impact fault coverage can have on quality. Seven different test sets and five different fault models were used for fault simulations of the 74181 16-function ALU [8]. These test sets were all generated so as to detect 100% of the stuck-at faults. Table 1 shows the expected defective parts per million (DPPM) shipped to the customer for each combination of fault model and test set using the Agrawal model with 95% yield and $n_0 = 3.5$. Figure 1 shows this data graphically. The shipped DPPM was found by multiplying $(1 - QL)$ by one million. Wired-AND, wired-OR, and the voting model are all bridging fault models. The shipped DPPM for each of the test sets was calculated assuming that only one type of fault could occur. For example, boxes in the figure represent the shipped DPPM that would be expected for each of the test sets if bridging faults were the only type of defect that could occur, and that when they did occur, they behaved as predicted by the voting model. (Stuck-open faults were

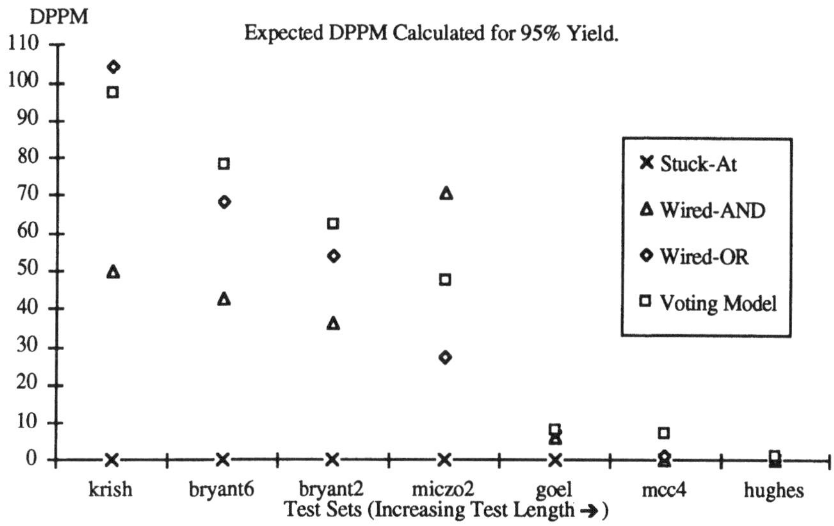

Fig. 1. Expected shipped DPPM for 74181 ALU for various 100% stuck-at test sets.

Table 1. Predicted shipped DPPM for seven test sets for the ALU.

Test Sets	Test Length	Fault Models				
		Stuck-At	Wired-AND	Wired-OR	Voting Model	Stuck-Open
krish	12	0	50	104	98	1576
bryant6	12	0	42	68	78	1288
bryant2	14	0	36	54	62	1633
miczo2	17	0	71	27	47	677
goel	35	0	6	7	8	582
mcc4	124	0	0	1	7	761
hughes	135	0	0	0	1	360

not included in Figure 1 as the predicted shipped DPPM for them was so large that the required scale would have led to little differentiation between the data points for the other fault models.)

It is clearly seen from Figure 1 that the expected quality level can vary greatly depending on what type of failure or failures are occurring and what type of failures were targeted by the test set. This means that in order to expect the predicted quality level to match reality, the fault model used must also match reality. It should be noted that more than one fault model may be necessary to model the behavior of all of the defects that occur.

Knowing how to model failures is only the first step. The second step is knowing which failures to model. For example, given an accurate bridging fault model, how are the bridging faults used during fault simulation selected? Simply selecting all possible pairs of nodes as candidate bridging faults is impractical since widely separated nodes are unlikely to short. Selecting only pairs of adjacent or crossing nodes is better, but it is still possible that some of these failures will not occur in practice. Clearly then, the type of failure to be modeled must be identified, as well as the conditions under which it occurs.

As a simple example, consider a chip for which 1,000 faults are to be modeled with a fault model that exactly matches the behavior of faulty chips. Assume that these 1,000 faults, or any subset thereof, are the only faults that occur on the actual chips. If these chips are tested with a test set that has 98% fault coverage of the 1,000 faults, what is the real fault coverage? If all 1,000 faults can occur with equal frequency, then the true fault coverage is 98%. But what if only 900 of the 1,000 faults modeled can actually occur? Then the true fault coverage could range from a best case of 100% if all of the undetectable faults cannot occur to a worst case of 97.8% if all of the undetectable faults can occur. And if only 500 of the modeled faults can occur, then the true fault coverage will range from 100% to 96%. Naturally, if the undetectable faults occur with greater frequency than the detectable faults, then the worst case numbers will be even lower; that is, the percentage of defects that actually occur and are detected will be lower than the percentage of the number of defects that are detectable.

Why then have tests generated for stuck-at fault models been so successful at detecting other types of failures? (If they had not been so successful, customers would have returned far more faulty parts than they have.) The answer can be seen in the results presented in Figure 1. The short test sets have been optimized for detecting stuck-at faults with as few vectors as possible, whereas the longer test sets have not; the hughes test is pseudorandom, the mcc4 test is pseudo-exhaustive, the miczo2 test is pseudorandom compressed, and the others are derived from more complex algorithms targeted at stuck-at faults. Note that the longer test sets predict quality levels that are much closer together for the different fault models than do the shorter test sets. Hence, past designs tested with long test sets tended to achieve the expected quality levels. But the figure shows that as test pattern generators create shorter, more efficient tests for stuck-at faults, the level of quality can be expected to significantly decrease if other types of failures predominate.

Based on this discussion, it should be clear that for test coverage to have any meaning, it must be based on fault models that accurately model how failures affect chips and where and how often those faults occur.

4. The Relationship Between Test Coverage, Yield, and Quality

Once a chip is being produced, the manufacturing process used to produce it can be measured to have a yield Y. (Of course, this number may not be known for certain as not all parts that pass the test but contain failures are returned by the customer. However, for the work presented below, it is assumed that a reasonable estimate of the yield is available.) The test set used on the chip has a test coverage T, which is based on fault models that accurately describe the way the chips actually fail. The combination of test coverage and yield allow the engineer in charge of the project to predict the quality level that will be seen by the customer. If that quality level is not high enough, which should be increased, test coverage or yield? The following analysis shows that for nearly all combinations of yield and test coverage that occur in practice, larger increases in quality can be achieved by increasing test coverage than by increasing yield.

Consider the Williams model shown in Equation 1. If test coverage is held constant while yield changes, how fast does the quality level change? This question can be answered by taking the derivative of quality level with respect to yield, which is shown in Equation 6. Taking the derivative of quality level with respect to test coverage while holding yield constant shows how fast quality level will change as a result of changes in test coverage. This derivative is shown in Equation 7.

$$\frac{dQL}{dY} = (1 - T)Y^{-T} \tag{6}$$

$$\frac{dQL}{dT} = -Y^{(1-T)}lnY \tag{7}$$

It should be noted that both of the derivatives are always positive on the range from 0 to 1, indicating that quality level is monotonic increasing with respect to yield and test coverage. If the derivative of quality level with respect to test coverage is greater than that with respect to yield, then test coverage should be increased rather than yield. Unfortunately, the relationship is not quite that simple; equal increases in the two parameters can require significantly different investments. For this reason, the inequality shown in Equation 8 will be solved, where K is the cost function used to relate the cost of increases in yield to the costs of increases in test coverage. Note that K is positive. How to estimate the value of K will be discussed in the next section. Solving Equation 8 for T results in Equation 9.

$$\frac{dQL}{dT} > K\frac{dQL}{dY} \tag{8}$$

$$-Y^{(1-T)}lnY > K(1 - T)Y^{-T}$$

$$-YlnY > K(1 - T)$$

$$T > 1 + \frac{1}{K}YlnY \tag{9}$$

Equation 9 indicates that if the test coverage is greater than $1 + (YlnY)/K$, then a larger increase in quality level will result from an increase in test coverage than from an equal increase in yield. Figure 2 shows a graph of Equation 9 for $K = 1$. For a given yield, if the test coverage falls above the black line, then an

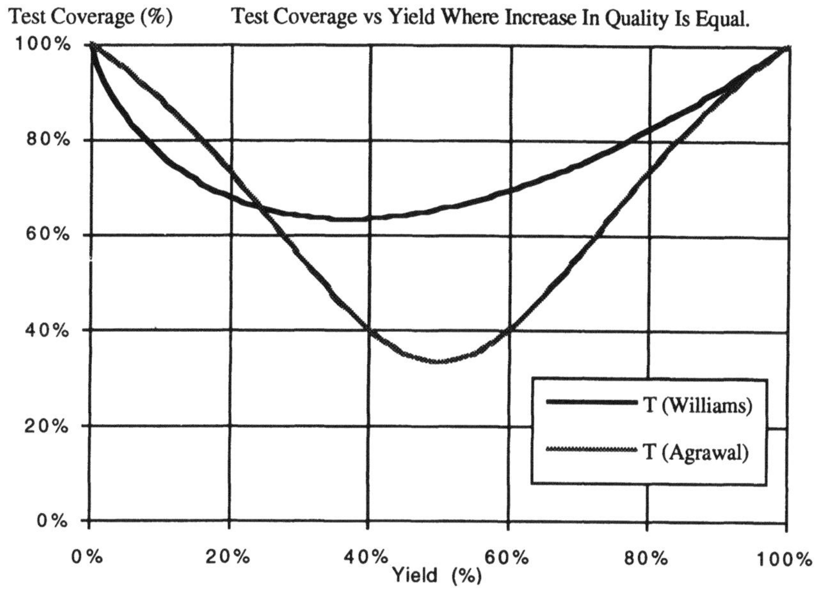

Fig. 2. Test coverage above which test coverage should be increased versus yield.

increase in test coverage will produce a larger increase in quality level than an equal increase in yield. Figure 3 shows how different values of K affect the graph of Equation 9. The right-hand side of Equation 9 is negative when $11\% < Y < 69\%$ and $K = 0.25$. Since T is constrained such that $0 \leq T \leq 1$, $T \geq 0$ is assumed for that range. This implies that for the range $11\% < Y < 69\%$ when $K = 0.25$, no matter what the test coverage is, increasing it will provide a greater increase in quality than an equal increase in yield.

The same analysis done above for the Williams model is now done for the Agrawal model. The derivatives of the quality level are shown in Equations 10 and 11.

$$\frac{dQL}{dY} = \frac{(1 - T)e^{-(n_0-1)T}}{[Y + (1 - T)(1 - Y)e^{-(n_0-1)T}]^2} \tag{10}$$

$$\frac{dQL}{dT} = \frac{Y(1 - Y)e^{-(n_0-1)T}[n_0 + T(1 - n_0)]}{[Y + (1 - T)(1 - Y)e^{-(n_0-1)T}]^2} \tag{11}$$

The inequality shown in Equation 8 is formed, leaving Equation 12, which is then rewritten in Equation 13.

$$Y(1 - Y)[n_0 + T(1 - n_0)] > K(1 - T) \tag{12}$$

$$T[K - Y(1 - Y)n_0 + Y(1 - Y)] > K - Y(1 - Y)n_0 \tag{13}$$

If $[K - Y(1 - Y)n_0 + Y(1 - Y))] > 0$, then the inequality shown in Equation 14 is true.

$$T > \frac{K - Y(1 - Y)n_0}{K - Y(1 - Y)n_0 + Y(1 - Y)},$$
$$\text{if } [K - Y(1 - Y)n_0 + Y(1 - Y)] > 0 \tag{14}$$

If $[K - Y(1 - Y)n_0 + Y(1 - Y)] < 0$, then the inequality shown in Equation 15 is true. However, note that the term $Y(1 - Y)$ in the denominator is always nonnegative for $0 \leq Y \leq 1$. Hence, if the denominator is negative, then the numerator, which is equal to the denominator minus $Y(1 - Y)$, must also be negative. In fact, it then follows that the absolute value of the numerator must be greater than that of the denominator. The fraction in Equation 15 is therefore constrained to be greater than 1. Therefore, when the denominator is negative, if the test coverage is less than 100%, an increase in test coverage will result in a larger increase in quality than an equal increase in yield. Since 100% is the maximum value for T, this is equivalent to the constraint $T \geq 0$.

$$T < \frac{K - Y(1 - Y)n_0}{K - Y(1 - Y)n_0 + Y(1 - Y)} > 1,$$
$$\text{if } [K - Y(1 - Y)n_0 + Y(1 - Y)] < 0 \tag{15}$$

Finally, if $[K - Y(1 - Y)n_0 + Y(1 - Y)] = 0$, then the right-hand side of Equation 13 must be negative. Hence, the result is $T > -\infty$, which is equivalent to $T \geq 0$.

Equations 16a and 16b, hereafter referred to as Equation 16, summarizes the results derived for the

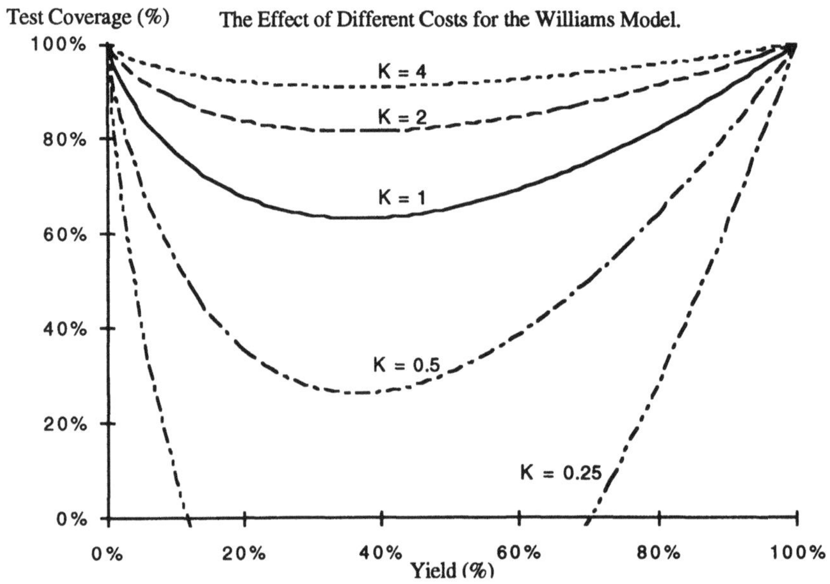

Fig. 3. Value above which to increase test coverage versus costs and yield (Williams model).

Agrawal model. Note that Equation 14 can be simplified as shown in Equation 16a.

$$T > \frac{1}{1 + \frac{1}{\frac{K}{Y(1-Y)} - n_0}}, \text{ if } [K + Y(1 - Y)(1 - n_0)] > 0 \qquad (16a)$$

$$T \geq 0, \text{ if } [K + Y(1 - Y)(1 - n_0)] \leq 0 \qquad (16b)$$

Figure 2 shows a graph of Equation 16 for $K = 1$ and $n_0 = 3.5$ in gray. In this case, the denominator is always positive. For lower values of yield, Figure 2 shows that the Agrawal model requires a higher test coverage than does the Williams model so that increases in test coverage increase quality level faster than equal increases in yield, whereas for $Y \geq 25\%$, the Agrawal model requires a lower test coverage than the Williams model. It is also interesting that Equation 16 is symmetrical with respect to $Y = 0.5$. Figure 4 shows how different values of K affect the graph of Equation 16.

From the information presented in Figure 4, it is clear that when $K \leq 1$ test coverage should be increased rather than similar increases in yield. Figure 5 shows a close-up of the graph of Equation 16 for cases of K ranging from 1 to 200. Note that the vertical scale has been changed to range from 98% to 100%. Consider an increase in test coverage of 0.1% at 50% yield on the curve for $K = 50$. Such an incraese will cause the expected shipped DPPM to go from 421 down to 337. In order to achieve the same improvement by changing yield, the manufacturing process would have to be improved to $Y = 55.5\%$, an 11% improvement. If the same analysis is done for the $K = 100$ curve, then the expected shippped DPPM would go from 208 to 125. Here, yield would have to improve to 63.35% in order to match the 0.1% improvement in test coverage. This would require a 26.7% improvement!

As a sanity check, a contour plot of quality with respect to yield and test coverage using the Agrawal model is shown in Figure 6. The ranges for test coverage and yield were chosen to reflect typical values for large chips. Be aware that the divisions along the T-axis represent much smaller changes than the divisions along the Y-axis. The figure clearly shows that, as predicted by Equation 16, increases in test coverage have a greater impact on quality than similar increases in yield.

The influence of n_0 on the analysis is shown in Figure 7. In order for a chip to fail, it has to contain at least one defect. Therefore, the smallest possible value for n_0 is 1. (Recall that n_0 represents the average number of faults per faulty device.) While n_0 does shift the curve up or down, the worst case does not affect quality as much as large values of K. The next section describes how to determine the cost function.

5. Computing the Cost Function K

The relationship derived above for determining when to change test coverage or yield depends upon the value of the cost function, K. Recall that K determines how much greater the improvement in quality needs to be

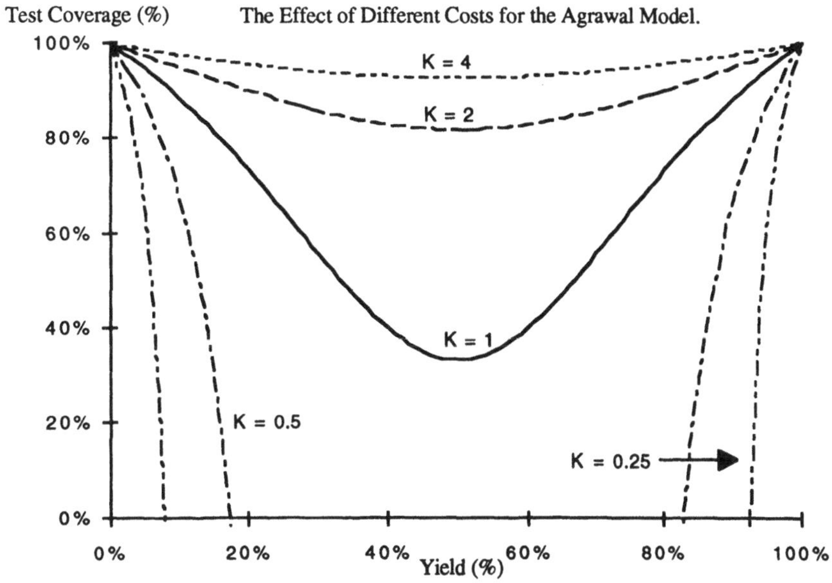

Fig. 4. Value above which to increase test coverage versus costs and yield (Agrawal model).

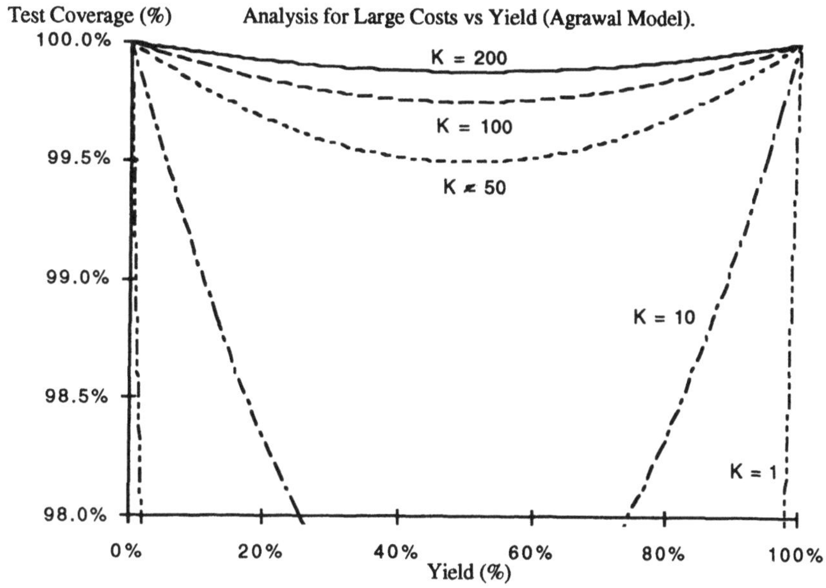

Fig. 5. Values where test coverage should be increased for large costs versus yield.

as a result of increasing test coverage than the improvement due to increasing yield a similar amount. While there seems to be no clear cut way to arrive at an exact value for K, there are several observations that can be made.

First, the cost of increasing test coverage a moderate amount is typically quite small, usually a matter of more simulation time. The use of structured design practices and built-in self-test techniques should make even higher test coverages achievable. However, applying the new set of test vectors may require longer socket times. If so, then either throughput will decrease or another tester must be purchased. Clearly, a decrease in throughput is undesirable. Hence, a ramification of increasing test coverage is that millions of dollars may have to be spent on test equipment, thereby increasing the value of K.

On the other hand, increasing yield a significant amount in a stable process is often extremely expensive, involving the retooling or replacement of millions

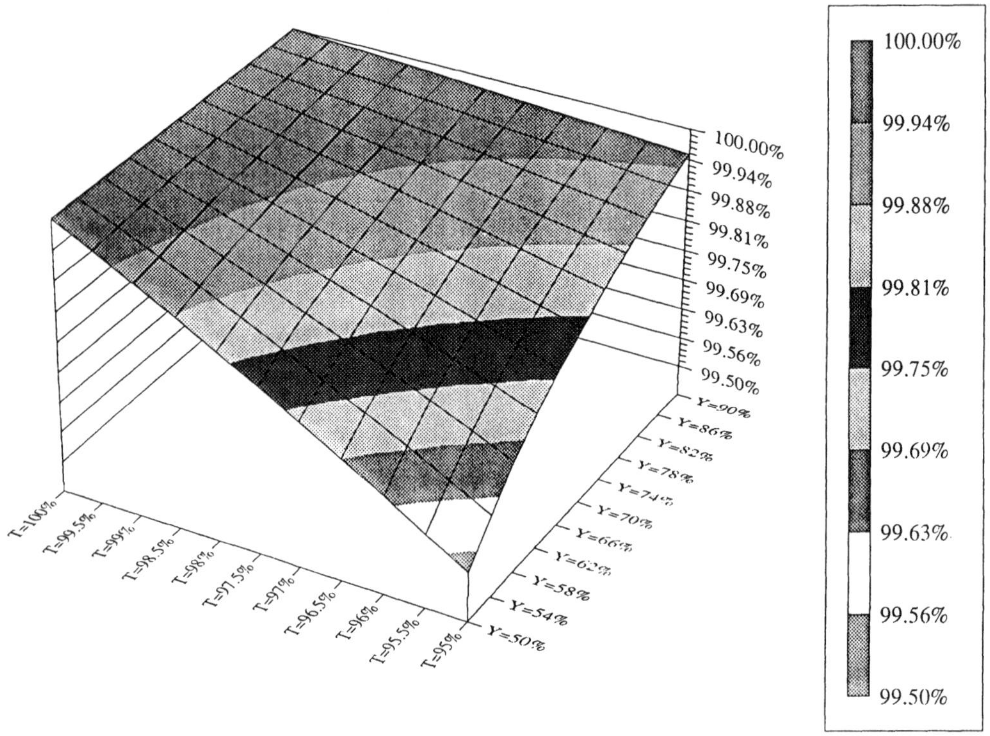

Fig. 6. Quality level versus T & Y [n_0 = 3.5 faults/faulty device].

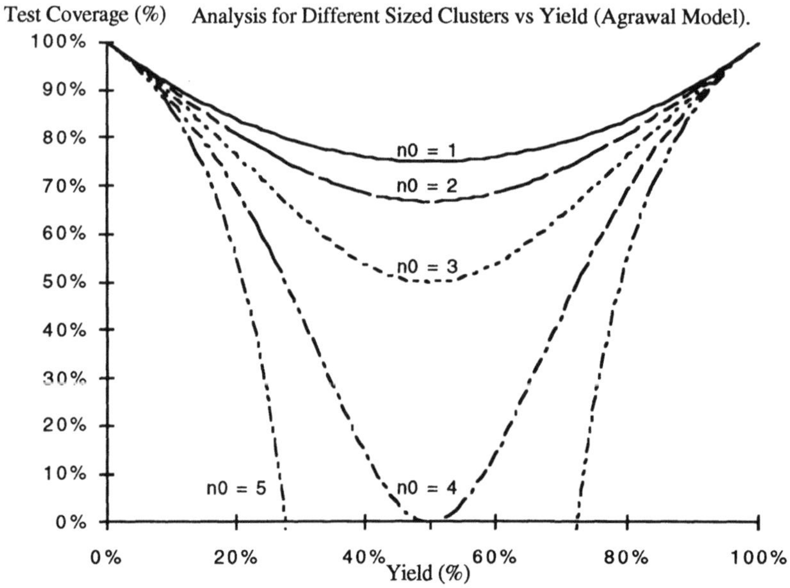

Fig. 7. Values for which test coverage should be increased versus cluster size and yield.

of dollars of manufacturing equipment. (This may not be true of immature or unstable processes that simply require minor adjustments.) In fact, small increases in yield are typically viewed as major victories by manufacturing teams. The larger the expense for a given increase in yield, the smaller K will be.

As discussed in the introduction, manufacturing defective parts costs money. If the yield is increased, fewer defective parts are made, resulting in cost savings. The larger the savings, the larger the value of K. To derive the amount of this savings, a projection over the life of the product must be made. Hence, the number of parts being manufactured and the cost of those parts must be considered.

Another factor that can strongly influence K is when the test is applied. For example, the cost of a defective chip discarded before packaging is often significantly less than the cost of a packaged chip. Typically, defective chips should be identified as early as possible to prevent wasting time and resources processing them further.

Clearly, these and other factors not mentioned influence K. The engineer in charge of quality must examine each of these factors and estimate a value for K. It was shown above that even for large values of K, increasing the test coverage will provide a larger increase in quality than expected from a similar increase in yield. For example, if the current yield is 60% and $K = 10$, the minimum test coverage where increases in test coverage cause larger increases in quality than increases in yield is 97%. Since it is expected that values of K will be much smaller than 10, and since high fault coverages are achievable through creating testable designs with built-in self-test techniques, it can be concluded that increasing test coverage will have a greater impact on quality than increases in yield.

6. Conclusions

This article shows that, for typical values of test coverage and yield, increasing the test coverage will have a greater impact on quality for a lower cost than similar increases in yield. In fact, this relationship often holds even when the increase in yield is much larger than the increase in test coverage. However, it must be ensured that the test coverage is based on fault models that accurately describe the behavior of faulty chips and that the simulated faults accurately represent the failures that actually occur. Utilizing this relationship between test coverage and quality, semiconductor manufacturers can increase quality at a lower cost than if only the process were improved.

References

1. T.W. Williams and N.C. Brown, "Defect Level as a Function of Fault Coverage," *IEEE Trans. Comput.*, vol. C-30, pp. 987–988, Dec. 1981.
2. D.V. Das, S.C. Seth, P.T. Wagner, J.C. Anderson, V.D. Agrawal, "An Experimental Study on Reject Ratio Prediction for VLSI Circuits: Kokomo Revisited," *Proc. Int. Test Conf.*, Washington, DC, pp. 172–720, Sept. 10–14, 1990.
3. P.C. Maxwell, R.C. Aitken, V. Johansen, and I. Chiang, "The Effect of Different Test Sets on Quality Level Prediction: When is 80% Better Than 90%?" *Proc. Int. Test Conf.*, Nashville, TN, pp. 358–364, Oct. 26–30, 1991.
4. V.D. Agrawal, S.C. Seth, and P. Agrawal, "Fault Coverage Requirement in Production Testing of LSI Circuits," *IEEE J. Solid-State Circuits*, vol. SC-17, pp. 57–61, Feb. 1982.
5. J. Galiay, Y. Crouzet, and M. Vergniault, "Physical Versus Logical Fault Models MOS LSI Circuits: Impact on Their Testability," *IEEE Trans. Comput.*, vol. C-29, pp. 527–531, June 1980.
6. J.P. Shen, W. Maly, and F.J. Ferguson, "Inductive Fault Analysis of MOS Integrated Circuits," *IEEE Design & Test*, vol. 2, pp. 13–26, Dec. 1985.
7. W. Maly, "Realistic Fault Modeling for VLSI Testing," *Proc. 24th DAC*, Miami Beach, FL, pp. 173–180, June 28–July 1, 1987.
8. S.D. Millman, *Nonclassical Faults in CMOS Digital Integrated Circuits*, Ph.D. thesis, Stanford University, Stanford, CA., Dec. 1989.

Steven D. Millman received an A.B. degree in Physics from Occidental College, Los Angeles, in 1984 while concurrently attending the California Institute of Technology. He then received M.S. and Ph.D. degrees in Electrical Engineering from Stanford University, Palo Alto, in 1985 and 1989, respectively. He has been with Motorola since 1989 and is the Manager of Large Core Development. His group writes synthesizable behavioral descriptions of previously intractable designs for Motorola's semiconductor sector and equipment divisions. Dr. Millman holds three patents and is a member of the Technical Program Committees of the Custom Integrated Circuits Conference and the Wafer-Scale Integration Conference.

JOURNAL OF ELECTRONIC TESTING: Theory and Applications, 5, 263-268 (1994)

Boundary Scan in Board Manufacturing

THOMAS A. ZIAJA
IBM, 11400 Burnet Rd., Austin, Texas 78758
EARL E. SWARTZLANDER, JR.
Department of Electrical and Computer Engineering, The University of Texas at Austin, Austin, Texas 78712

Abstract. This article discusses the development of a board level manufacturing test for a surface mount board implemented with boundary scan. The board examined is a composite of several actual products. Methods for effectively developing a boundary scan test are examined along with some of the advantages of approaching the development in unique ways. Additionally, the criteria for using these methods are developed. Results for test development time and the resulting test coverage show that with two weeks of test development using boundary scan it is possible to increase the rate of solder opens detection from 80% to 99% for a large ball-grid-array module.

Keywords: Board and system test, boundary scan, boundary scan description language, design-for-test.

1. Introduction

The electronic circuit card manufacturing environment has changed substantially in the past several years. At the chip level, fabrication and packaging technologies are creating increasingly dense integrated circuits. In addition, emerging technologies and assembly processes for electronic circuit cards are producing boards with greater solder joint densities. These influences challenge the current methods of mechanical interfacing for board test resulting in longer test development time or reduced test quality.

Increasing density will eventually make traditional mechanical accessibility unachievable, requiring new test strategies. Complex components requiring many thousands of test patterns may exceed the resources available to adequately test them as part of an assembled board. With these aspects of board manufacturing in mind, a test strategy is developed to illustrate the problems and the approaches needed to solve them.

The test strategy focuses on a fictional composite of several actual boards. In order to preserve confidentiality for the individual products and to provide a concise example for the points to be made, such a composite board is appropriate. The composite board is a hybrid surface mount and pin-through-hole (PTH) board with components on both sides and contains approximately 1500 nets and twenty components. One large pin-grid-array module with 400 pins and one large ball-grid-array surface mount module with 625 pins together account for over half of the total nets. The other components consist of buffers and other miscellaneous logic. The leads of the PTH connectors provide access to many of the signal nets of the board buses.

The traditional manufacturing test strategy for such a board consists of a manufacturing verification test followed by a functional verification test. The manufacturing verification test insures that all components have been assembled and soldered onto the board correctly. Defects such as solder shorts, incorrect components, and other manufacturing defects are detected during the manufacturing verification test. The functional verification test confirms that the assembled board functions correctly. Thus, the following test strategy reflects the manufacturing verification test.

2. Test Strategy

The traditional approach to test development for a hybrid surface mount board depends on the physical characteristics of the board. Since the goal is to detect all assembly related defects the tester needs access to all signal nets on the board. A bed-of-nails test fixture typically provides this access by interfacing with one side of the board and thus requires that all signal nets are accessible from one side. Tests may then be performed for shorts between nets, incorrect or missing components, etc.

Two problems arise, however, when confronted with the composite board described above. Some signal nets are not physically accessible by the tester interface from only one side. This problem frequently occurs from board physical design efforts such as minimizing signal distances between modules that hide nets underneath modules. A fixture that accesses both sides of a board is significantly more complex so a board with hidden or inaccessible nets may suffer reduced test quality. With limited physical access to the board signal nets, an alternative is to use some form of electronic access. The preferred method is the IEEE 1149.1 boundary scan standard [1], [2]. The second problem arises from component complexity when a single VLSI module may exceed tester capabilities. Normally this is resolved by logically partitioning the module test into subtests so that each subtest covers a portion of the physical module and together forms a complete module test. Since the goal for manufacturing verification is to insure correct assembly, partitioning alone need not degrade the quality of the achievable module test. However, if the module is difficult to partition then test quality degradation may occur.

For the composite board, signal nets are fully accessible as a result of their placement and routing, thus enabling tester access through the fixture bed-of-nails probes. Also, the large pin-grid-array module provides access to many signals since its pins preclude module placement on the opposite side. This inherently decides the side of the board probed by the test fixture. Since all pin-through-hole modules are generally placed on the same side of a board, including components such as connectors, a large number of signals are automatically available for physical access by the test fixture probes. Complete physical access to all signal nets allows application of a non-powered shorts test and forms the foundation for board test since the majority of defects are solder defects.

With shorts testing covered through the use of complete probe access, the solder opens defect category is the next most important category to consider. While this board contains a mixture of SMT and PTH modules, we focus on the two VLSI modules since chip level tests are available for the other modules. The large pin-grid-array module contains over 400 pins and implements a complex function. But, because of its use on other products currently in production an existing test is available that toggles 95% of the module signal pins. Since it is a PTH module on a standard .100 inch grid visual inspection of the pins for solder opens is possible. The board also contains a large surface mount ball-grid-array module with over 600 leads and over 400 signal nets. A ball-grid-array (BGA) is a surface mount version of a pin-grid-array module, but with a denser placement of pins [3]. Because the package body hides the solder connections, the assembly process is critical. It is also important to thoroughly test the BGA solder connections since visual inspection is impossible. X-ray inspection is of limited usefulness due to package X-ray opacity, so the electrical test for solder shorts and solder opens must detect those defects.

The tester used for the composite board manufacturing verification test is an in-circuit board tester configured to drive and receive signals in parallel. Partitioning is necessary to test all of the BGA module's signal nets for solder opens since the module size exceeds the tester resources. Due to the complexity of such a module the prospect of obtaining usable board level test patterns from the chip manufacturer is low. Generally, chip level tests are rarely usable at the board level due to module interactions and require modification although this is not feasible for complex modules. Also, chip level tests normally contain many redundant patterns unnecessary during manufacturing verification testing. Considering these factors, the decision was made early in the product design cycle to implement the IEEE 1149.1 boundary scan structure for the ball-grid-array surface mount module.

The result is a combination of boundary scan and direct probe access that provides advantages over using boundary scan alone. With limited probe access, boundary scan is suitable because it requires only four pins to test any module in the boundary scan chain for solder opens and shorts. This requires that testing of the modules in the chain exhibit closure so that modules outside the chain are not involved. Shorts testing between bounday scan and non-boundary scan modules requires additional probes for all non-boundary scan nets. Non-boundary scan modules require physical probing when their signal nets do not connect to a boundary scan module. It is sufficient if every signal net on the board connects to at least one boundary scan cell although this still requires powered shorts testing. By using complete probe access and testing shorts without the board powered there is no risk of component damage due to shorts during powered testing.

Another advantage to complete probe access is the diagnostics achievable since isolation to a pair of shorted nets or even multiple nets is a direct byproduct of a shorts test using full physical access. Testing for shorts with boundary scan techniques requires a more complex algorithmic approach to isolate the short to its component nets [4]. It may be difficult to integrate a boundary scan shorts test with a standard shorts test that uses signal nets that interconnect non-boundary scan modules. Boundary scan shorts test techniques may rely on physical locality, so that shorts between nets that do not have module leads physically close may require significantly more time to diagnose. Though there is little published data to indicate the relationship between the probability of a short and board assembly processess, current experience indicates that shorts are most probable between the closely spaced leads of surface mount modules. But increasing wiring densities and component placements combined with the increasing use of double-sided assembly techniques may lead to a more random mix of board level shorts in the future.

The biggest advantage to implementing boundary scan is the higher initial test quality for solder opens on surface mount modules. This can approach 100% detection of solder opens on a module in contrast to optimistic levels of 80% for a standard 244 pin quad flat-pack initially and 95% after several months of test development. Initial levels of 80% require extensive supplementary inspection, either visually or automated through the use of X-ray or other equipment, to bridge the gap to the necessary 100% coverage. It is vital that some method be used to detect all possible solder opens on surface mount modules as these defects contribute to failures after shipment. Complete probe access is advantageous for solder opens testing also. A solder opens test that toggles all leads of a complex module is simpler using parallel inputs and outputs driven and received by a tester through physical probes rather than through other boundary scan modules. Also, shifting the test patterns and results through the boundary scan chain to the tester is slower than parallel application. Some increased efficiency and diagnostics (to the failing module lead) are achievable with parallel test application.

3. Test Development

To develop the boundary scan tests for solder opens, the first step is to obtain the boundary scan description language file (BSDL file) for each boundary scan module on the board and input the BSDL files into a program that generates tester-specific tests. In this case the desired test is a pins-toggle test to detect solder opens on the component leads.

The use of a BSDL file may appear straightforward, but often may require a surprising amount of test development time because the BSDL file may not exist, (often the case in new product development). To most effectively use boundary scan the information contained in a BSDL file is essential. A BSDL file describes the specific implementation for a module's boundary scan circuitry along with the legal instructions, physical pinouts and any other information necessary to generate a board level module test [5]. In order to use boundary scan to test the large ball-grid-array module a BSDL file was desirable and since none existed one was developed specifically for manufacturing board in-circuit test. Thus, inclusion of only the mandatory instruction set and cell designs was necessary in the BSDL file. For instance, even though the module had a built-in self-test (BIST) instruction, it was not included in the BSDL file; this allowed the use of the boundary scan circuitry for board level testing even before the BIST circuitry design was complete. Figure 1 shows the relevant portion of the BSDL file.

Writing a BSDL file to describe the target BGA module resulted in some unexpected problems due to insufficient tester resources. A BSDL file does not provide the test generation software with information such as tester resource limitations, but instead assumes unlimited driver and receiver tester resources. Tester resource limitations arise due to the requirement to drive all bidirectional pins controlled by a single control latch cell. If that number of pins exceeds the tester resources then it cannot be performed, and the test developer must partition the module test. In essence, the test developer must manually correlate the boundary scan test to the tester resources. However, the nature of boundary scan facilitates partitioning. Testing a boundary scan module using partitions can be accomplished by simply modifying the complete boundary scan test so that only a portion of the drivers and receivers in the test are active at any one time, within the limits of the tester hardware. This requires several partitions to test the module completely. This task, however, may itself be extremely complex since a complex module requires a large set of test patterns to toggle pins even when tested via boundary scan. A simpler approach is to implicitly specify the critical tester resource limitations through modification of the BSDL file so that the test generation software can automatically generate partitioned tests.

Boundary scan allows the generation of partitioned tests with the test generation software through modification of the BSDL file. Whereas modifying hundreds

```
attribute Instruction_Length of BGAmodule: entity is 8;
attribute Instruction_Opcode of BGAmodule: entity is
   "EXTEST  (00000000),  SAMPLE  (00000001),  BYPASS  (11111101, 11111111)," &
   "SECRET  (00000010, 00000110, 00001010, 00001110," &
--  The preceding is actually an INTEST but is labeled SECRET to BC_4 for inputs
--  The following opcodes are true SECRET instructions.
-- All instructions listed as secret are not to be used.
   "01111100, 01111101, 01111111, 11111100)"
```

Fig. 1. The portion of the boundary scan description language file showing the instruction description. Only the mandatory instructions are used in the board test, all others are described as secret to prevent their use by the tester.

or thousands of test patterns to partition a boundary scan pin-toggle test for a complex VLSI module is unrealistic, modifying the corresponding BSDL file is limited in complexity to the number of boundary scan cells. The procedure is to divide the module logically into partitions by creating a unique BSDL file for each partition so that each module test does not exceed tester resources. The BGA module contained nearly two boundary scan cells for every signal pin. Thus, with 425 signal pins the resulting boundary scan chain was 800 cells long. This approach partitions the BGA module by describing it as four separate modules, each implementing boundary scan. Figure 2 illustrates a portion of a BSDL file describing the first partition.

Generating a BSDL file such as that in Figure 2 requires several steps:

1. The maximum number of concurrent drives and receivers should be determined. In this case, 288 drivers and receivers could be used at one time.
2. The set of boundary scan cells to include in any partition is selected by dividing the boundary scan chain into equal lengths, so that the number of cells in each subchain is equal or less than 288. The 800 cells are divided into four sets of 200 cells each.
3. In each resulting BSDL subfile, all cells not included in that partition are described as internal cells.
4. All necessary control cells should be included; for bi-directional pins associated cells are included to avoid splitting input, output and control cells for the same physical pin.

The resulting BSDL subfiles each describe a single logical partition of the physical module. The tester sees four modules each with a single boundary scan chain but sharing a common test access port (TAP) interface. To perform a complete test for solder opens each partitioned module test is run consecutively. The segment in Figure 2 illustrates the BSDL file for the first 200 cells. Note the control cells included outside of the cell target range.

4. Evaluation

The effect on test quality is now illustrated through analysis of two approaches. The first case considers composite board test without using boundary scan for the ball-grid-array surface mount module, while the second case assumes that boundary scan is used. Table 1 summarizes both cases.

Case 1 results in a 91% detection rate for solder opens, while case 2 shows a 98% detection rate, using the formula for defect rate [6]:

$$D_{out} = D_{in}(1 - C_t)$$

where D_{out} is the outgoing defect rate, D_{in} is the incoming defect rate and C_t is the tester coverage for opens. In particular, the escape rate from the tester is lower for case 2 using boundary scan. The defects assumed are solder opens, with the defect rate indicating the percentage of boards with at least one solder open on a signal pin. Non-signal pins include power and ground and are excluded from the analysis. This analysis method is similar to that typically used for board test since it considers the test effectiveness for a specific defect category for a specific component type [7]. In this study, the component signal pins are equivalent to the number of solder joints.

A summary of the most important costs and resulting benefits due to implementing boundary scan in this instance is presented in Table 2. This analysis illustrates the quality considerations of implementing boundary scan. A complete economic analysis would include daily production rates, total production volume, time to market and other specific costs but must include the increase in quality to balance the other economic considerations.

There are basically three options to be considered when developing a test for a board containing boundary scan. The first is to provide complete probe access to all nets along with implementing boundary scan in all modules. This allows the greatest flexibility and

```
-- Cells included:  0 to 200, and 303;    file name: BGALINK1
-- Control cells: 0 disables outputs, result is tristate for output
--num cell port         type             safe  [ccell, disval, result]
"0 (BC_1,  *,          control,         X),"&
"1 (BC_1,  OUT1,       output3,         X,       0,      0,       Z),"&
"2 (BC_4,  IN1,        input,           X),"&
"3 (BC_4,  BIDIR1,     input,           X),"&
"4 (BC_1,  *,          control,         X),"&
"5 (BC_1,  BIDIR1,     output3,         X,       4,      0,       Z),"&
--  cells 6 to 198 similar.
"199 (BC_1,  BUS1(14), output3,         X,     303,      0,       Z),"&
"200 (BC_4,  BUS1(15), input,           X),"&
"201 (BC_1,  BUS1(15), output3,         X,     303,      0,       Z),"&
"202 (BC_4, *,         internal,        X),"&
"203 (BC_1, *,         internal,        X),"&
-- cells 204 to 301 internal
"302 (BC_4, *,         internal,        X),"&
"303 (BC_1, *,         control,         X),"&
"304 (BC_1, *,         internal,        X),"&
-- cells 305 to 798 internal
"799 (BC_4, *,         internal,        X)";
end BGAmodule;
```

Fig. 2. Illustration of a boundary-scan description language file to partition a module with 800 boundary scan cells so that only the first 200 cells are recognized by the board tester. This file is the first of four, each of which the tester interprets as being for a separate module in one boundary scan chain.

Table 1. Comparison of tester effectiveness on solder opens using boundary scan on a ball-grid-array module. A typical value for D_{in} might be .20 for a newly introduced surface mount process.

Module	Signal Pins	Case 1: C_t	Case 2: C_t
pin-grid array	300	.95	.95
miscellaneous	475	.99	.99
ball-grid array	425	.80	.99 with bscan
Total:	1200	.91 = 91%	.98 = 98%
D_{out}	—	$D_{in} \times (1 - .91) = D_{in} \times .09$	$D_{in} \times (1 - .98) = D_{in} \times .02$

Table 2. Cost/benefit comparison for implementing boundary scan on a complex surface mount module.

Costs	Benefits
Silicon overhead	Shorter test development
Increased chip development time	Lower manufacturing costs —repair, scrap, etc.
Extra module pins	Increased quality

allows non-powered shorts testing. Another option is full probe access with only certain, strategic, modules implemented with boundary scan. This was the choice for the composite board and allows non-powered shorts testing while maintaining the benefits of boundary scan for complex surface mount modules. The third option is to provide only limited probe access to the board's signal nets. This should be used only when full probe access is unachievable, since it requires that shorts testing be performed while the board is powered. Boundary scan should be implemented in all modules with physically inaccessible signal nets. Additionally, passive components (resistors, etc.) require probes. Full probe access to all signal nets on the board allows non-powered shorts testing. This provides a diagnostic advantage and reduces the risk of component damage due to excessive current on component signal leads. For these reasons boundary scan should not be viewed as simply a method of reducing test fixture complexity.

The best case for implementing boundary scan can be made for complex VLSI surface mount modules.

Due to the nature of surface mount modules it is difficult to insure that their pins are soldered onto the circuit board, and boundary scan provides a way to confirm the solder joint integrity. The most efficient solution is to implement boundary scan only for surface mount modules too large or complex to be tested with traditional methods. Similarly, implementation of boundary scan in pin-through-hole modules is justifiable for complex modules; however, visual inspection of PTH leads lessens the need for boundary scan. Additional applications for boundary scan at the functional and self-test levels are also justifiable.

Several aspects of test development for boundary scan compliant modules have been illustrated. First, there is no unique BSDL file for a module. Manufacturing verification testing requires only a subset of the complete BSDL file adhering to tester resource limitations. This can be more cost effective than buying extra hardware to enhance the tester resources. Second, by using the tester to drive and receive signals through physical probes the boundary scan test may be less complex and exhibit better diagnostic capabilities than if other boundary scan modules are used as electronic probes. This approach also requires that fewer modules be implemented with boundary scan. Third, the use of boundary scan provides a tremendous potential increase in the percentage of possible defects covered. Specifically, solder open defects at the leads of surface mount devices are completely detectable with little effort compared to traditional tests using chip level patterns.

Test development time for the BGA module using boundary scan was two weeks and achieved 100% testing of the leads for solder opens compared to approximately two months to achieve 80% coverage using chip level patterns. An estimate of six months to achieve 95% coverage using chip level patterns is reasonable, although this would require investing in additional tester hardware.

5. Conclusion

This study illustrates a method for manufacturing to develop BSDL files for complex modules specifically for board level manufacturing verification testing. The boundary scan cell ordering, instruction definitions, and pin identifications are required, but as soon as this information is available, the BSDL file for the board test may be written and tailored to the needs of manufacturing test. This offers a simple method of decreasing the lead time for board test development and for increasing the quality of the board level manufacturing test.

References

1. IEEE Standard 1149.1-1990, "IEEE Standard Test Access Port and Boundary-Scan Architecture," *IEEE Standards Board*, 345 East 47th Street, New York, NY 10017, May 1990.
2. R. Stevens and T. Ziaja, "A Strategy for Effective Board Test," *Proceedings of the 6th Workshop on New Directions for Testing*, pp. 25–35, Montreal, Canada, May 1992.
3. J. Tuck, "The Next Great Package?" *Circuits Assembly*, vol. 4, pp. 22–26, Feb. 1993.
4. N. Jarwala and C. Yau, "A New Framework for Analyzing Test Generation and Diagnosis Algorithms for Wiring Interconnects," *Proceedings of the International Test Conference*, pp. 63–70, 1989.
5. *HP Boundary-Scan Tutorial and BSDL Reference Guide*, Hewlett-Packard, PN E1017-90001, 1990.
6. C. Pynn, *Strategies for Electronics Test*, McGraw-Hill Book Company, New York, 1986.
7. M. Tegethoff, T. Figal, and S. Hird, "Board Test DFT Model for Computer Products," *Proceedings of the International Test Conference*, pp. 367–371, 1992.

Thomas A. Ziaja received the B.S. degree from the University of Missouri, Columbia and the M.S. degree from the University of Texas, Austin, both in Electrical Engineering. He has been employed at IBM since 1982 and is currently pursuing the Ph.D. degree in Electrical Engineering from the University of Texas, Austin. His research interests include design for testability, testability analysis and the economics of testing.

Earl E. Swartzlander, Jr. received the B.S.E.E. degree from Purdue University, the M.S.E.E. degree from the University of Colorado, and the Ph.D. degree in Electrical Engineering from the University of Southern California. At the University of Texas, Professor Swartzlander holds the Schlumberger Centennial Chair in Engineering in the Department of Electrical and Computer Engineering. He is conducting research on application specific processor technology and design, including high-speed computer arithmetic, systolic signal processor architecture, VLSI and WSI technology, and system prototyping.

He was a member of the Board of Governors for the IEEE Computer Society (1987-1991) and is the Editor-in-Chief of the *IEEE Transactions on Computers*, the hardware area editor for *ACM Computers Reviews*, and the founding Editor-in-Chief of the *Journal of VLSI Signal Processing*.

He obtained his doctorate in computer design with the support of a Howard Hughes Doctoral Fellowship. He has written the book *VLSI Signal Processing Systems* (Kluwer, 1986) and edited five books including two collections of reprints on Computer Arithmetic. He has written or cowritten over 100 papers in the fields of computer arithmetic, signal processing, and VLSI implementation. He is a Fellow of the IEEE and is a registered professional engineer in Alabama, California, Colorado, and Texas. He has received Outstanding Electrical Engineer and Distinguished Engineering Alumnus awards from Purdue University.

JOURNAL OF ELECTRONIC TESTING: Theory and Applications, 5, 269-272 (1994)

Comparing Quality Assurance Methods and the Resulting Design Strategies: Experiences from Complex Designs

C. v. REVENTLOW
SICAN GmbH, Garbsener Landstr, 10, 30419 Hannover, Germany

Abstract. Design quality assurance methods are compared in terms of cost and error detection efficiency using statistical data from several complex designs (Complexity > 100.00 gates). A suitable strategy for error detection as a function of system complexity is identified, and the resulting design flow is described. The statistical control of the design process as a feedback loop to achieve error prevention is demonstrated. Using data from complex projects practical advice for managers is given.

Keywords: Inspections, quality assurance methods, VLSI design methodology.

1. Characterization of the Problem

For systems of low complexity a simple design strategy is typically used: Each designer implements and debugs his modules. As shown by Aas [1], designers are capable of detecting 80% of their own errors in this debugging phase. The remaining 20% of module errors have to be found during system simulations together with the problems of module interactions. This conventional strategy often fails for complex systems (complexity > 100.00 Gates) due to the high system simulation times, the resulting decrease of the designers productivity by a factor of 2.4 [2], the increasing number of members and communication interfaces in the design team and the overall complexity of the design goal and solutions. Thus improved quality assurance methods are required.

2. Method of Investigation

Using a method of W. Deming [3] following Lord Kelvin's statement: "...but when you cannot measure it, when you cannot express it in numbers, your knowledge is of unsatisfactory kind..." the design process has been analyzed using data of several complex designs with design teams of 5 .. 20 designers. Both error detection and error prevention has been investigated.

3. Error Detection Methods

Several error detection methods have been investigated.

- System simulations
- Module simulations performed by an independent test team
- Inspections [4](The quality assurance team reads the material, the designer only answers questions of the team; the working speed is regulated by the error detection efficiency of the team.
- Walk-Throughs [5] (The designer explains the material to the quality assurance team.)

A first advantage of the manual methods (Inspection, Walk-Through) is that they can be used in the early phase of the development cycle during system specification and architecture and architecture development (in opposition to current simulation oriented methods). Defining the efficiency of a quality assurance method as the percentage of errors in a module found by the method, a combination of inspections followed by simulations has been found to be most effective for RT-Level models and schematics (Figure 1). Interesting to see is the low error detection efficiency for reused modules, which makes reuse a dangerous strategy. The Walk-Through has been omitted in the diagram, as it only detects 50% of all errors. The reason for this low error detection efficiency is the control of the speed of the Walk-Through by the designer of the material, who already knows all details. This explains the advantage of the Inspection where the speed is controlled by the inspection team, which still has to understand the product to be analyzed.

Figure 2 shows the cost to detect and remove an error depending on the strategy used. From the cost point of view the Inspections are far superior to simulation

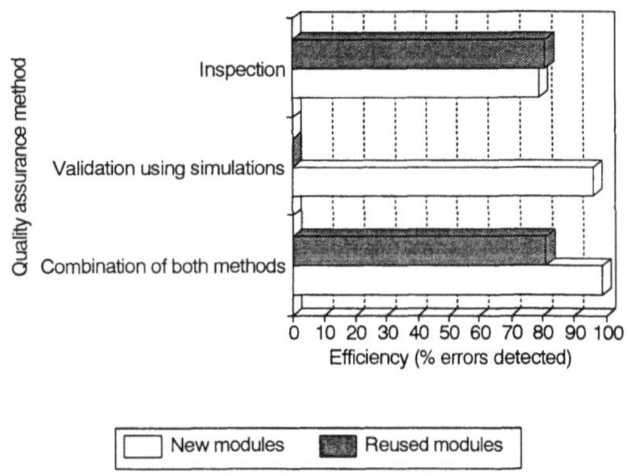

Fig. 1. Error detection efficiency.

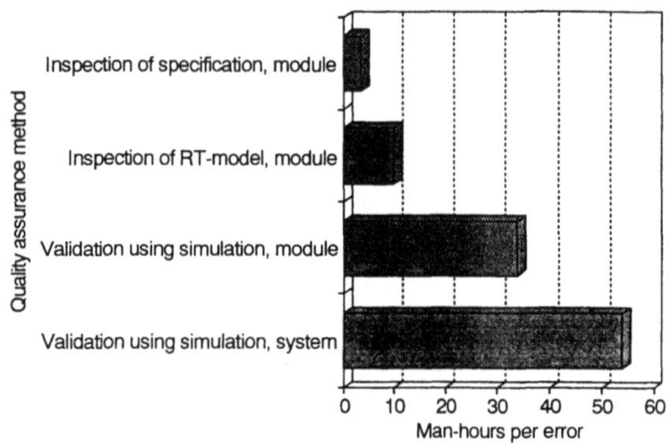

Fig. 2. Error detection and removal costs.

methods. If we define the cost effectiveness of a quality assurance method as the quotient of the effort saved due to early error detection compared to the effort to detect and remove the error by the method studied, again the inspection turns out to be a far superior strategy compared to all other methods (Table 1).

Table 1. Cost effectiveness of quality assurance methods.

Method	Cost Effectiveness
Inspection of module specification	16
Inspection of RT—level model	5
Validation using module simulations	1.6
Validation using system simulations	1

The difference between validation using system and module simulations can be explained by the high system simulation times for complex systems. It has been observed, that designers productivity decreases by a factor of 2.4 for long simulation times (hours or more). Thus no difference should be observed for systems of low complexity.

A further advantage of inspections as an error detection method is that it works as a communications tool in the team. The effort for rework is drastically reduced shortly after the introduction of this method (Figure 3).

3. Quality Assurance Methods and the Resulting Design Flow

The quality assurance methods strcuture the design flow. Four different resulting development styles have been investigated:

- a conventional strategy without structuring the design flow in phases. Quality assurance is only performed at the end of the development cycle. Products: specification/architecture. RT-model, schematic diagram.

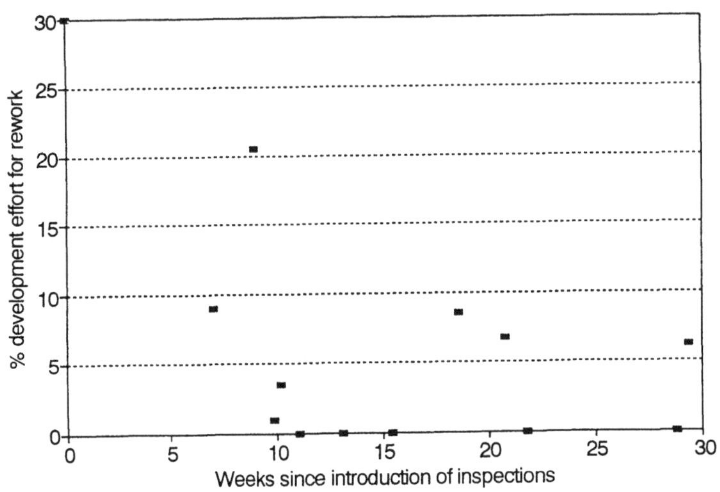

Fig. 3. Influence of introduction of inspections on rework.

- a structured three phased approach with quality assurance before the beginning of the next phase. Phases: specification/architecture, RT-modeling, development of schematic diagrams.
- a structured two phased approach again with quality assurance at the end of each phase. Phases: specification/architecture and development of schematic diagrams.
- reuse. Modified modules from previous projects are reused.

The resulting average number of errors per module is shown in Figure 4 for the case of using both inspections and simulations as quality assurance methods. The structured approaches lead to best quality, as error propagation into the next development phase is minimized. The conventional strategy and reuse gives the worst results.

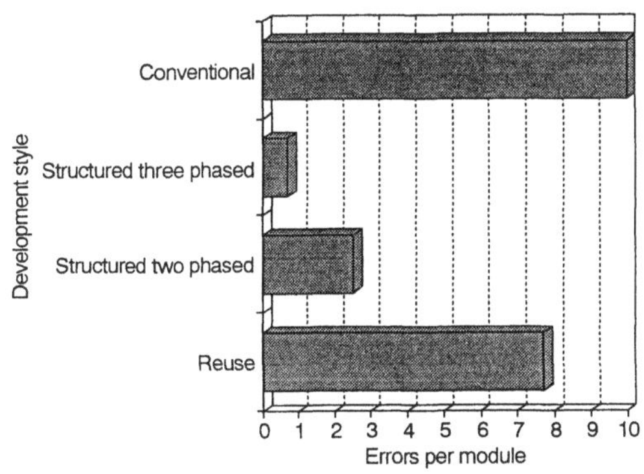

Fig. 4. Development style vs. error density.

From the cost point of view no difference in development costs between the conventional approach and the structured three phased approach has been observed. Using the structured two phased approach resulted in 25% savings. The effort for reuse depend strongly on the size of the changes required, but was typically below the effort for the three phased structured approach.

This leads to the following results for a project manager:

- The conventional approach should not be used in our environment out of quality and economical reasons.
- Due to the high error density in reused modules and the low error detection efficiency of the quality assurance methods for these modules, reuse should be avoided out of quality reasons.
- For practical purpose only the two phased and the three phased structured approaches remain. As the simulation time on gate level typically exceeds the simulation time on RT-level by at least a factor of two, the two phased approach is only appropriate for low system complexities (in our environment typically below 50.000 gates), where simulation times on gate level are still reasonable and/or for systems, where the number of system level simulations required is low.

4. Error Prevention

Statistical analyses allows the optimization of the design process in terms of quality and cost effectiveness. The central concept is the feedback loop. The selection of the best error detection strategy and the control of its efficiency is a first example. As a further example, Figures 5 and 6 show the average error densities (errors per complexity unit) of specifications for a design team and for one of its members. This information is fed back to the designer. He can identify his strength (documentation) and his weakness (control flow).

5. Summary

Using statistical analyses and control of the design flow to select the best error detection method and design strategy has been described. Establishing a feedback loop allows error prevention. Statistical data from complex projects are used to illustrate the method and to give practical advice for project managers.

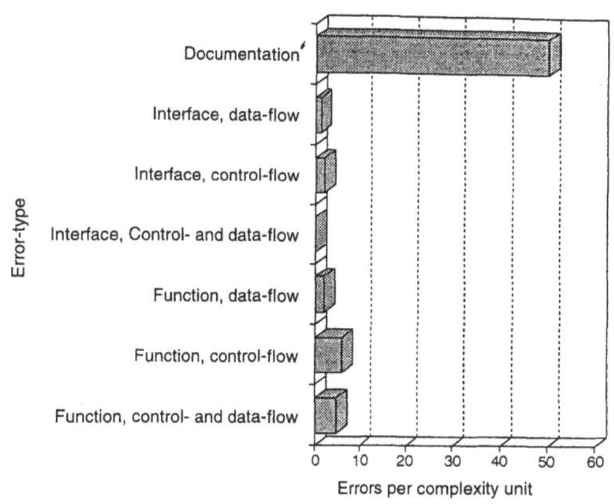

Fig. 5. Average team error densities.

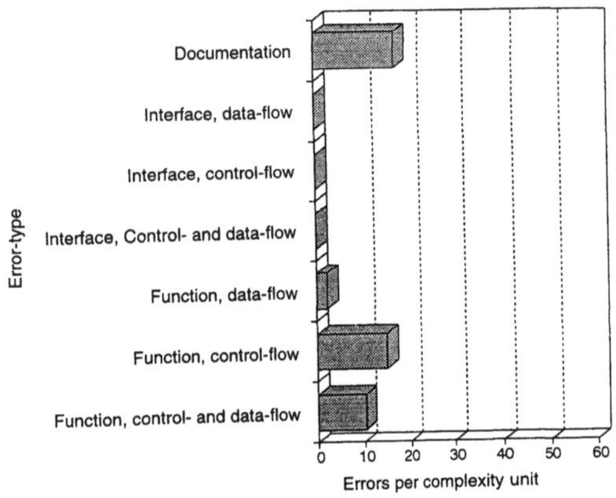

Fig. 6. A designer's error profile.

References

1. E.J. Aas, K. Klingsheim, and T. Steen, "Quanitfying Design Quality: A Model and Design Experiments," *Proceedings EURO ASIC'92*, Paris, 1-5, June 1992.
2. F.P. Brooks Jr., *The Mythical Man-Month*, Addison-Wesley, Reading, MA, 1982, pp.
3. W.E. Deming, "Quality, Productivity and Competitive Position," MIT Center for Advanced Engineering Study, Cambridge MA, 1982.
4. M.E. Fagan, "Design and Code Inspections to Reduce Errors in Program Development," *IBM Sys. J.*, vol. 15, 1976.
5. C.L. McClure, *Managing Software Development and Maintenance*, Van Nostrand Reinhold, New York, 1981.

Christian v. Reventlow received an M.S. in Physics from the Ruhruniversität Bochum, Germany, and a Ph.D. in Electrical Engineering from Technical University Berlin, Germany. From 1984 he worked as a system engineer in the PC, networking, and video processing area. Currently he is the manager of the Full Custom Projects department at SICAN GmbH, Hannover, Germany. His research interests include the management of the VLSI design process and the development of VLSI architectures for the above mentioned areas.

JOURNAL OF ELECTRONIC TESTING: Theory and Applications, 5, 273-283 (1994)

Trade-offs in Scan Path and BIST Implementations for RAMs

M. NICOLAIDIS, O. KEBICHI, AND V. CASTRO ALVES
Reliable Integrated Systems group, TIMA/INPG, 46 avenue Félix Viallet, 38031 Grenoble Cédex, France

Abstract. In this article we propose efficient scan path and BIST schemes for RAMs. Tools for automatic generation of these schemes have been implemented. They reduce the design effort and thus allow the designer to select the more appropriate scheme with respect to various constraints.

Keywords: Aliasing, BIST, coupling faults, RAM test algorithms, scan path, signature analysis.

1. Introduction

In this article we first present a new scan path technique for RAMs that applies one test pattern per clock cycle. Then, we consider several techniques for address generation, test data generation, and output response verification related to scan path and BIST implementations for RAMs. The advantages and the drawbacks of each technique in terms of area overhead, fault coverage, and test length have been discussed in order to derive the best scan path and the best BIST scheme. Tools allowing automatic implementation of these techniques have been developed. They eliminate the design effort and thus allow the designer to select the scheme satisfying various constraints, such as low area overhead, high fault coverage for both RAM block and BIST circuitry, and low test time. Based on these results, it turns out that for embedded RAMs the proposed scan path and BIST techniques are more cost effective than the traditional testing, while in most situations BIST is more effective than scan path.

2. Fault Models and Test Algorithms for Static RAMs

Stuck-at faults, transition, and coupling faults are the most important classes of faults occurring in static RAMs [1], [2]. The more efficient algorithms detecting these faults are the marching algorithms [2], as those proposed in [1], [3], [4]. In this article, we consider these kinds of algorithms, and we illustrate our technique by means of the algorithm B proposed in [3], which offers a very high fault coverage: all faults affecting the decoders, all stuck-at faults in the read/write logic, all single and multiple cell stuck-at faults, all single and multiple transition faults, and all single and multiple idempotent coupling faults are detected.

The algorithm is given in Table 1. It performs $17n$ read and write operations (n being the number of words in the RAM).

2.1. Marching Test for Word Oriented RAMs

The above algorithm considers bit-oriented RAMs. In word oriented RAMs, one writes simultaneously into several RAM cells, and thus, some new faults can appear. These faults are similar to the ones occurring in multiport RAMs [5]. For instance, when we perform simultaneous transitions into two cells *Ci* and *Cj* these transitions can modify the value of a third cell Ck, resulting on a fault called complex coupling [5]. Also, when we perform simultaneous transitions into two cells Ci and Cj the transition on one cell can prevent the transition on the other cell, resulting in a fault called concurrent coupling [5]. Such faults can appear in single port RAMs when each RAM row implements a single word. In this case a cell of a word can be a neighbor of three cells of another word. Thus we must test the RAM for the eight possible transitions (i.e., 000 → 111, 111 → 000, 001 → 110, 110 → 001, 010 → 101, 101 → 010, 011 → 100, and 100 → 011) of any three consecutive cells of each word. This can be done by performing the algorithm of Table 1 four times, using one of the following test data pairs (000000...000..., 111111...111...), (001001...001..., 110110...110...), (010010...010..., 101101...101...), or (011011...011..., 100100...100...) each time. If the RAM uses a 1-out-of-2 column multiplexing, then

Table 1. MARINESCU's B algorithm.

Init.	S1	S2	S3	S4
W0	R0W1W0W1	R1W0R0W1	R1W0W1W0	R0W1R1W0
W0	R0W1W0W1	R1W0R0W1	R1W0W1W0	R0W1R1W0
W0	R0W1W0W1	R1W0R0W1	R1W0W1W0	R0W1R1W0
*	*	*	*	*
*	*	*	*	*
W0	R0W1W0W1	R1W0R0W1	R1W0W1W0	R0W1R1W0

→ time

each row implements two words. In this case, a cell of a word can be a neighbor of at most two cells belonging to another word. Thus, we must test the RAM for the four possible transitions of any two consecutive cells of each word. This can be done by performing the algorithm of Table 1 twice, using one of the test data pairs (000...0..., 111...1...), or (0101..01..., 1010...10...) each time. If a 1-out-of-4 column multiplexing is used then, each row implements four words. In that case, a cell of a word cannot be a neighbor to more than one cell of another word. Thus, we need to test the RAM only for the two possible transitions (i.e., 0 → 1, 1 → 0) of each cell. Thus we will perform the algorithm of Table 1 once using the test data pair (0000...00..., 1111...11...). This is also true for greater order multiplexing. RAMs are generally implemented using column multiplexing of 1-out-of-4 or of greater order. Thus, the algorithm of Table 1 can be used by replacing the write 1 and write 0 of a cell by write 1 or write 0 to all the word cells.

3. Address Generation

The addressing order of march algorithms like the one of Table 1, and consequently the addressing order of the march algorithms derived in the previous section for word oriented RAMs, require to address all the RAM cells in an increasing order (sequences S1 and S2 in Table 1) and in a decreasing order (sequences S3 and S4). The exact addressing order is not significant (the marching algorithms are functional). However, the decreasing order must be the reverse of the increasing one. Usually in RAM testing, the binary addressing order is used. In BIST this addressing can be implemented by an Up/Down binary counter. However, using Up/Down LFSRs (introduced in [6]) is more efficient since they are more compact, faster and if a RAM is embedded, they can also be used as signature analyzers for other blocks. For more details concerning the design of Up/Down LFSRs, see [6], [8], [13].

3.1. *Address Generation for Scan Path*

A popular DFT technique uses scan paths for accessing the inputs and outputs of embedded blocks. It has the drawback: if a block has *n* inputs then, one needs *n* clock cycles (*n* shifts) in order to apply each test vector. This will happen in the case of march algorithms, if one uses addressing based on the binary order. A new technique which copes with this problem has been proposed independently by us [9], [10] and by Maeno et al. [11]. We first present this technique for scan BIST and then for external scan path testing.

An efficient BIST scheme is the scan BIST [12]. It uses a single pseudorandom test pattern generator and a signature analyzer to test all the blocks via the scan paths, as shown in Figure 1.

Generally, the pseudorandom test ensures a good fault coverage for the combinational blocks and also for the sequential ones that (via the scan path) are tested

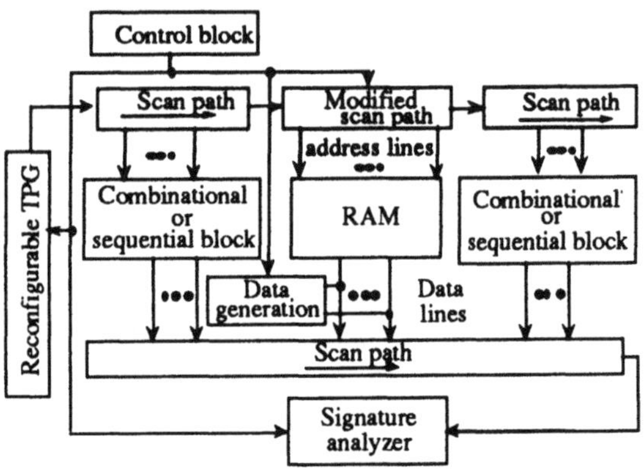

Fig. 1. Global BIST with scan path.

as combinational blocks. On the other hand, deterministic BIST for RAMs can be implemented efficiently and ensures higher fault coverage. In such a scheme, it is important to have an address generator compatible with the scan path in order to reduce the test time. If one injects in the scan path input, the output of a cell of an n-bit LFSR, we will need n shifts for the first address and only one shift for each other address. So, if one uses an Up/Down LFSR then during the Up operation, one just needs to inject the output of one cell (for instance the nth cell) into the scan path input. A particular implementation of the RAM scan path part extends this property for the whole (i.e., Up and Down) LFSR operation. The critical point is the transition from the Up to the Down function and vice versa. We exemplify the implementation for the external XOR LFSRs:

- The scan path part corresponding to the RAM address lines is implemented to perform both shift-left and shift-right.
- When the LFSR is in the Up state, we inject the input of the first cell of the LFSR into the scan path, and when it is in the Down state, we inject the input of the nth cell of the LFSR.
- The scan path part corresponding to the RAM address lines changes from shift-right to shift-left m clock cycles after the Up/Down LFSR changes from Up to Down operation and vice versa.

For more details see [10].

The above scheme can be applied to any RAM block of a system. A single control block can be used for all the BISTed RAMs since this block is independent of the RAM size (see [8]). Only one LFSR will be used for both: test pattern generation for combinational (or sequential) blocks and RAM address generation. During the testing of a RAM block, the test pattern generator will be reconfigured to give an Up/Down LFSR with the appropriate length.

In the case of external scan path testing, the addresses will be injected into the scan path through an input pad of the chip. The new scan path scheme will be as follows: similar to the scheme of Figure 1, the scan path part corresponding to the RAM address lines will be implemented to perform both shift-left and shift-right. Then, the Up/Down LFSR used in the scheme of Figure 1 will be simulated in order to determine the address sequence to be injected into the scan path. This implementation will lead to a slight increase in the scan path complexity, but it divides the test length by n.

4. Test Data Generation

Two techniques can be used for test data generation. The first uses a dedicated block to generate the test data locally to the RAM block. The second injects the test data via a scan path.

Due to the simplicity of the test data in RAM test algorithms, the test data generation block is quite simple. As we have seen previously for RAMs using a multiplexing of 1-out-4 or of greater order, the marching test algorithms require the generation of the data 000000...00 and 111111...11. In that case, the test data generator is given in Figure 2a. For a 1-out-of-2 column multiplexing the march test algorithms require to generate the data pairs (000000...000, 111111...11) and (01010...01,10101...10), and the test data generator is presented in Figure 2b. Finally, for RAMs without column multiplexing, we need four test data pairs and the test data generation block is given in Figure 2c. In all cases, the test data generator is a compact block. The signals VTi are generated by the BIST controller.

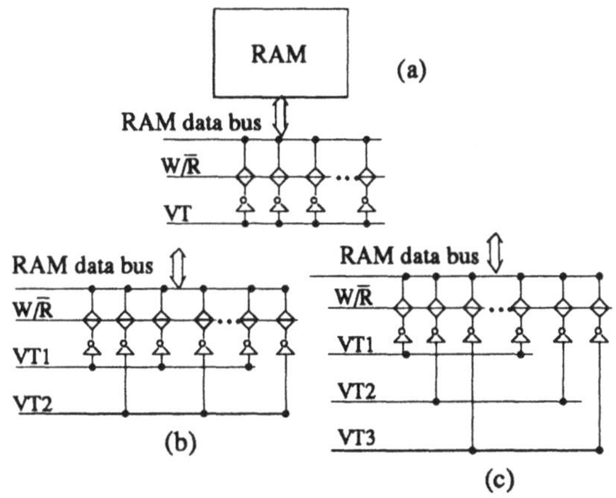

Fig. 2. Test data generators for RAMs with (a) MUX 1/4, (b) MUX 1/2, (c) no MUX.

If the test data are applied via a scan path, then we will need m shifts to apply each data. For reducing these shifts the test data will be transformed: If the test algorithm uses the pair of test data (00000...00, 11111...11) (column multiplexing of 1-out-of-4 or of greater order) then, we will replace this pair by (01010 01, 10101...10). Then, the one data of the pair is obtained from the other by means of one shift. Furthermore, with these data the RAM is tested for both the $1 \rightarrow 0$ and $0 \rightarrow 1$ transitions of each cell.

If the algorithm uses the data pairs (000000...000, 111111 11) and (101010...10, 010101...01), they will be replaced by the pairs (001100...11, 110011...00) and (100110...01, 011001...10). In that case, we need two shift operations to transform one vector of a pair into the second one. Furthermore, using these pairs the RAM is tested for all the possible transitions of each pair of consecutive cells.

Similarly, if the initial pairs are (000000...000, 111111...111), (001001...001, 110110...110), (010010...010, 101101...101) and (100100...100, 011011...011), the new pairs will be (000111...000, 111000...111), (001110...001, 110001...110), (010101 010, 101010...101) and (100011...100, 011100...011) (in that case we need three shift operations to transform one vector of a pair into the other). Furthermore, using these pairs, the RAM is tested for all the possible transitions of each triplet of consecutive cells. In the usual case (i.e., column multiplexing of 1-out-of-4 or of greater order), only one shift will be needed to apply each test data.

5. Output Response Verification

We consider three schemes allowing to verify the output responses of a RAM. The first is the popular scheme of parallel signature analysis. The second consists of collecting the output responses via a scan path, and the third performs the output response verification by using a comparator.

The advantage of signature analysis is that it requires an acceptable amount of hardware and verifies the output responses by checking a small amount of information. The drawback is that it involves aliasing which reduces thc fault coverage.

The marching test algorithms for RAMs involve few expected output vectors (e.g., two vectors for RAMs having a column multiplexing of 1-out-of-4 or of greater order). Thus the output response verification can be performed efficiently by a comparator (see [6]). The comparator requires significantly lower hardware than the signature analyzer and it does not involve aliasing. But, it has the drawback that it will not detect many of the faults affecting the comparator and the BIST circuitry (for instance faults under which the sequence control block jumps from the sequence S1 to the sequence S4, or the address generator generates only a subset of the RAM addresses, or stuck-at faults in the signals VTi or in the inverters of Figure 2, etc....). Thus the signature analysis offers a more uniform fault coverage for the faults affecting the RAM and the BIST circuitry (same aliasing). To detect the faults in the BIST circuitry in the scheme using the comparator, one can duplicate this circuitry. A solution which moderates this overhead is presented in Section 5.1.

On the other hand, if the comparator is used in a scheme where the addresses and the test data are applied via scan paths, then, these scan paths are tested easily by shifting a few values. Furthermore the comparator can be tested easily via the data scan path. Therefore in this case, the comparator is more advantageous than the signature analyzer from both fault coverage and area overhead view points.

Finally, one can verify the output responses via a scan path. This can be done in two ways. The first (to be called *compaction free scan out*) loads in the scan path each read data, shifts out this data, and then, loads a new data. This technique does not involve any aliasing since there is no information compaction, but requires m shifts for each read data.

The other technique (to be called *space compaction scan out*) loads the read data in the scan path, performs one shift, and loads the next data in the scan path. In this case, we mix $m - 1$ bits of the new data with $m - 1$ bits of the previous data using XOR gates. This space reduction involves some aliasing called *error cancellation* [15]. Of course, the aliasing is much lower than in signature analysis which involves both error cancellation and error masking [15].

5.1. Output Response Comparison with BIST Circuitry Checking

In the following, we present techniques that detect the faults in the BIST circuitry without using duplication.

Address generator testing:

The Up/Down LFSR can be checked by using the parity code [6]. This implementation requires an extra (parity) cell and a parity checker. Next, we propose another technique (testing and not checking) that detects the faults in the Up/Down LFSR by means of test sequence modification. To do that, we use the sequences of Table 2 instead of those of Table 1.

The detection of faults in the Up/Down LFSR is based on the following proposition:

Proposition 1. If the NOR functions in the Up and Down feedback loops are realized by independent gates

Table 2. Modified MARINESCU's test algorithm.

Init.	S0	S1	S2	S3	S4	S5
W0	R0	R0W1W0W1	R1W0R0W1	R1W0W1W0	R0W1R1W0	R0
W0	R0	R0W1W0W1	R1W0R0W1	R1W0W1W0	R0W1R1W0	R0
W0	R0	R0W1W0W1	R1W0R0W1	R1W0W1W0	R0W1R1W0	R0
*	*	*	*	*	*	*
*	*	*	*	*	*	*
W0	R0	R0W1W0W1	R1W0R0W1	R1W0W1W0	R0W1R1W0	R0

→ time

then, under any single fault, the set of addresses generated by the Up/Down LFSR during the Up sequence is different from the one generated during the Down sequence.

Proof. (see [7])

From the above proposition there is at least:

a. one address generated during the Down sequence and not generated during the Up one, or,
b. one address generated during the Up sequence and not generated during the Down one.

Case a. is detected by the read operations of sequences S0 and S3 and case b. is detected by the read operations of sequences S2 and S5.

Comparator testing:

We will use the two-output totally self-checking double-rail checker proposed in [14]. This comparator (Figure 3.b) is implemented as a tree of double rail-checker cells as the cell of Figure 3.a. This checker is easily testable, since it requires only 4 test vectors in order to test all the double-rail checker cells. If we use a standard (i.e., single-output) comparator, we need $2m + 2$ test patterns.

For the double-rail checker there are 4^{k-2} test sets of 4 vectors. They can be obtained from the test vectors of an XOR tree. These vectors will be applied on one side of the double-rail checker (say direct input data) and their complements will be applied on the other side (say complemented data). For instance a set of test vectors for the 8 inputs double-rail checker is:

T0	0 0 0 0 0 0 0 0	1 1 1 1 1 1 1 1
Ta	0 1 0 1 1 1 0 1	1 0 1 0 0 0 1 0
Tb	1 1 1 0 0 1 1 0	0 0 0 1 1 0 0 1
Tc	1 0 1 1 1 0 1 1	0 1 0 0 0 1 0 0
	Direct data.	Complemented data.

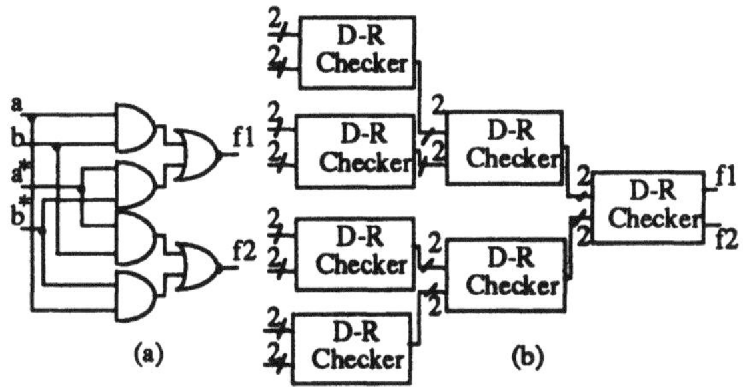

Fig. 3. Double rail checker for (a) 2 inputs; (b) 8 inputs.

The test algorithm will be extended to add the new vectors. We will use a test set which includes the all 0's vector (see above the vector *T0*). Since this data appears during the RAM test, we need to add only three additional vectors (e.g., the vectors *Ta*, *Tb* and *Tc*). A compact test data generator is derived as follows: let *S* be the set or the RAM inputs, *Sa* be the set of bit positions in which *Ta* has the value 1, define similarly *Sb* and *Sc*. $Sab = Sa \cap Sb$, define similarly *Sbc* and *Sac*. Within any set of four vectors testing an XOR tree, each bit position takes twice the value 0 and twice the value 1. Since the vector *T0* is the all 0's vector, then within the vectors *Ta*, *Tb* and *Tc* each bit position has twice the value 1 and once the value 0. Then we find $Sa = Sab \cup Sac$, $Sb = Sab \cup Sbc$ and $Sc = Sac \cup Sbc$. We also find $Sab \cap Sac = Sab \cap Sbc = Sac \cap Sbc = \emptyset$, and $Sab \cup Sac \cup Sbc = S$. Thus, a compact test data generator is designed as in Figure 4.

This generator works as follows: $VTab = VTac = 0$ and $VTbc = 1$ for applying *Ta*, $VTab = VTbc = 0$ and $VTac = 1$ for applying *Tb*, $VTbc = VTac = 0$ and $VTab = 1$ for applying *Tc*, $VTab = VTbc = VTac = 1$ for applying the all 0's data, $VTab = VTbc = VTac = 0$ for applying the all 1's data (required for RAM testing). We note that this generator requires the same number of gates (i.e., *n* inverters) as the generator of Figure 2.

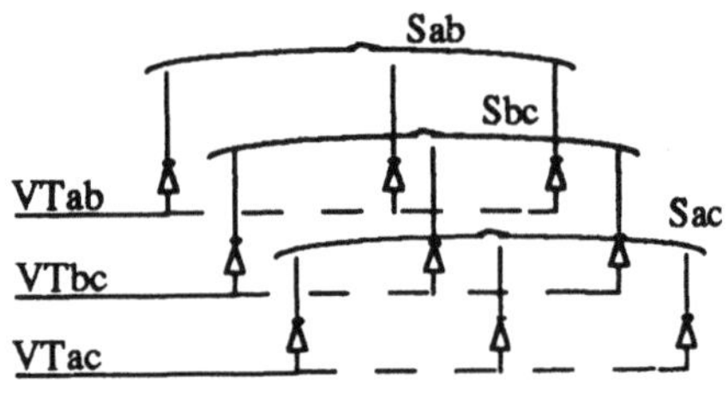

Fig. 4. The test data generator.

A second generator will be used for applying the data on the complemented inputs of the double rail-checker. This generator is derived from the one of Figure 4 by using the signals *VTab**, *VTbc**, *VTac**, which are the complements of the signals *VTab*, *VTbc*, and *VTac*.

Checking the sequence generator:

The new test algorithm uses 8 sequences (7 sequences shown in Table 2 and 1 sequence for testing the double-rail checker). Thus, the 3-cells LFSR of Figure 5 can be used as sequence generator. To check this LFSR we add a parity prediction cell *P* as proposed in [6]. Signals $P1$ and $P2$ check the parity of the LFSR and give the error indication (i.e., $(P1, P2) = 01$ or 10 for correct operation and $(P1, P2) = 00$ or 11 for error detection). One can check that the first error produced by any single fault occurring in the LFSR modifies either $P1$ or $P2$ but not both, and thus it is detected.

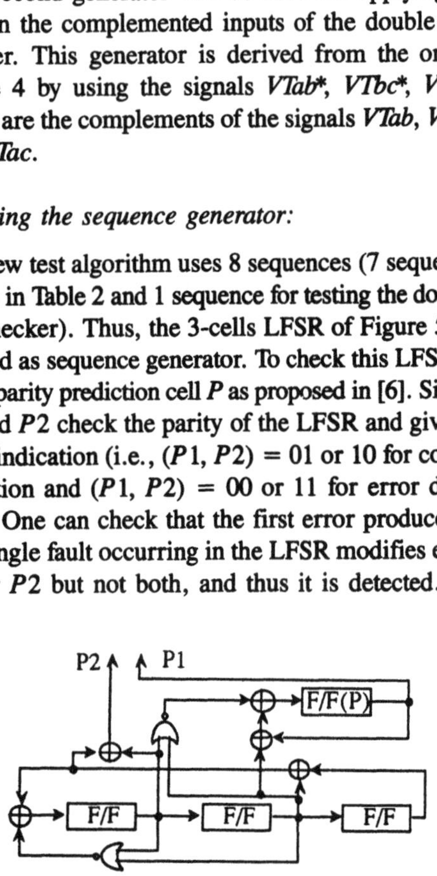

Fig. 5. Checking the sequence generator.

Checking the logic:

The remaining logic of the BIST circuitry is checked by means of duplication. The resulting BIST scheme is given in Figure 6. We note the following:

- The error indications $f1$, $f2$ and $P1$, $P2$ are compacted into the single error indication $g1$, $g2$ by using a double-rail checker cell.
- Half of the address generator cells receive the Up/Down signal from the block "logic" and the other half from the block "logic.*"[4] This allows detection of errors on the Up/Down signal, since these errors involve the LFSR with a behavior described in Proposition 1.

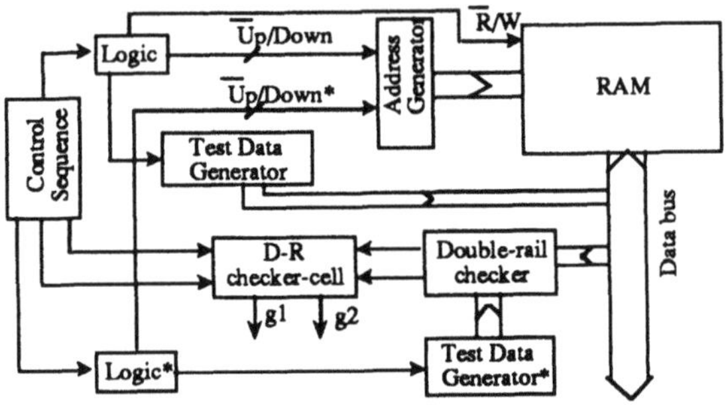

Fig. 6. The BIST structure.

Finally, errors on the R/W are detected by the fact that the values applied on the double-rail checker by the block "Test Data Generator*" do not correspond to the ones coming from the data bus.

6. Combination and Comparison of the Different Schemes

The schemes for address generation, test data generation and output response verification can be combined in different ways to obtain several BIST and scan path architectures. For instance, Figure 7 presents a BIST scheme which uses a comparator for output response verification and Figure 8 presents a BIST scheme using a MISR. Figure 9a presents a scheme that uses a scan path for addressing the RAM, another scan path for applying the test data, and a comparator for output response verification.

Figure 9b presents another scheme that uses a scan path for addressing the RAM, and another scan path for both applying the test data and collecting the output responses. It requires *m* shifts for applying each test data and *m* shifts for transferring out each read data. To avoid test length increase, we can use one shift for

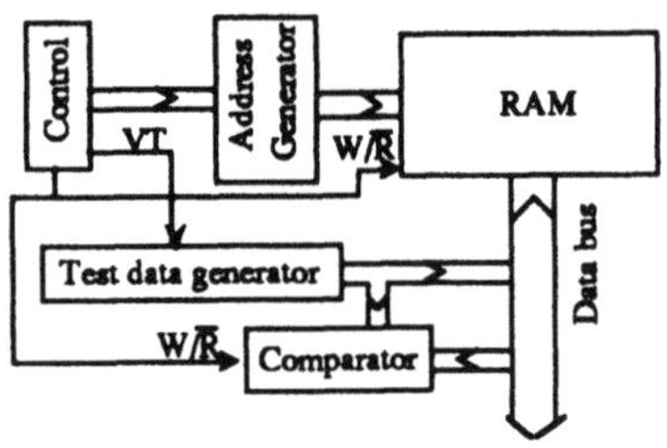

Fig. 7. BIST with comparator.

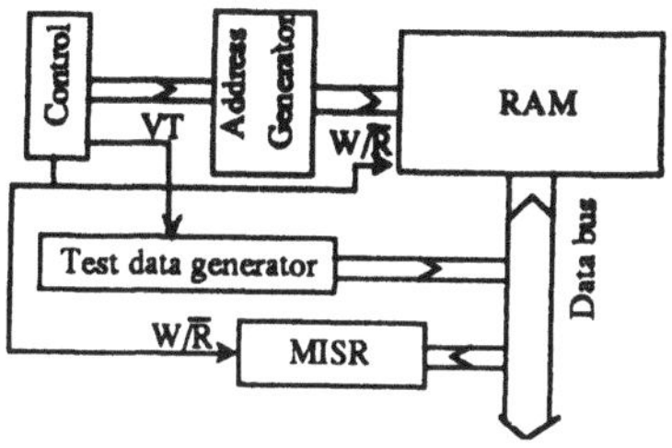

Fig. 8. BIST with MISR.

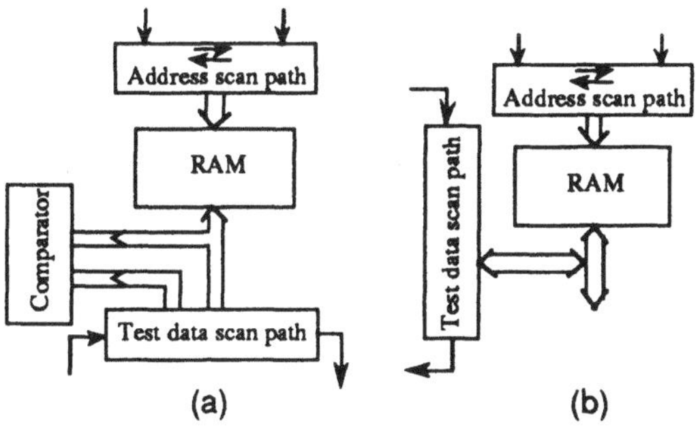

Fig. 9. (a) Scan path using comparator for output response verification, (b) single scan path for both test data application and output response verification.

applying each modified test data (according to the technique of Section 4), and one shift for each read data (according to the space compaction scan out technique of Section 5). These two techniques require the use of two scan paths, otherwise test data and output responses will be merged. In order to reduce the overhead, in Figure 10 each cell of the double scan path is composed of a master and two slave latches. The scheme works as follows: The master latch is used for both the test data and the output responses, while slave 1 latch is used for the output responses and slave 2 latch for the test data. When we inject the output response in the scan path, the test data are saved in slave 2 latch, and when we shift the test data, the output response is saved in slave 1 latch. This scheme is less efficient than the one of Figure 9b since it involves larger overhead and requires the external verification of a large amount of information. With respect to Figure 9a this scheme is definitely worse (higher overhead and very high amount of information to be verified externally).

The area overheads are presented in Figure 11 (Figure 11a for 8 bit word RAMs, Figure 11b for 16 bit word RAMs). The BIST area overhead has been obtained by using the RAM BIST generator presented in [8]. Another generator developed recently around the CADENCE CAD tools and the ES2 SOLO 2030 standard cell library is used for the BIST scheme using the comparator plus BIST circuitry checking (Figure 6).

A third tool is used for two scan path schemes, the first corresponds to Figure 9a and the second to Figure 9b (using a single scan path for data). When the scan path plus comparison technique (Figure 9a) is applied, there is a saving in test length compared to the compaction free scan path technique (Figure 9b), but the area overhead is larger. For large and medium sized RAMs, the difference in area overhead is insignificant, and the scan path plus comparison scheme (requiring *m* times less test of length) is more attractive. For small RAMs, the area overhead difference increases (but remains low) and the test length difference is not critical (17408 clock cycles instead of 2176 clock cycles, for 128 by 8 bit RAM). Thus the advantage of the scheme of Figure 9a is reduced.

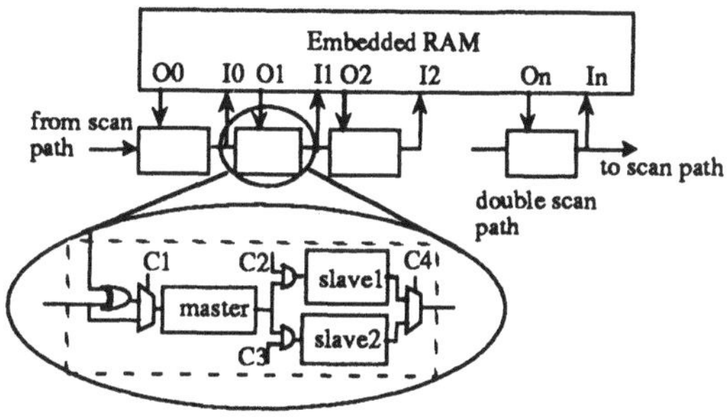

Fig. 10. Modified double scan path.

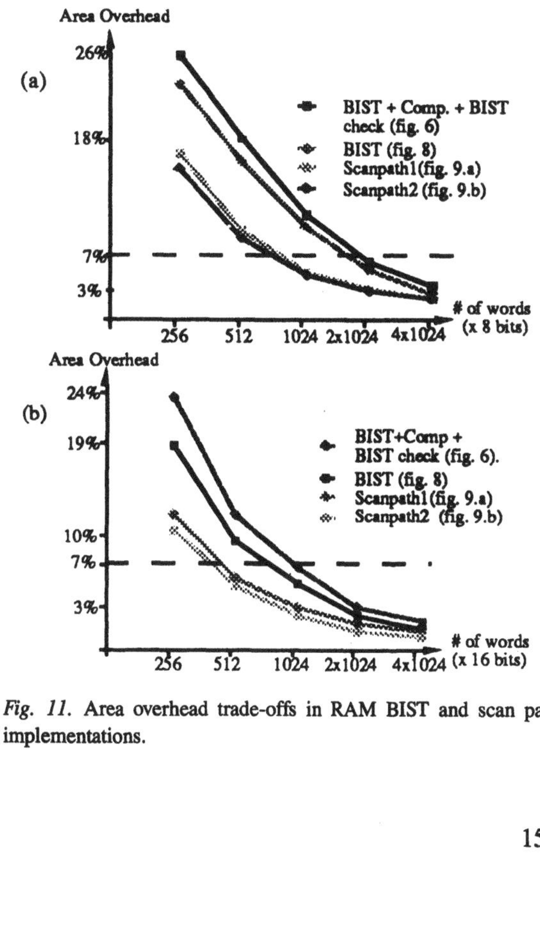

Fig. 11. Area overhead trade-offs in RAM BIST and scan path implementations.

The designer can make choice according to the area overhead that he accepts to pay. For instance, for 8-bit words, if this overhead is 7%, all the schemes are acceptable for RAMs with more than 1024 words. Both the scan path schemes are acceptable for RAMs with size between 512 and 1024 words, and no scheme is acceptable for RAMs with less than 256 words. For a more accurate choice, an economic analysis taking into account of several other cost factors has to be given [16].

Tables 3 and 4 present the advantages and the drawbacks of the different schemes for addressing the RAM, applying the test data and verifying the output responses. To summarize these tables, for BIST implementations:

- Using an Up/Down LFSR as address generator is better than using an Up/Down counter, since the Up/Down LFSR is less complex, faster, and in embedded RAMs it can be reconfigured into a signature analyzer for verifying the output responses of other blocks.
- Although the comparator does not involve any aliasing, using a MISR for output response verification (Figure 8) is better than using a comparator (Figure 7), since the comparator offers a low coverage for faults affecting the BIST circuitry. For large RAMs, using a comparator and checking the BIST circuitry (Figure 6) is more interesting since the fault coverage is high and the area overhead increase is not significant.

For scan path oriented implementations:

- Addressing order based on Up/Down LFSRs does not increase the test length, while the one based on Up/Down binary counters multiply the test length by n.
- If the test data are applied via a scan path, then using standard test data multiplies the test length by m, while using the modified test data proposed in this article does not increase this length.
- If the output response verification is performed by a comparator, there is no aliasing and also the coverage of faults occurring in the DFT circuitry is high if the test data are applied via a scan path.

Therefore, the best BIST implementation consists of:

- address generation based on Up/Down LFSR
- output verification based on MISR for small RAMs, or on comparator combined with BIST circuitry checking for large RAMs

The best scan path based technique consists of:

- using a bidirectional scan path for addresses combined with addressing order based on Up/Down LFSRs
- using a comparator for output response verification

In this scheme there is no aliasing, the coverage of the DFT circuitry is high, and the test length is not increased, while the area overhead remains close to the other scan path schemes.

The comparision of the best BIST scheme against the best scan path scheme gives:

1. both the schemes require the same test length.
2. both the schemes ensure high fault coverage for the BIST/DFT circuitry.
3. the BIST allows the circuit to be tested itself but not the scan path scheme.
4. the area overhead involved by the BIST scheme is low for medium and large RAMs, but due to the area of the BIST sequencer (which is independent of the RAM size) the BIST area overhead becomes significant for small RAMs. The area overhead for the scan path remains low even for small RAMs.

7. Economic Viability

Integrating testability in the design process leads to a higher quality product. This has well recognized cost advantages; it is cheaper to discover faults in the early stages, for example at component test, rather than later, when a faulty chip has been incorporated into a system and shipped to a customer. Increasing the reliability of high quality products has definite advantage for the manufacturer's reputation. Thus, the expenditure in the component test stage may drastically reduce field maintenance costs, and therefore be a justifiable expense [17], [18]. These arguments in combination with the fact that DFT techniques decrease significantly the test generation cost and the test program cost, make DFT technique attractive.

In order to make an accurate choice between conventional testing and DFT techniques like scan path and BIST, several parameters concerning the fabrication cost and the test cost at different stages of the product fabrication have to be considered. Such parameters are for instance yield, manufacturing costs (e.g., processing cost, masks cost, packaging cost, board assembling cost...), test generation cost, ATE cost, labor rates

Table 3. Address generation and output response verification for BIST.

address generation	silicon area	speed
Up/Down counter	low	slow for large RAMs
Up/Down LFSR	lower than Up/Down counter	fast

output response verification	aliasing	BIST circuitry fault coverage
MISR	yes	high
Comparator	no	high if BIST checking

Table 4. Address generation, test data generation, and output response verification for scan path and mixed approaches.

address generation using scan path	test length	silicon area
Up/Down LFSR based address ordering	unchanged	low
Up/Down counter based address ordering	multiplied by n	low

test data generation	test length	silicon area	DFT circuitry fault coverage
scan path using standard test data	multiplied by m	low	high
scan path using modified test data	unchanged	low	high
test data generation block	unchanged	very low	high if with MISR, low if with comparator

output response verification	test length	silicon area	DFT circuitry fault coverage	aliasing
MISR	unchanged	low	high	yes: error cancellation + error masking
compaction free scan path	multiplied by m	low	high	no
space compaction scan path	unchanged	low	high	yes: error cancellation
scan path + comparison	unchanged	very low	high if with test data scan path	no
BIST + comparison	unchanged	increased if checking BIST	high if checking BIST	no

of test programmers, board, system, and field diagnosis and repairing costs, system down time cost... [19], [16]. Most of these parameters are known to the manufacturer and do not depend upon the choice of the test technique. On the other hand, the choice of the test technique determines some parameters very sensible for the economics model like test length, fault coverage, test pattern generation cost, test programming cost, additional silicon area, and additional design effort for the DFT techniques (for instance see [16]). It has been shown that for low production volumes, BIST is economically more attractive than conventional test and scan path, while for high production volumes, scan path takes the advantage due to the lower reliability of BISTed circuits involved by the additional hardware. However if extensive burn-in is used, BIST is more attractive even if it requires high area overhead like 30% to 40% [16]. In these estimations it was considered that BIST requires lengthy test patterns (pseudorandom testing), and so for the scan path (use of several shifts per test vector). Also test pattern generation and fault simulation effort have been considered for scan path. Concerning the RAM BIST and scan path techniques presented in this article, the test length is not increased with respect to the deterministic test patterns, while for conventional testing of embedded RAMs, the test length should be increased due to the lower controllability and observability. In addition, conventional testing will involve test generation (and eventually fault simulation) cost for backward and forward propagation of the deterministic test algorithm. In addition, reduced controllability and observability will involve reduction of the fault coverage for conventional testing of embedded RAMs.

Furthermore, tools like the one developed in the framework of the present project allow automatic generation of BISTed and scan path RAM designs, so that the additional design effort is eliminated. Taking into account that the area overhead required for the proposed RAM BIST schemes is much lower than the pessimistic estimation of 30% to 40% considered in [16], we can conclude that for embedded RAMs the BIST and scan path techniques are definitely more advantageous than the conventional test techniques.

As concerning the choice between BIST and scan path, the results of this article show that for large and medium sized RAMs, the BIST implementation is definitely more advantageous, since it ensures self-test by means of slightly higher overhead with respect to scan path. For small RAMs, exact cost estimation has to be given for making a decision. This article gives parameters like area overhead, fault coverage, test length, design effort, etc., and makes this estimation easier. According to the values of the other parameters, the volume of the production and the eventual use of burn-in, scan path or BIST can be more interesting. However, high area overhead for small RAMs is not significant for the total area of a VLSI circuit. Thus, if the other blocks are BISTed, small RAMs has also to be BISTed to ensure a complete self-test of the circuit. Furthermore, if the circuit includes several RAMs, then the sequencer of the BIST (involving high overhead for small RAMs) can be shared between these blocks, resulting in a considerable reduction of the area overhead. Finally, in case of applications requiring periodic testing, BIST becomes even more attractive if it is implemented using the transparent BIST technique [8], [13]. Such a BIST preserves the contents of the RAM and avoids the use of spare RAMs for saving the RAM contents before testing, it does not reduce the fault coverage [13] and requires only 1% to 2% more area overhead with respect to standard BIST [8].

8. Conclusion

In this article we have presented efficient techniques for designing scan path and BISTed RAMs. The new scan path scheme applies one test pattern per clock cycle and reduces test length drastically. Then, we have studied various techniques for applying addresses and test data and verifying the output responses in BIST and scan path designs for RAMs. The more efficient of these techniques have been selected to derive the best scan path and BIST schemes. Tools for automatic generation of these schemes have been implemented and experimented over various RAM sizes. These experiments show that we can satisfy all the following constraints: low area overhead, high fault coverage for both the RAM block and the DFT circuitry and low test length.

A preliminary analysis shows that for embedded RAMs, the proposed scan path, and BIST techniques are definitely more cost effective than the conventional test techniques. On the other hand, for medium and large RAMs, BIST is found to be more effective than scan path, while for small RAMs the choice will depend upon the production volume, the use of burn-in etc., and an exact cost computation is needed in each case to make a definite choice. However, since for small RAMs high area overhead is not significant with respect to the total area of VLSI circuit, BIST implementation can be chosen even for small RAMs.

References

1. S.M. Thatte and J.A. Abraham, "Testing of Semiconductor Random Access Memories" *Proc. IEEE 7th Intern. Conf. on Fault-Tolerant Comput.* (FTCS), USA, 1977, pp. 81–87.
2. A.J. Van de Goor and C.A. Verruijt, "An Overview of Deterministic Functional RAM Chip Testing," *ACM Computing Surveys*, vol. 22, no. 1, March 1990, pp. 5–33.
3. M. Marinescu, "Simple and Efficient Algorithms for Functional RAM Testing," *Proc. IEEE Intern. Test Conf.* (ITC), USA, Nov. 1982, pp. 236–239.
4. D.S. Suk, and S.M. Reddy, "A March Test for Functional Faults in Semiconductor Random Access Memories," *IEEE Trans. on Comput.* vol. C-30, no. 12, Dec. 1981, pp. 982–985.
5. C.V. Alves, M. Nicolaidis, P. Lestrat, and B. Courtois, "Built-in Self-Test for Multi-Port Rams," *Proc. IEEE Intern. Conf. on Computer-Aided Design* (ICCAD), Santa Clara, USA, Nov. 1991, pp. 248–251.
6. M. Nicolaidis, "An Efficient Built-In Self-Test for Functional Test of Embedded RAMs," *Proc. IEEE 15th International Symp. on Fault-Tolerant Comput.*, Ann Arbor, USA, June 1985, pp. 118–123.
7. M. Nicolaidis, O. Kebichi and V.C. Alves, "Trade-Offs in Scan Path and BIST Implementation for RAMs," TIMA Research Report No. RR 93. 2-I-Sept. 1993.
8. O. Kebichi and M. Nicolaidis, "A Tool for Automatic Generation of BISTed RAMs," *Proc. IEEE Intern. Conf. on Computer Design*, (ICCD), Cambridge, USA, Oct. 1992, pp. 570–576.
9. M. Nicolaidis, "Transparent Testing of Integrated Circuit," Greek Patent 920100088 5/3/92, Institute of Industrial Property, Athens, Greece, March 1992.
10. M. Nicolaidis, V.C. Alves, and O. Kebichi, "Trade-Offs in Scan Path and BIST Implementation for RAMs," *Proc. IEEE European Test Conf.* (ETC), Rotterdam, April 1993, pp. 169–178.
11. H. Maeno, K. Nii, S. Sakayanagi, and S. Kato, "LSSD Compatible and Concurrently Testable RAM," *Proc. IEEE Intern. Test Conf.* (ITC), Baltimore USA, Sept. 1992, pp. 608–614.
12. P. Bardel, W. McAnnoy, J. Savir, *Built-In Test for VLSI: Pseudorandam Techniques*, Wiley Interscience, New York, USA, 1987.
13. M. Nicolaidis, "Transparent BIST for RAMs," *Proc. IEEE Intern. Test Conf.* (ITC), Baltimore USA, Sept. 1992, pp. 598–607.
14. W.C. Carter, and P.R. Schneider, "Design of Dynamically Checked Computers," *Proc. IFIP 68 Cong.*, Amsterdam, 1968, pp. 878–883.
15. V. Zorian and A. Ivanov, "EEODM: An Effective BIST Scheme for ROMs," *Proc. IEEE Intern. Test Conf.* (ITC), Washington DC, USA, Sept. 1990, pp. 871–879.
16. P. Varma, A.P. Ambler, and K. Baker, "An Analysis of the Economics of Self-Test," *Proc. IEEE Intern. Test Conf.*, USA, Oct. 1984, pp. 20–30.
17. I.D. Dear and A.P. Ambler, "Predicting Cost and Quality Improvement as Result of Test Strategy Planning," *IFIP Workshop on fast prototyping of VLSI*, March 1987, pp. 225–237.
18. A.P. Ambler, M. Paraskeva, D.F. Burrows, W.L. Knight and I.D. Dear, "Economically Viable Automatic Insertion of Self-Test Features for Custom VLSI," *IEEE Intern. Test Conf.* (ITC), Sept. 1986, pp. 232–243.
19. R.E. Huston, "An Analysis of ATE Testing Costs," *Proc. IEEE Intern. Test Conf.* (ITC), Oct. 1983, pp. 396–411.

Michael Nicolaidis is a "chargé de Recherche" of the CNRS (French National Research Center) and leader of the Reliable Integrated Systems group in the TIMA Laboratory of INPG-Grenoble. He received the degree in electrical engineering from the Polytechnical School of the University of Thessaloniki (Greece) and the engineer Doctorate Thesis from the INPG, National Polytechnical Institute of Grenoble, France. His main research interests include testing of VLSI systems, DFT, BIST, self-checking and fail-safe systems. He has served on the Technical Program Committees of the IEEE Fault Tolerant Computing Symposium, the VLSI Test Symposium and the European Design and Test Conference, and he is presently the vice chairman of technical activities of the IEEE Test Technology Technical Committee (TTTC).

Omar Kebichi received the B.E.E. degree from Blida University (Algeria) in 1989 and the DEA degree from the Insitut National Polytechnique de Grenoble in 1990. He is currently working towards the Ph.D. degree on microelectronic in the Institu National Polytechnique de Grenoble. His research topics are VLSI design, BIST, DFT, self-checking circuits and synthesis for testability.

Vladimir Castro Alves obtained his electronic engineering degree from the Federal University of Rio de Janeior in 1985. He obtained his DEA and Ph.D. degrees from the Institut National Polytechnique de Grenoble in 1989 and 1992 respectively. He holds presently a Eurochip Lecturer position in the University of Aveiro where he is also responsible for the VLSI design Laboratory. His interest are VLSI design, self-checking circuits, memory testing and synthesis for testability.

JOURNAL OF ELECTRONIC TESTING: Theory and Applications, 5, 285-297 (1994)

Techniques for Estimating Test Length Under Random Test

AMITAVA MAJUMDAR*
Crosscheck Technology Inc., 2833 Junction Ave., Suite 100, San Jose, CA 95134

SARMA B.K. VRUDHULA
(a.k.a. Sarma Sastry)
Department of Electrical and Computer Engineering, University of Arizona, Tucson, AZ 85721

Abstract. When a circuit is tested using random or pseudorandom patterns, it is essential to determine the amount of time (test length) required to test it adequately. We present a methodology for predicting different statistics of random pattern test length. While earlier methods allowed estimation only of upper bounds of test length and only for exhaustive fault coverage, the technique presented here is capable of providing estimates of all statistics of interest (including expected value and variance) for all coverage specifications.

Our methodology is based on sampling models developed for fault coverage estimation [1]. Test length is viewed as a *waiting time* on fault coverage. Based on this relation we derive the distribution of test length as a function of fault coverage. Methods of approximating expected value and variance of test length are presented. Accuracy of these approximations can be controlled by the user. A practical technique for predicting expected test length is developed. This technique is based on clustering faults into equal detectability subsets. A simple and effective algorithm for fault clustering is also presented. The sampling model is applied to each cluster independently and the results are then aggregated to yield test lengths for the whole circuit. Results of experiments with several circuits (both ISCAS '85 benchmarks and other practical circuits) are also provided.

Keywords: Circuit testing, test quality, urn models, waiting time distribution.

1. Introduction

A popular methodology for testing large circuits is Built-In Random Self-Test (BIRST) where a given circuit is partitioned into smaller combinational logic blocks and each block is tested using random or pseudorandom patterns [2]. The main advantage of applying random or pseudorandom patterns is that all combinational blocks can be tested at the same time. However, different blocks have varying testability properties and covering the same proportion of faults in different blocks may (and usually do) require different amounts of test time. Thus, the total amount of time spent in testing a given circuit depends on the desired level of fault coverage and the amount of time required to test the worst block. It is therefore, important to determine how long a combinational block needs to be tested in order to obtain a specified level of fault coverage. This is called *test length estimation*.

*This work was done while the author was with the Department of Electrical Engineering, Southern Illinois University, Carbondale, IL 62901.

Apart from this application, test length estimation also finds extensive use in 1) design for testability tasks (such as test point insertion and partitioning) in deciding which designs can be tested in a reasonable amount of time and for 2) on-line test scheduling in fault tolerant systems [3], to determine how much time should be spent in testing different modules so that a circuit functions correctly with a specified confidence (this also influences the number of spares included in a fault tolerant system).

Most research efforts in the past have focused on the problem of predicting expected fault coverage as a function of test length. Among them are [4]-[7]. The question of test length estimation has been addressed in [7], [8] and are discussed below.

Wagner et al. [7] use a probabilistic model to examine both random and pseudorandom testing and derive formulae for average fault coverage and test length. Their model is based on detectabilities of different faults (where detectability is defined as the number of patterns that detect a fault). This information is used to form a detectability profile H of a circuit that simply associates with a value k the number h_k of faults with detectability k. The authors present a detailed analysis for estimating expected fault coverage of a circuit for a given test length. For test length estimation as well, the same underlying model is used. The expression derived for expected test length (Equation 12 in [7]) gives the average amount of time required to determine whether a circuit is faulty (i.e., given a fault has occurred). In doing so the authors assume that all faults can occur with equal probability. However, the problem of estimating test length for a specified fault coverage was not addressed.

Savir and Bardell [8] present a methodology for computing upper bounds on test length (for a given confidence ϵ) for 100% coverage of a set of hard to detect faults. In order to simplify analysis, they assume that all faults in this set have a detectability p (that is usually taken as the detectability of the least testable fault). The technique provides an estimate of the amount of time required to ensure that the escape probability is below a given threshold (escape probability being the probability that at least one fault in the set is untested).

One disadvantage of this methodology is that due to the inherent nature of escape probability as a parameter, we can predict test lengths only for 100% fault coverage in a fault set. Furthermore, the obtained upper bounds on test length (see Equation 15 in [8]) approach infinity as the specified escape probabilities approach 0. In general, the upper bound for a given escape probability may grossly overestimate the expected test length for 100% coverage.

For a better understanding of the feasibility of random testing and also for effective test scheduling, it is necessary to estimate the distribution of test lengths for different fault coverage values (not only for 100% coverage). This allows computation of not only expected values of test length but also of other important statistics such as variance and other higher moments. This problem was addressed briefly in [1] where an important relation between test length and fault coverage was established but was not formally proven.

In this article we present some new results on fault coverage which facilitate our characterization of the distribution of test length. Furthermore, we define simple techniques that allow us to estimate different test length statistics.

The rest of the article is organized as follows: In Section 2 we introduce the notation used in our analysis. A sequential sampling model for fault testing, first introduced in [1], is described in Section 3. Based on this model we derive the distribution of test length as a function of fault coverage and the detectabilities of faults in a given circuit. Since the test length results derived in this article are based on the distribution of fault coverage, we first present some key theorems on fault coverage. Next, in Section 4.2 we present the main result of this article, namely the distribution of test length under very general conditions. The distribution of test length under the special case where the number of faults tested by a random pattern is binomially distributed, is derived in Section 4.3. Approximations for expected value and variance of test length are given in Section 4.4. In Section 5 we define an algorithm for fault clustering by detectability and present results of experiments with several circuits in order to validate our models. Finally we offer some concluding remarks in Section 6.

2. Notation

In this section we define the key concepts considered in the following analysis. Most of the notation is adopted from [1].

Definition 2.1. The detectability p_f of a fault f is the probability that given f is present in the circuit it is detected by a randomly selected vector.

Definition 2.2. The fault coverage X_r is a random variable that denotes the number of distinct faults that are tested by the application of r randomly selected test vectors.

Definition 2.3. The fault uncoverage Y_r is a random variable that denotes the number of distinct faults that are untested after the application of r randomly selected test vectors. Thus $Y_r = M - X_r$ where M is the total number of faults in the set under consideration.

Definition 2.4. Test length R_n is a random variable that denotes the number of test patterns required to obtain a fault coverage of at least n.

$\binom{n}{k}$ = binomial coefficient n choose k.

$x^{\underline{r}} = x(x-1)\cdots(x-r+1)$.

$\Delta f(x) = f(x+1) - f(x)$.

$\Delta^n f(0) = [\Delta^n f(x)]_{x=0}$

3. Model of Fault Testing

Let C denote a combinational circuit with n inputs. The set of all $N = 2^n$ possible test patterns is denoted by $\mathfrak{I}$ and the set of all possible *ssa* faults in C is denoted by $\mathfrak{F}$, with $M = |F|$. Associated with each test pattern $t_i \in \mathfrak{I}$, is a subset $f_{t_i} \subset \mathfrak{F}$ that are detected by t_i. Let $w_i = |f_{t_i}|$.

Let us first consider how one could experimentally obtain an estimate of the number of random test patterns that are required to detect a given number n or more faults. The sequence of steps required to do this is given below.

1. $i = 1$.
2. $D = \phi$ /* D is the set of detected faults*/.
3. Generate a test pattern t_i randomly.
4. Perform fault simulation to determine f_{t_i}.
5. $D \leftarrow D \cup f_{t_i}$.
6. if $|D| < n$ {
   $i = i + 1$
   go to 3
   }
7. Stop.

The value of i after the program's execution is the number of patterns required to test for n faults. If the above procedure is repeated a sufficiently large number of times, say k, then the average of the k observations $i_1, i_2, \ldots, i_k$ can be taken as an estimate of the average number of test patterns needed to detect n faults. We now describe a two stage sampling scheme that models the experimental procedure described above.

The application of a sequence of random test patterns may be viewed as a random selection of the corresponding subset of faults. Thus, if we apply a sequence of r test patterns, $t_1, t_2, \ldots, t_r$ chosen from $\mathfrak{I}$, we would observe r numbers $w_1, w_2, \ldots, w_r$, $0 \leq w_i \leq M$. Let W be the random variable that denotes the number of faults tested by the application of a randomly selected test pattern. The quantities $w_1, w_2, \ldots, w_r$ are independent realizations of W. Since a test pattern t_i tests for a subset f_{t_i} of fauls, we can represent the application of t_i as sampling the set $\mathfrak{F}$ of faults, w_i number of times, without replacement. Now, for any two test patterns, t_i and t_{i+1}, the set $f_{t_i} \cap f_{t_{i+1}}$ is not necessarily empty. Therefore, the faults detected by t_i are marked as being detected and replaced back in the population before the application of t_{i+1}. After applying a sequence of r test patterns, the total number of faults that are marked as being detected is the fault coverage and is denoted as X_r. Similarly, let R_n denote the number of test patterns that are required to attain a fault coverage of at least n.

An implicit assumption made in this model is that each fault in the set $\mathfrak{F}$ is equally likely to be detected. Clearly, this is not true in practice. The results based on this assumption are still very useful since we can partition the fault set into clusters based on estimates of their detectabilities. This is similar to estimating the detectability profile as was done in [7] or partitioning the fault set into the so-called easy and hard to detect faults, as was done in [8]. In Section 5 we present a procedure for clustering faults based on their detectabilities. Once the fault set has been partitioned, we can apply the results based on the two stage sampling model to the individual clusters and aggregate them to obtain estimates of fault coverage and test length for the complete fault set.

In the following section we examine the distribution functions of X_r and R_n under this two stage sampling scheme. The approach taken here is as follows. The distribution functions of X_r and R_n are expressed in their most general form, i.e., in terms of the random variable W. Recall that W represents the number of faults that are detected by the application of a random test pattern. We then derive expressions for the factorial moments of all orders of X_r and R_n. Again these are expressed in terms of the corresponding moments of W. Once this is done, we proceed to examine special cases for the distribution of W. In particular, after the fault set has been partitioned into equal-detectability classes, the number of faults from a given cluster that are detected by a random test pattern follows a binomial distribution. Thus, by assuming that W for a given cluster is binomial, we can obtain highly simplified expressions for the moments of fault coverage and test length.

Note: In the interest of clarity and conciseness, most results are presented here without proof (barring a few important ones). Complete details may be found in [9].

4. Distribution of Test Length

Most of the results presented in the remainder of this article are expressed in terms of two new random variables, Z and Y_r, instead of W and X_r. $Y_r = M - X_r$ and $Z = M - W$. The quantity Z represents the number of faults that are untested by a randomly selected pattern. Similarly, Y_r represents the number of faults that remain untested after application of r random test patterns. The reason for using Z and Y_r is that the resulting expressions are algebraically simpler. It is for this same reason that in many problems involving *urn models* one finds that dealing with the number of empty urns often leads to simpler expressions than when dealing with the number of occupied urns.

The following lemma shows a simple and useful relation between fault uncoverage of r test patterns, Y_r, and the test length R_n for a coverge of n or more faults.

Lemma 1. The *pdf* of test length R_n for coverage of n or more faults out of M faults is given by

$$P\{R_n \le r\} = P\{Y_r \le M - n\}.$$

Proof. The two events $\{R_n \le r\}$ and $\{Y_r \le M - n\}$ are such that

$$\{R_n \le r\} \Rightarrow \{Y_r \le M - n\}$$
$$\{Y_r \le M - n\} \Rightarrow \{R_n \le r\}.$$

These relations imply that

$$P\{R_n \le r\} \le P\{Y_r \le M - n\}$$
$$P\{R_n \ge r\} \ge P\{Y_r \le M - n\}.$$

□

Equation (1) further implies that

$$\begin{aligned} P\{R_n = r\} &= P\{R_n \le r\} - P\{R_n \le r - 1\} \\ &= P\{Y_r \le M - n\} - P\{Y_{r-1} \le M - n\} \end{aligned} \quad (2)$$

Equation 1 states that test length can be viewed as a waiting time on fault uncoverage. This property is used to simplify the task of constructing the distribution of R_n by first deriving the distribution of fault uncoverage Y_r and then using Equation (1) to derive an expression for test length.

4.1. Distribution of Fault Uncoverage

Fault uncoverage Y_r is the number of faults that remain untested after the application of r random test patterns. The number of faults tested by each of these r test patterns is assumed to be an independent realization of the random variable W. Consider the case where the r realizations of W are known, i.e., $w_1, w_2, \ldots, w_r$ are given. the conditioning on the knowledge of $w_1, w_2, \ldots, w_r$ describes a hypothetical situation where one draws r random numbers $w_1, w_2, \ldots, w_r$ from the distribution of W and the on each trial k, draws w_k faults without replacement from the fault set $\mathfrak{F}$, marks them as being tested and returns them to the population before the next draw. We first derive an expression for the probability that the fault uncoverage $Y_r = j$, given $w_1, w_2, \ldots, w_r$.

Theorem 1.[1] The *pmf* and *pgf* for Y_r conditioned on w_i, $i = 1, 2, \ldots, r$, are respectively given by (see Section 2 for notation).

$$P\{Y_r = j \mid w_1, w_2, \ldots, w_r\} = \frac{1}{\Pi_{i=1}^{r} M^{\underline{w_i}}} \binom{M}{j} \Delta^{M-j}\left[\prod_{i=1}^{r} (0)^{\underline{w_i}}\right] \quad (3)$$

$$G_{Y_r}(s \mid w_1, w_2, \ldots, w_r) = \frac{1}{\Pi_{i=1}^{r} M^{\underline{w_i}}} (s + \Delta)^M \left[\prod_{i=1}^{r} (0)^{\underline{w_i}}\right] \quad (4)$$

□

Some interesting special cases such as $j = M$ and $r = 1$ can be easily checked for the above *pmf*. Thus

$$P\{Y_r = M \mid w_1, w_2, \ldots, w_r\} = \begin{cases} 1 & w_1 = w_2 = \cdots = w_r = 0 \\ 0 & \text{otherwise} \end{cases}$$

since $0^{\underline{w_i}} = 1$ if and only if $w_i = 0$ and equals 0 otherwise. Similarly, when $r = 1$,

$$P\{Y_1 = j \mid w_1\} = \begin{cases} 1 & j = M - w_1 \\ 0 & \text{otherwise} \end{cases}$$

This follows from the fact that $\Delta^{M-j}[0^{\underline{w_i}}] = w_1!$ if and only if $w_1 = M - j$ and equals 0 otherwise.

Equations 3 and 4 allow us to examine the *pmf* of Y_r for general (and unrestricted) distributions of W. It is found that these general formulae are more easily derived using the factorial moments (see Section 2) of Y_r conditioned on the w_i's. Expressions for these quantities are easily obtained by differentiating G_{Y_r} w.r.t. to s. This is given in the following lemma.

Lemma 2. The jth factorial moment of Y_r conditioned on w_i, $i = 1, 2, \ldots, r$, is given by

$$\begin{aligned} E(Y_r^{\underline{j}} \mid w_1, w_2, \ldots, w_r) &= M^{\underline{j}} \prod_{i=1}^{r} \left[\frac{(M - w_i)^{\underline{j}}}{M^{\underline{j}}}\right] \\ &= M^{\underline{j}} \prod_{i=1}^{r} \left[\frac{(z_i)^{\underline{j}}}{M^{\underline{j}}}\right] \end{aligned} \quad (5)$$

where $z_i = M - w_i$. □

Although, the above equations are general, they are of little practical use, since if the w_i's $(1 \leq i \leq r)$ are known, we can assume that fault simulations were carried out for all r vectors and therefore the actual fault uncoverage is also known.

In order to exploit the utility of these formulae as well as to obtain simplifications we remove the conditioning on w_i. We assume that each w_i is an independent realization of W. Similarly the numbers $z_i = M - w_i$ are independent realizations of the random variable $Z = M - W$. The following theorem expresses the unconditional factorial moments of Y_r in terms of the factorial moments of Z.

Theorem 2. The unconditional j^{th} factorial moment of Y_r is given by

$$E(Y_r^{\underline{j}}) = M^{\underline{j}} \left[\frac{E(Z^{\underline{j}})}{M^{\underline{j}}} \right]^r. \tag{6}$$

□

Equation 6 expresses the factorial moments of fault uncoverage in terms of the corresponding moments of W, the number of faults tested on any trial. Thus, the mean and variance of the number of faults that are not tested by r random test patterns are given by

$$E(Y_r) = M \left(1 - \frac{E(W)}{M}\right)^r \tag{7}$$

$$Var(Y_r) = E(Y_r^{\underline{2}}) + E(Y_r) - E^2(Y_r) \tag{8}$$

The *pmf* of Y_r can be obtained from its factorial moments. Using this and the relation between the test length R_n and Y_r given in Equation 1, we derive the distribution function of R_n.

4.2. Distribution of Test Length

Factorial moments and *pmf* of a discrete random variable U (which ranges in value between 0 and M) are related in the following manner (see [10], page 73).

$$P\{U = n\} = \frac{1}{n!} \sum_{s=0}^{M-n} \frac{1}{s!} E[U^{\underline{s+n}}].$$

Substituting from Equation 6 into the above expression we get the unconditional *pmf* of fault uncoverage Y_r for any distribution of Z (or W) as follows. Let

$$\alpha_j = \frac{E[Z^{\underline{j}}]}{M^{\underline{j}}}. \tag{9}$$

The *pmf* of Y_r is then given by

$$P\{Y_r = n\} = \binom{M}{n} \sum_{s=0}^{M-n} (-1)^s \binom{M-n}{s} (\alpha_{s+n})^r. \tag{10}$$

We can now use Equation 1 to obtain a distribution of R_n. This is stated below.

Theorem 3. The *pmf* and *pgf* of R_n can be expressed as follows.

$$P\{R_n = r\} = \begin{cases} 0 & \text{for } r = 0 \\ \sum_{j=M-n+1}^{M} (-1)^{j-M+n-1} \binom{M}{j} \binom{j-1}{M-n} \alpha_j^{r-1}(1-\alpha_j) & \text{for } r \geq 1 \end{cases} \tag{11}$$

and

$$G_{R_n}(s) = \sum_{j=M-n+1}^{M} (-1)^{j-M+n-1} \binom{M}{j} \binom{j-1}{M-n} \left[\frac{s(1-\alpha_j)}{1-s\alpha_j} \right]. \tag{12}$$

Proof. The fact that $P\{R_n = r\} = 0$ when $r = 0$ needs no proof. When $r > 0$, we use Equation 1 to get

$$P\{R_n \leq r\} = P\{Y_r \leq M - n\}$$
$$= \sum_{i=0}^{M-n} \binom{M}{i} \sum_{s=0}^{M-i} (-1)^s \binom{M-i}{s} \alpha_{s+i}^r$$

Summing on $s + i = j$, we get

$$\begin{aligned} P\{R_n \leq r\} &= \sum_{j=0}^{M} \alpha_j^r \sum_{t=0}^{M-n} (-1)^{j-t} \binom{M}{t} \binom{M-t}{j-t} \\ &= \sum_{j=0}^{M} (-1)^j \alpha_j^r \binom{M}{j} \sum_{t=0}^{M-n} (-1)^t \binom{j}{t} \\ &= \sum_{j=0}^{M} (-1)^{j-M+n} \binom{M}{j} \binom{j-1}{M-n} \alpha_j^r. \end{aligned} \tag{13}$$

Equation 11 now follows directly from the above equation and the relation of Equation 2. Further Equation 12 is a straightforward consequence of Equation 11. □

The statistics of interest in this case are the expected test length and its variance. Expressions for these quantities can be derived easily from Equation 12. We state these below. Let

$$\beta_j = (-1)^{j-M+n-1} \binom{M}{j} \binom{j-1}{M-n}. \qquad (14)$$

Then

$$E(R_n) = \sum_{j=M-n+1}^{M} \frac{\beta_j}{1-\alpha_j}. \qquad (15)$$

$$Var(R_n) = \sum_{j=M-n+1}^{M} \frac{\beta_j(1+\alpha_j)}{(1-\alpha_j)^2} - E^2(R_n) \qquad (16)$$

The above equations show that if statistics of Z or W are known, we could easily define an empirical method for estimating the *pmf* of test length. This can be achieved by conducting a small number of simulations and empirically estimating $E[Z^{\underline{j}}]$ for $j = 1, \ldots, M$. Then we can substitute these values in the above equations to obtain the *pdf* of R_n.

However, for certain special cases of the distribution of W the above expressions simplify considerably. We focus on one such special case that has been known to provide excellent predictions of fault coverage (see [1]) even for large circuits. This is discussed in the next section.

4.3. Binomially Distributed W

In this case, we assume W is binomially distribued with parameters M and p where M is the size of the fault set and p is the detectability of a fault in that set. The implicit assumption in this case (as has been discussed earlier) is that the set of all faults has been partitioned (clustered) into equal detectability subsets (detectability profile).

This yields a much simpler form for the unconditional distribution of fault uncoverage. The binomial assumption is particularly useful for large circuits where there may be a large variance in detectabilities of faults. In such situations, we carry out approximate controllability and observability analysis using STAFAN [11]. This allows us to partition the fault set $\mathfrak{F}$ roughly into a number of classes, each class having a different but approximately constant detectability of faults. Fault clustering is discussed in Section 5.

It is clear that if W is binomially distributed with parameters M and p then Z is also binomially distributed with parameters M and $(1 - p)$. Its factorial moments can be expressed as

$$E(Z^{\underline{t}}) = M^{\underline{t}}(1-p)^t. \qquad (17)$$

This follows from the fact that the *pgf* of Z is $G_Z(s) = (p + s(1-p))^M$ and its tth factorial moment is given by $\left.\frac{d^tG_Z(s)}{ds^t}\right|_{s=1}$. Substituting this in Equation 6 we get

$$E(Y_r^{\underline{j}}) = M^{\underline{j}} \left[\frac{M^{\underline{j}}(1-p)^j}{M^{\underline{j}}}\right]^r = M^{\underline{j}}(1-p)^{rj}. \qquad (18)$$

Examining Equations 17 and 18, we can see that when W is binomially distributed with parameters M and p, Y_r is also binomially distributed with parameters M and $(1 - p)^r$.

Having obtained the distribution function of Y_r, we can now use Lemma 1 to obtain the *pdf* of R_n. For the case when Z is binomially distributed with parameters M and $1 - p$, Equations 9 and 17 tell us that $\alpha_j = (1 - p)^j$. Then

$$E(R_n \mid Z \text{ Binomial}) = \sum_{j=M-n+1}^{M} \frac{\beta_j}{1-(1-p)^j} \qquad (19)$$

and

$$Var(R_n \mid Z \text{ Binomial}) = \sum_{j=M-n+1}^{M} \frac{\beta_j(1+(1-p)^j)}{(1-(1-p)^j)^2} - E^2(R_n \mid Z \text{ Binomial}). \qquad (20)$$

Based on these relations we define some techniques to estimate expected test length (Equation 19). But first we turn to the problem of approximating these quantities.

4.4. Approximating Expected Value and Variance of Test Length

While Equations 15 and 16 express expected value and variance of test length exactly, they are not amenable to easy computation. The root of this intractability is the rapid loss of precision due to alternating sums of large coefficients. In order to obtain estimates of these statistics we need accurate approximations where the error can be controlled by the user. In this article we restrict ourselves to approximating expected value and variance for binomially distributed W.

Let $D_n = E[R_{n+1}] - E[R_n]$ $(1 \le n \le M - 1)$. Thus $E[R_n] = \Sigma_{i=1}^{n-1} D_i$. Substituting for β_j (from Equation 14) in Equation 19 we express D_n as follows.

$$D_n = \sum_{j=M-n}^{M} \frac{(-1)^{j-M+n}\binom{M}{j}\binom{j-1}{M-n-1}}{1-(1-p)^j} - \sum_{j=M-n+1}^{M} \frac{(-1)^{j-M+n-1}\binom{M}{j}\binom{j-1}{M-n}}{1-(1-p)^j}$$

$$= \sum_{j=M-n}^{M} \frac{(-1)^{j-M+n}\binom{M}{j}}{1-(1-p)^j}\left[\binom{j-1}{M-n-1} + \binom{j-1}{M-n}\right]$$

$$= \sum_{j=M-n}^{M} \frac{(-1)^{j-M+n}\binom{M}{j}\binom{j}{M-n}}{1-(1-p)^j}$$

$$= \binom{M}{M-n}\sum_{z=0}^{n} \frac{(-1)^z\binom{n}{z}}{1-(1-p)^{z+M-n}}$$

where $z = j - M + n$.

Expanding the denominator in the above fraction as an infinite series we get

$$D_n = \binom{M}{M-n}\sum_{z=0}^{n}(-1)^z\binom{n}{z}\sum_{r=0}^{\infty}(1-p)^{(z+M-n)r}.$$

Now, changing the order of summation results in

$$D_n = \binom{M}{M-n}\sum_{r=0}^{\infty}(1-p)^{(M-n)r}\sum_{z=0}^{n}(-1)^z\binom{n}{z}[(1-p-p)^r]^z$$

$$= \binom{M}{M-n}\sum_{r=0}^{\infty}(1-p)^{(M-n)r}(1-(1-p)^r)^n. \tag{21}$$

Truncating the above infinite series at an appropriate value of r (say at $r = r^*$) yields excellent approximations for D_n.

In order to define r^* we need to specify an allowable upper bound on the error. We solve the problem of determining r^* in the following manner: For a given n, the term $(1 - (1 - p)^r)^n$ approaches 1 as r increases. Let $\delta \le (1 - (1 - p)^r)^n \le 1$ for all $r \ge R_\delta$. By approximating Equation 21 as

$$D_n \approx \binom{M}{M-n}\sum_{r=0}^{R_\delta}(1-p)^{(M-n)r}(1-(1-p)^r)^n$$

the error λ satisfies the relation

$$\binom{M}{M-n}\frac{\delta(1-p)^{(M-n)(R_\delta+1)}}{1-(1-p)^{M-n}} \le \lambda \le \binom{M}{M-n}\frac{(1-p)^{(M-n)(R_\delta+1)}}{1-(1-p)^{M-n}}.$$

For a specified upper bound Λ on error, we can use the above relation to determine r^* as

$$r^* \ge \frac{ln\left[\dfrac{\Lambda(1-(1-p)^{M-n})}{\binom{M}{M-n}}\right]}{(M-n)ln(1-p)} - 1. \tag{22}$$

For a typical case with $M = 40$, $n = 30$, $p = 0.1$ and $\Lambda = 0.01$, $r^* = 24$ satisfies the error bound requirements.

For variance computation, we carry out a similar exercise for approximating second moments of R_n. Let $C_n = E[R^2_{n+1}] - E[R^2_n]$. Thus $E[R^2_n] = \Sigma_{j=1}^{n-1} C_j$. Expanding terms in $E[R^2_n]$, in a manner similar to that shown above, yields

$$C_n = \binom{M}{M-n}\sum_{r=0}^{\infty}\sum_{t=0}^{\infty}(1-p)^{(M-n)(t+r)}[(1-(1-p)^{t+r})^n + (1-p)^{M-n}(1-(1-p)^{t+r+1})^n].$$

For a specified upper bound on error we can truncate the above series at appropriate values of r and t. Since r and t appear only in the form of a sum $(r + t)$ in the above equation, we need only find a value R^* such that by setting $r + t = R^*$ we satisfy our error bounding requirements.

Since computation of C_n requires evaluation of a double summation it is expected that estimating variance will take more time than estimating expected values. From our experience we find that the amount of time required even for variance computation is far less than carrying out actual fault simulations to empirically estimate average values and variance of test length.

5. Techniques and Experimental Results for Expected Test Length Estimation

We first give a brief description of the algorithms used to gather experimental data. Then we present a comparison with the theoretical results.

5.1. A Methodology

The analysis presented in Section 4 assumes that the set of faults in a circuit has been clustered (partitioned) into equal detectability subsets and the analysis applied to each subset. Based on this we can divide our estimation technique into three subtasks, namely

T1: fault detectability estimation,
T2: fault clustering by detectability, and
T3: test length estimation for a specified coverage from each cluster.

It should be noted that in addition to assumptions made in our sampling model, the factors that have strong influence on the accuracy of test length predictions are 1) accuracy of fault detectability estimates (Task T1) and 2) the criteria used for clustering (Task T2).

T1) Detectability Estimation: It is well known that exact fault detectability estimation (Task T1) is an intractable problem. Due to this reason we need to use a detectability estimation program (such as STAFAN [11]) to first estimate signal probability c_i and observability o_i of each line i in the circuit. Then the detectability of a sutck-at 0 fault on line i is computed as $c_i \times o_i$ and the detectability of a stuck-at 1 fault is computed as $(1 - c_i) \times o_i$. Signal probability estimates obtained by STAFAN are known to be quite accurate. However, the accuracy of observability estimates is not known and may differ significantly from exact values. A methodology for exact detectability computation using binary decision diagrams is described briefly at the end of this section.

T2) Fault Clustering: Our aim is to find clusters of faults such that all faults within a cluster have approximately same detectability. In other words, we need to find subsets of faults such that the variation of detectabilities within each subset is small. This would allow us to use Equation 19 for estimating test length for a specified fault coverage within that cluster. The clustering algorithm defined by us is based on the following observations and reasoning:

- We can quantify the above notion of variation by using the sample variance or equivalently sample standard deviation (denoted by σ) of fault detectabilities in a cluster. Thus, our criteria for clustering translate to restricting the standard deviation of detectabilities within each cluster. A consequence of such restriction is that the number of clusters tends to increase, and the number of faults within each cluster decreases.
- A factor that opposes this tendency to decrease the number of faults in a cluster is the fault sampling model adopted by us. In general, the accuracy of our model increases with an increase in the number of faults in a cluster. Therefore, while clustering the set of faults, we need to balance these two needs.
- From our experience we have found that test length predictions are more sensitive to detectability variations for clusters containing low detectability faults than for clusters with high detectability faults. Therefore, in clusters with higher detectability faults, we can tolerate larger standard deviation.

Based on these observations it is clear that a ceiling on standard deviation (σ) in each cluster may not be a good criterion. Instead, we use another metric, namely the coefficient of variation γ defined as $\gamma = \sigma/\mu$, μ being the mean of detectabilities of faults in a cluster. Using a ceiling Γ on γ we find that we can balance the two opposing factors discussed above. The effect of this is that standard deviation of detectabilities in low detectability clusters is restricted to smaller values than in high detectability clusters. This is just the effect that is desired since test length predictions are more sensitive to variations in low detectability clusters than in high detectability clusters.

A simple algorithm FAULT_CLUSTER has been defined and implemented [12]. For a user specified ceiling Γ, FAULT_CLUSTER starts by clustering faults with the lowest detectabilities. A *greedy* approach is used in determining whether a set of faults (with detectability p) can be included in a given cluster without violating Γ. If not, a new cluster is started with this set of faults. Evidently this algorithm is not optimal and the problem of optimal clustering fo test length estimation is open.

Our observation has been, however, that even such a simple procedure yields good clustering. Obviously, sizes of clusters increase with an increase in the ceiling for γ. Clusters obtained for a circuit using this procedure is illustrated in Table 1. Clusters for other larger circuits were also quite compact (18 clusters for c880

Table 1. Fault clustering by detectabilities for a Viterbi Decoder chip with a ceiling of 0.1 on γ.

	Viterbi Decoder				$\Gamma = 0.15$		
Cluster	Size	Lowest Detectability	Highest Detectability	μ	σ	η
1	9	0.027	0.03625	0.032528	0.002639	0.081132
2	9	0.05775	0.065	0.061139	0.002998	0.049035
3	112	0.11275	0.199	0.129645	0.016262	0.125432
4	205	0.234	0.396	0.265916	0.037537	0.141162
5	277	0.46425	0.75075	0.522961	0.072995	0.139580

with $\Gamma = 0.15$, 13 clusters for C1908 with $\Gamma = 0.2$, 10 clusters for a multiplier circuit with $\Gamma = 0.15$, 22 clusters for c3540 with $\Gamma = 0.11$).

T3) Expected Test Length Estimation: For a specified fault coverage within a cluster, the approximation method of Section 4.4 is used to obtain an estimate of expected test length. M in this equation is substituted with the cluster size N (column 2 in the above table). All faults in that cluster are assumed to have a detectability p whose value can be selected in one of several ways depending on whether a pessimistic, optimistic, or a midpoint approximation is desired. For a pessimistic (optimistic) estimate of test length, the lowest (highest) detectability from each cluster is used as p. For a midpoint approximation, we can either set $p = \mu$, i.e., the mean value of detectabilities in that cluster or set p to be their median.

Suppose we are given a fault set F which has been partitioned into K clusters with cluster i having N_i faults ($1 \leq i \leq K$). Then there are two equally interesting ways of framing the test length estimation question.

Q1 A fault coverage of n_i ($\leq N_i$) is required from cluster $i (1 \leq i \leq K)$. What are the expected test lengths T_i for each cluster?

Q2 For a cluster $i*$ we want a coverage of n_{i*}. What is the expected test length T_{i*}? Further, using T_{i*} as test length for each cluster $j (\neq i*)$ what is the maximum coverage n_j such that $T_j \leq T_{i*}$? Then the test length for covering $\Sigma_{j=1}^{K} n_j$ faults from the complete fault set is taken as T_{i*}. The cluster $i*$ is called the *anchor* for this test length computation. This formulation is illustrated by the following example.

Example 1. Let $i* = 1$, i.e., cluster 1 is the anchor for our computations. Further, let us assume that $T_1 = 20$ patterns are required to test for $n_1 = 50$ faults from cluster 1. In cluster 2, testing for 60 faults requires 19 patterns but that for 61 faults requires 21 patterns. Then we assume n_2 to be 60. Similarly, we can find n_j based on T_1 for all other clusters j.

Test length estimates are often required for a specified fault coverage from the complete fault set and not from individual clusters. Examining the above formulations of the test length question, we see that answers to Q2 better suit our needs.

We define a simple procedure that allows the user to specify an anchor cluster $i*$ and the coverage n_{i*} desired. The procedure first determines T_{i*} and next finds n_j for each cluster $j(\neq i*)$ according to Q2 above. For a given total coverage value C from the complete fault set F, both $i*$ and n_{i*} can be adjusted through iterations such that $\Sigma_{i=1}^{K} n_i = C$. Although these adjustments seem complex, in practice however, we find that they are quite easy to perform, whether manually or using a simple adjustment program (as an outer loop) that converges rapidly to the desired coverage value C. These adjustments are guided by a set of simple rules summarized below.

R1 For a large increase (decrease) in total coverage, increase (decrease) the desired coverage in low detectability clusters.

R2 While increasing (decreasing) total coverage iteratively to a specified value C, if for $i* = j$, an increment (decrement) of the desired coverage n_j by 1 results in larger (smaller) total coverage than C, then set $i*$ to the next higher detectability cluster and recompute after incrementing (decrementing) its desired coverage.

These rules are based on the observation that a small perturbation in coverage specification from a low detectability cluster results in larger perturbations from higher detectability clusters in order to obtain similar test length estimates.

5.2. Experimental Results

We now present results of our experiments with several circuits including a Viterbi decoder network, a multiplier and four ISCAS '85 benchmark circuits. We compare two sets of theoretical estimates of expected test length for each circuit with those obtained by running fault simulations. The two sets are 1) pessimistic estimates using the lowest detectability value from each cluster as p and 2) midpoint estimates by setting $p = \mu$ (the mean detectability) for each cluster.

Our aim is to show that the midpoint estimates can be used to approximate actual test lengths quite accurately and that the pessimistic estimates can be used as an upper bound for expected test length. With this in mind we also provide an estimate of one standard deviation above and below the midpoint estimates, shown in the plots as an error-bar. Standard deviation (computed as the square-root of the variance) is estimated using approximations given in Section 4.4.

Faults considered for our experiments were those for which we were able to obtain reasonably accurate estimates of detectabilities. This was done to ensure some level of confidence in our test length predictions. Test length predictions for faults with inaccurate detectability estimates, especially for low detectability faults, can show large deviations from observed values. While STAFAN is our detectability estimation program of choice, for experimental purposes we also used fault simulated detectability estimates to verify STAFAN estimates. Results of our experiments are illustrated in Figure 1 through 6.

Discussion: Since test length prediction for high coverage specifications is usually of more interest to the test engineer (and also for the sake of clarity), we have shown test length predictions only for these cases. It should be noted however, that prediction errors are negligible for small coverage specifications.

It is clear from the plots that even for large coverage specifications, the midpoint theoretical estimates of expected test length can serve as excellent approximations for observed values. Furthermore, we find that all the observed values lie well within one standard deviation of corresponding midpoint predictions. In addition to this we find that our pessimistic predictions (using lowest detectability from each cluster) in all cases can serve as an upper bound on test length. Similarly, we found that our optimistic predictions (using highest detectability from each cluster), although not shown here, can serve as a lower bound. This illustrates the capability of the methodology in predicting test length.

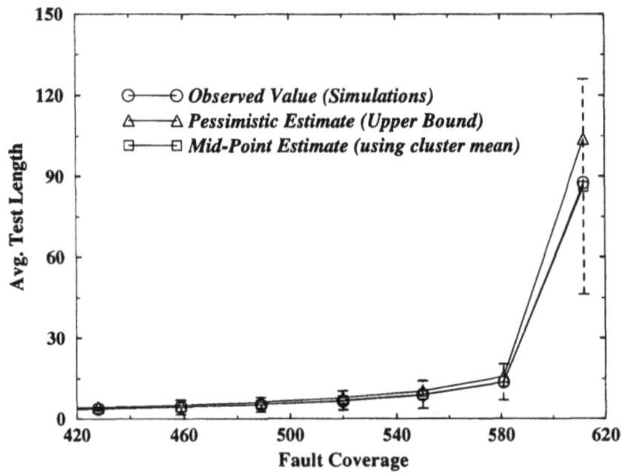

Fig. 1. Average test length estimates for a Viterbi Decoder circuit.

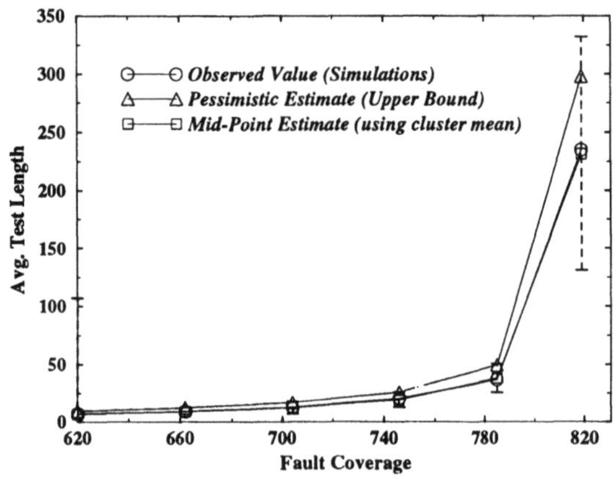

Fig. 2. Average test length estimates for C880.

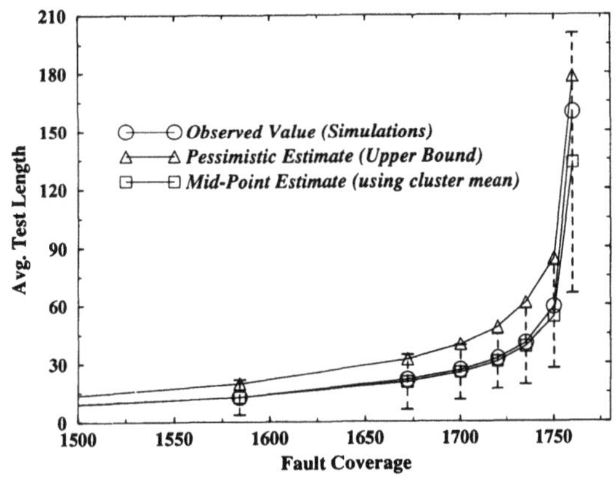

Fig. 3. Average test length estimtes for a multiplier circuit.

One should note however, that in our experiments we dealt exclusively with faults for which we had accurate detectability estimates. The effect of inaccurate detectabilities can be quite pronounced and was verified by the following simple experiment. It is well known

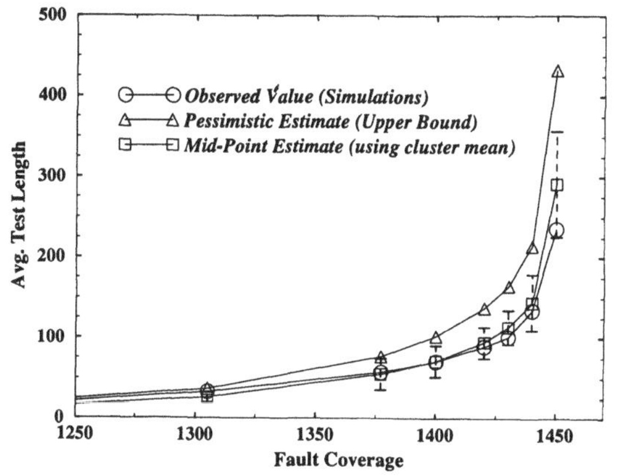

Fig. 4. Average test length estimates for C1908.

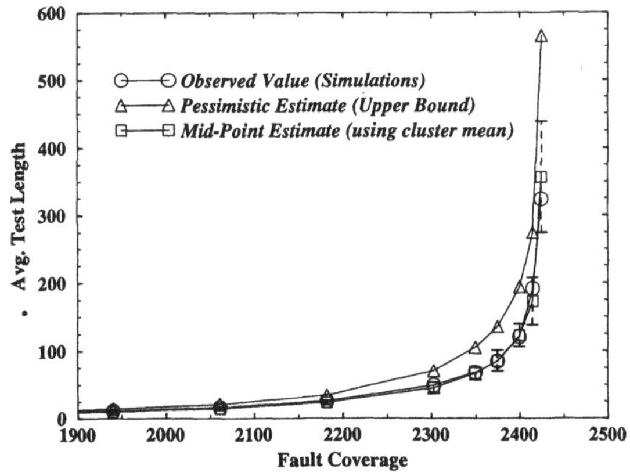

Fig. 5. Average test length estimates for C2670.

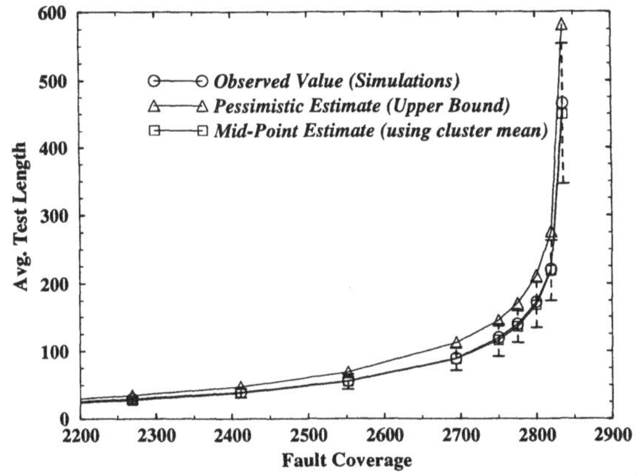

Fig. 6. Average test length estimates for C3540.

that STAFAN provides better estimates of signal probabilities than does COP. Their respective accuracies in observability computation however, suffer from similar effects caused by similar simplifing assumptions (although STAFAN has the advantage of better signal probability estimates for its observability computation). Based on this logic we expect STAFAN to give better estimates of detectabilities than COP. When applied to test length estimation, we found that our predictions using COP numbers were far worse than those obtained using STAFAN detectabilities.[2] This corroborates our intuition that test length estimates are quite sensitive to fault detectability values and their accuracy depends strongly on that of detectability estimates.

5.3. Exact Detectability Computation using BDDs

Ideally we would like exact detectability values for all faults. Towards this end we explore an avenue common to several areas in VLSI design, that of employing binary decision diagrams (BDD). While this approach does not change the complexity of the general problem of detectability computation (which is known to be NP-complete), we are encouraged by the volume of favorable results where BDDs were successfully used in keeping complexities at manageable levels.

The first step in our approach is a test generation procedure which produces, for a given fault *f*, not just a single test pattern for *f*, but a BDD whose on set consists of all its test patterns. Such a BDD is called a *test BDD* for fault *f*. The process of determining a test BDD involves 1) determining the BDD for the faulty function and 2) a method of traversing the "good" and "fault" BDDs. This process has been developed and is documented in [13]. The complexity of this procedure is $O(G_1G_2)$ where G_1 and G_2 are the number of nodes in the "good" and "fault" BDDs respectively. Given the test BDD T_f for *f*, its detectability can be computed as the 1-probability of T_f. Simple procedures for performing this task have been studied and can be implemented easily [14]. Future work in this direction involves extensive experimentation with different circuits in generating test BDDs and in finding ways of improving its efficiency.

6. Conclusion

Estimation of test length for a specified fault coverage is the focus of this article. A sequential fault sampling model is used to represent the process of testing faults by a sequence of random patterns. Using this model, we obtain several novel results that offer a significant improvement over earlier techniques for test length prediction.

These improvements are due mainly to the derivation of exact distributions of fault uncoverage and test length. Whereas, earlier techniques yielded either expected values for fault coverage and upper bounds for test length to achieve 100% coverage, the availability of the actual probability distribution functions presented in this paper, allows engineers to estimate other statistics of interest as well. In this respect the analytical resuts presented here are unique.

We establish an important relation between the random variables test length and fault uncoverage (and equivalently fault coverage). The process of testing faults using a sequence of random patterns is viewed as a process of sequentially sampling faults from a fault set. Theoretical results for the distribution of fault uncoverage of r random patterns, conditioned on the number of faults tested by each vector, are obtained. These conditions are removed for a completely general case of the distribution of Z, the number of faults untested by a random vector.

Next we use the theoretical resuts for fault uncoverage for obtaining expressions for the distribution function of test length. We also derive the probability generating function of test length which further allows us to express all its moments.

Based on these results we define some simple techniques for estimating the average test length required for a specified coverage. These techniques require prior estimation of fault detectabilities. We define a simple algorithm for clustering the set of faults by their estimated detectabilities. Next we device a scheme for obtaining expected test length predictions from each fault cluster and aggregating these results to obtain a test length prediction for the complete fault set. Finally we present the results of experiments with several circuits to illustrate the usefulness of the above models and analytical results.

It was observed that test length predictions are very sensitive to the accuracy of detectability estimates. A brief description of a methodology for finding exact fault detectabilities is also given. This methodology is based on determining the test BDD for a given fault. Future work involves finding efficient ways of obtaining test BDDs and also for optimal clustering of faults based on detectabilities. Work in both these directions is currently underway.

Acknowledgment

This research was supported by a grant from the National Science Foundation MIP-9111206.

Notes

1. Proofs of Theorems 1 and 2 and of Lemma 2 are given in [9].
2. Initially all test lengths were obtained using COP detectability estimates. Later estimates computed by STAFAN were used.

References

1. A. Majumdar and S. Sastry, "On the Distribution of Fault Coverage and Test Length in Random Testing of Combinational Circuits," *Proc. of 29th ACM/IEEE Design Automation Conf.*, pp. 341-346, 1992.
2. K. Kim, D.S. Ha, and J.G. Tront, "On Using Signature Registers as Pseudorandom Pattern Generators in Built-in Self-Testing," *IEEE Trans. on Computer-Aided Design*, vol. 7, pp. 919-928, Aug. 1988.
3. M.A. Breuer and A.A. Ismaeel, "Roving Emulation as a Fault Detection Mechanism," *Trans. on Comp.*, vol. C-35, pp. 933-939, Nov. 1986.
4. V.D. Agrawal, "Sampling Techniques for Determining Fault Coverage in LSI Circuits," *Journal of Digital Systems*, vol. V, pp. 189-202, 1981.
5. T.W. Williams, "Test Length in a Self-Testing Environment," *IEEE Design & Test*, April 1985, vol. 2, pp. 59-63.
6. W.K. Huang, M. Lightner, and F. Lombardi, "Predicting Fault Coverage for Random Testing of Combinational Circuits," *Proc. IEEE Int. Test Conf.*, pp. 843-848, Sept. 1987.
7. K.D. Wagner, C.K. Chin, and E.J. McCluskey, "Pseudorandom Testing," *IEEE Trans. on Computers*, vol. C-36, pp. 332-343, Mar. 1987.
8. J. Savir and P.H. Bardell, "On Random Pattern Test Length," *IEEE Trans. on Computers*, vol. C-33, pp. 467-474, June 1984.
9. A. Majumdar, *Stochastic Models for Testability Analysis of Digital Circuits*, Ph.D. thesis, University of Southern California, Dept. of Electrical Engineering-Systems, 1992.
10. F.N. David and D.E. Barton, *Combinatorial Chance*, Charles Griffin & Company Limited, London, UK, 1962.
11. S.K. Jain and V.D. Agrawal, "STAFAN: An Alternative to Fault Simulation," *Proc. 21st Design Automation Conference*, pp. 18-23, June 1984.
12. A. Majumdar and S.B.K. Vrudhula, "Random-Test Length Estimation: Analysis and Techniques," Tech. Rep. SIUC/DEE/TR-93-2, Dept. of Electrical Engineering, Southern Illinois University at Carbondale, 1993.
13. S. Jayaraman, "Test Generation for Single Stuck-At Faults in Combinational Circuits Using Binary Decision Diagrams," Masters Thesis, Dept. of Electrical Engineering, Southern Illinois University-Carbondale, 1993.
14. R. Krieger, "PLATO: A Tool for Computing Exact Signal Probabilities," *Proc. 6th Int. Conf. on VLSI Design*, Bombay, pp. 65-68, Jan. 1993.

Amitava Majumdar received his B.E. (Hons) degree in Electrical and Electronics Engineering from BITS, Pilani in India, in 1983, the M.S. degree in Electrical and Computer Engineering from the University of Massachusetts, Amherst, in 1986 and a Ph.D. degree in Electrical Engineering from the University of Southern California, Los Angeles, in 1992.

Since 1992 he has been an assistant professor in the Department of Electrical Engineering at Southern Illinois University, Carbondale. His research interests include VLSI design automation, testability analysis of digital circuits, design for testability, fault tolerant computing, reliability modeling and performance evaluation of computer systems.

Sarma B.K. Vrudhula (a.k.a. Sarma Sastry) is an Associate Professor of Electrical and Computer Engineering at the University of Arizona in Tucson. He received the B.Math (Honors) degree from the University of Waterloo, Ont., Canada, in 1976 and his M.S. and Ph.D degrees in electrical engineering from USC in 1980 and 1985, respectively. During 1985–1992 he was on the faculty of the EE-Systems department at USC. His current research interests span a number of areas in computer-aided design and design automation of digital systems. These include various optimization problems that arise in VLSI chip layout, stochastic models for testability analysis of circuits, partitioning for built-in self-test, automatic synthesis of asynchronous designs, and logic level and RT level synthesis for low power. His teaching interests cover undergraduate and graduate courses in logic design, VLSI design, digital systems testing, computer architecture, and discrete mathematics.

JOURNAL OF ELECTRONIC TESTING: Theory and Applications, 5, 299–305 (1994)

Fuzzy Optimization Models for Analog Test Decisions

MOUNIR FARES AND BOZENA KAMINSKA
Electrical Engineering Department, Ecole Polytechnique de Montréal, P.O. Box 6079, Station A, Montréal, Québec, Canada, H3C 3A7

Abstract. Test decisions still constitute one of the most difficult and time-consuming design tasks. This is particularly true in the analog domain where some basic test questions have not yet been completely resolved. Since the gap between a good and a bad analog circuit is not always well-defined, extensive tests may result in the rejection of many fault-free ICs. The objective of this article is to propose fuzzy optimization models that can help in the more realistic formulation and resolution of the analog test problem. The set of good or fault-free ICs is considered as a fuzzy set. Each performance test is represented by a membership function. A global test measure is obtained by aggregating all the performance tests. An illustrative example using these concepts is provided.

Keywords: Analog circuits, fuzzy decision-making, parametric optimization, test selection.

1. Introduction

Current trends in VLSI technology are toward the integration of whole systems on a single chip. Mixed-signal ICs are occupying a growing portion of the electronics market. The design and testing time of such mixed-signal ICs is primarily dominated by the analog parts. Since analog testing is very time-consuming, the time to market of such circuits is becoming a critical economical parameter. It is widely admitted today that test considerations have to be included in the earliest stages of the design process, and test decisions are still one of the most difficult and time-consuming design tasks.

In the digital domain, the test space is characterized by the large number of DFT techniques that have evolved over the years [1], [2], starting from the early ad hoc techniques for designing easily testable circuits. More structured techniques have gradually been introduced and developed, leading to the large selection of DFT techniques now available including the BIST, Figure 1. Each of these techniques can be associated with a set of measures reflecting the various costs and gains associated with a specific technique. It is nevertheless extremely difficult for the designer to select and apply the appropriate DFT technique [2]–[5]. Furthermore, the application of a single specific DFT technique may result in several implementation alternatives, depending on the resources available [3]. A large part of digital test decisions can be considered as selection problems. This type of test decision has been addressed by the authors in [6].

Analog test decisions are of a different nature and are more difficult than digital test decisions. The basic test question of how to define a good or a faulty analog circuit has not yet been completely resolved. In fact, the boundary between a good analog circuit and a bad one is still very fuzzy. A fault-free analog circuit is generally described by a set of performance constraints. The question here is what performances need to be tested, since the list of performances may be very large and not all of them need testing. An extensive test that includes all performances is very costly, and many fault-free circuits may fail such tests. To obtain more realistic test results and to reduce test costs, a minimal set of performance tests has to be selected. The next question is, what tests should be applied to verify each performance? Do we need to fulfill each performance 100%?

Many practical test decisions are complicated by the existence of multiple, and generally conflicting, objectives and constraints, and there is considerable uncertainty about what the consequences of the various possible alternatives will be. Much of the decision-making in the design process takes place in an environment in which the design goals, constraints and consequences of possible actions are not precisely known. The human nature of the decision-making process makes its formulation very difficult. Designers often do not like to give precise target values to the expected performances.

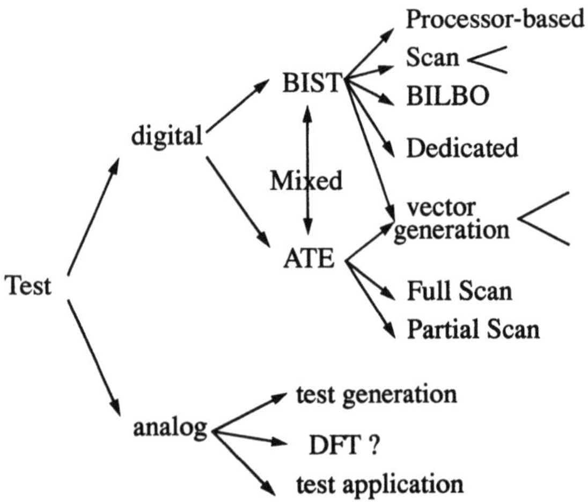

Fig. 1. Test space exploration.

They use objectives, like minimize power dissipation, maximize gain, that are difficult to tanslate into precise numbers. To deal quantitatively with this imprecision, we propose a testing approach based on fuzzy set theory. Since most analog tests are typically specification-dependent, we define the concept of a fuzzy specification constraint [7], [8] and explore the application of this concept to the formulation of the analog testing problem. By a fuzzy specification constraint, we mean a design specification that can be characterized as a fuzzy set. The following statement illustrates an example of a fuzzy specification constraint: "The output resistance should be *essentially smaller* than 100 Ω." The italicized words are the source of fuzziness. As pointed out in [9], [10], fuzzy set theory makes it possible to quantify and manipulate qualitative statements, vagueness, or subjectivity of opinion, because the elements in human thinking are not numbers but labels of fuzzy sets. Humans are more efficient in qualitative evaluation than in quantitative analysis.

This article is organized as follows: After a review of previous efforts, in Section 3 we discuss analog test decisions. In Section 4, we will summarize some basic concepts of fuzzy set theory. Section 5 describes how performance tests are formulated as fuzzy sets. The aggregation of the various test results is presented in Section 6. Section 7 presents a detailed example and Section 8 concludes the article.

2. Previous Efforts

In the test domain, the issue of automating the DFT technique selection was first reported in [3] by Abadir and Breuer, where they pointed out the need for mechanisms for evaluating and making choices among the numerous existing DFT techniques. They presented a knowledge-based approach for modifying the design in order to make it easily testable. The proposed approach contains mechanisms for evaluating and selecting the best DFT techniques. The authors also described how to explore the various alternative implementations for a single DFT technique. Another promising approach that can be used to select DFT methodologies has been proposed by Zhu and Breuer [4]. They developed a knowledge-based system that helps in the selection of test methodologies for a particular type of logic structure. Their prototype system deals with testing PLAs. Zhu and Breuer covered the case where the requirements for the various attributes can be stated precisely by numbers. The user is asked to specify the requirements as a vector of the desired attribute values. The designer needs to change his requirements several times before an optimal solution is found. As a number of selection attempts end in some form of failure, this system may not be able to give any solution at all.

Another research project in the test domain was carried out by Bhawmik, Narang and Chaudhuri [11]. They presented two knowledge-based expert systems (KBESs) for selecting test methodologies for PLAs and random logic circuits. The knowledge-based approach has been used in all known works, resulting in very problem-specific solutions that are difficult to apply to other DFT selection problems. In addition, less attention has been given to the cases where the requirements cannot be specified precisely by numbers.

It should be noted, too, that in the design process many of the design objectives and constraints are based essentially on the subjective judgments of a designer and are therefore inherently imprecise. The designer may not clearly specify a desired value, especially when dealing with qualitative data. At best, the objectives may be described using linguistic labels like *very low, high*, *acceptable*, etc.

3. Analog Test Decisions

A fault in an analog circuit usually means that at least one of the circuit responses is outside the design specifications. Most analog test researchers describe a fault-free analog circuit by a set of specification constraints like:

$$F_i(x) \geq spec_i$$

where $F_i(x)$ is a performance function and $spec_i$ is a specified limit value on that performance function.

This type of formulation for describing a fault-free circuit generally results in extensive tests and tighter tolerances than are necessary in reality. The following problems arise:

1. The number of performances that can describe circuit functionality may be large, even though the designer may be able to eliminate many unnecessary performance tests for that particular application. We have to identify a small set of critical performances to be tested, in order to speed up the manufacturing test and therefore reduce test costs.
2. In the minimal test set retained, some performance tests may be more critical than others. The admissible tolerances may vary from one performance to another.
3. The above formulation divides the space of the circuits under test (CUTs) into two categories: good, or fault-free, CUTs and bad, or faulty, CUTs. There is a rough transition between the two categories. Such a formulation has to be relaxed, however in order to obtain more realistic test results, Figure 2.

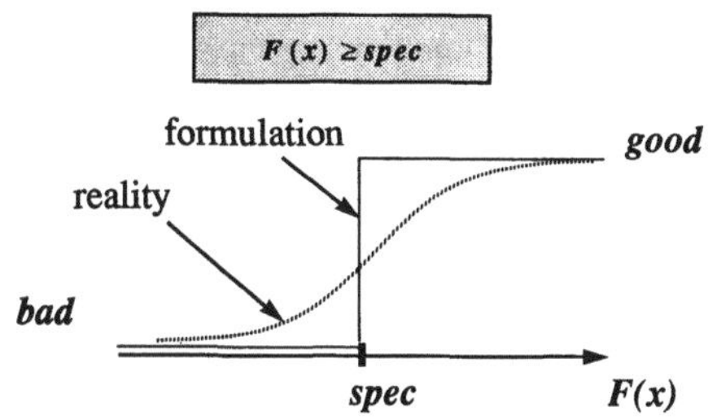

Fig. 2. The definition of fault-free and bad analog circuits.

When the deviation of the circuit response is far from its specified value, the circuit is definitely considered faulty. However, when the deviations are within a certain interval around the specified value, they are usually accepted, with varying degrees of acceptability. We therefore have to allow for varying degrees of "goodness" and "badness." Our basic idea is to model each specification test as a fuzzy set, but first we have to summarize some basic definitions of fuzzy set theory.

4. Fuzzy Sets and Terminology

Fuzzy set theory was first introduced by Zadeh [9] to deal with the vagueness and uncertainty inherent in natural language and human thinking. Let S be a collection of objects. A fuzzy subset A in S is characterized by its membership function μ_A, which associates every element x with its grade of membership $\mu_A(x)$ in the set, and $\mu_A(x)$ is a real number in [0, 1]:

$$\mu_A: S \to [0, 1]$$
$$x \to \mu_A(x)$$

Thus a fuzzy set A in S may be represented as a set of ordered pairs:

$$A = \{(x, \mu_A(x)) | x \in S\}$$

Example 1. Let us assume that we have a set S of identical analog circuits CUTs to be tested, for example op amps:

$$S = \{CUT_i\} \quad i = 1, \ldots, n$$

The class H of CUTs that have a high dc gain is a fuzzy set, since ***high*** does not define sharp boundaries for this class. This set can be characterized by the membership function shown in Figure 3.

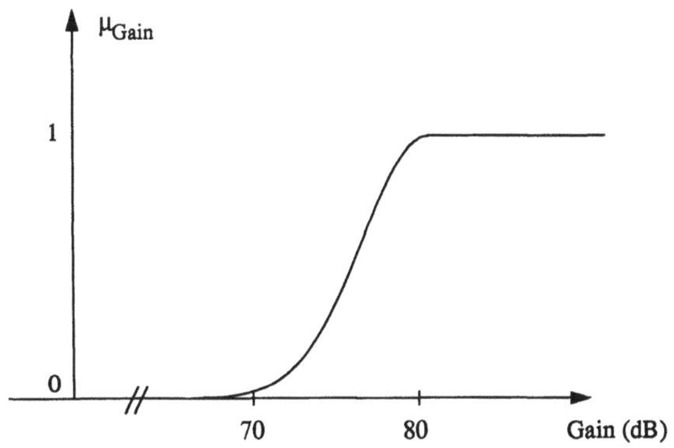

Fig. 3. Membership function for the fuzzy set of CUTs with high dc gain.

Suppose we have the dc gain values for the CUTs shown in Table 1.

Table 1. A simple example.

CUTs	CUT1	CUT2	CUT3	CUT4	CUT5
dc gain A (dB)	80	75	79	83	70

CUT_1 has dc gain $A = 80$ dB, so it belongs to the fuzzy set H with a grade of membership equal to 1. It has full membership in the set. The grade of membership of CUT_2 is 0.25. For all CUTs, we obtain the following fuzzy set H:

$$H = \{CUT_1/1,\ CUT_2/0.25,\ CUT_3/0.73,\ CUT_4/1,\ CUT_5/0.83\}$$

Basic Operations on Fuzzy Sets

Let A and B be two fuzzy sets in S with membership functions μ_A and μ_B respectively. The operations of union and intersection for fuzzy sets are defined via their membership functions:

Union

The fuzzy set $A \cup B$ is characterized by the membership function:

$$\mu_{A \cup B}(x) = Max(\mu_A(x), \mu_B(x)) \quad x \in S$$

Intersection

The membership function for the intersection $A \cap B$ is defined as:

$$\mu_{A \cap B}(x) = Min(\mu_A(x), \mu_B(x)) \quad x \in S$$

5. Test Performances as Fuzzy Sets

In general, analog testing has two major objectives: functional verification and fault isolation. Functional verification determines whether or not the CUTs performances (or responses) are within specification constraints [12]. The task in fault isolation is to identify the faulty element in the CUT. Here, we are concerned with functional verification. Each performance test T_i concerning the ith performance parameter is formulated as a fuzzy set. This means that we allow varying degrees of goodness and badness with respect to the performance to be tested. A membership function μ_{T_i} is defined to characterize the test T_i. This membership function measures the degree of goodness of the CUT as a function of the CUT responses. To illustrate, suppose that we are testing a specification constraint of the type:

$$F_i(x) \geq spec_i$$

With respect to the performance F_i under consideration, a fault-free circuit is a circuit that verifies the above equation. The set of such fault-free circuits is a fuzzy et that can be characterized by the membership function shown in Figure 4.

When a CUT response F_i is much smaller than the specified limit value $spec_i$, the circuit is considered faulty, and the degree of badness of such a circuit increases with the amount of the response deviation from the specified limit value, part I in Figure 4. On the other hand, as soon as the CUT response verifies the above equation, the CUT is considered to be fault-free. In this case, too, we may have varying degrees of goodness, part II in Figure 4. In Figure 4, a transition region is shown, in the membership function, between good and bad CUTs.

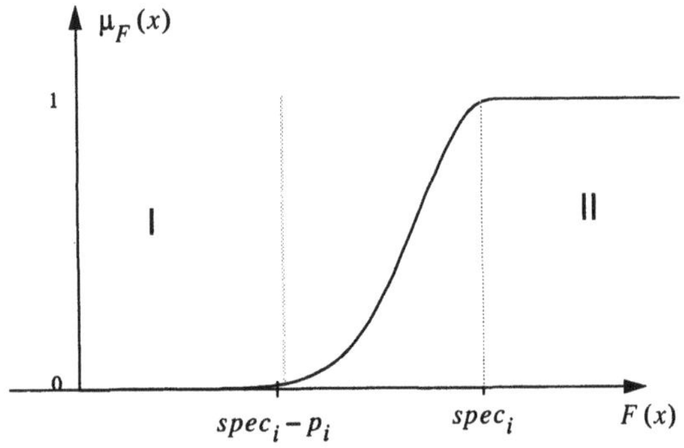

Fig. 4. Membership function characterizing a response test.

6. Aggregating Several Performance Tests

Note that a fault-free analog circuit is generally described by more than one specification constraint. In fact, we usually have to test several performance parameters before stating that the CUT is fault-free or functionally good. Note that these test requirements are considered simultaneously through the use of the connective *and* i.e., a fault-free circuit must verify all the T_i simultaneously.

$$T_1 \text{ and } T_2 \ldots \text{ and } T_n$$

As pointed out in [7], *and* corresponds to the intersection of fuzzy sets. Since each test T_i is formulated as a fuzzy set, this implies that the combined effect of all the tests may be represented by the intersection:

$$T_1 \cap T_2 \ldots \cap T_n$$

This intersection defines the set of fault-free circuits, with respect to all performances taken simultaneously, as a fuzzy set G:

$$G = T_1 \cap T_2 \ldots \cap T_n$$

According to the definition of intersection, the membership function of G becomes:

$$\mu_G(CUT) = min(\mu_{T_1}(CUT), \mu_{T_2}(CUT), \ldots, \mu_{T_n}(CUT))$$

where $\mu_G(CUT)$ indicates the degree of goodness of the CUT with respect to n tests simultaneously. If the CUT succeeds in all the tests, the degree of membership will be $\mu_G(CUT) = 1$. This will occur only if we have 1 as the degree of membership in each of the μ_{T_i}, which means that the circuit is completely within the specifications. $\mu_G(CUT)$ may be interpreted as the global functionality measure of the circuit.

Note that the aggregation seen as the intersection of fuzzy sets reflects one possible interpretation of *and*. In this interpretation the global functionality measure is mainly influenced by the weak performances. This may be suitable in situations where all the tests are critical, and there is no room for compromise.

However, under other circumstances, other interpretations may be more convenient, for example a weighted sum that takes into account all the test results with weights indicating the relative importance of the test. More explicitly, we may express $\mu_G(CUT)$ as

$$\mu_G(CUT) = \sum_{i=1}^{n} \omega_i \mu_{T_i}(CUT)$$

In this formulation, all the performances contribute, depending on their relative importance, to the final decision. This formulation is particularly useful when we want to do some trade-offs by accepting the degradation of a less important parameter in order to improve a more important one.

7. Examples and Comments

The fuzzy technique introduced here may apply to several selection problems encountered in the design of both digital and analog circuits. This example demonstrates how it is used in functional verification of an analog circuit. Functional verification determines whether an analog network's functionality falls outside design specifications [17]. However, design specifications are often expressed in a fuzzy manner and a precise quantification may restrict the space of good circuits, resulting in the rejection of functionally good circuits. The analog circuit to be tested in this example is the two-stage CMOS operational amplifier, Figure 5.

Let us assume that a good circuit is defined by the following specifications:

Gain	$> G_{min}$
Gain-bandwidth	$> GB_{min}$
Power dissipation	$< P_{max}$
Slew rate	$> SR_{min}$

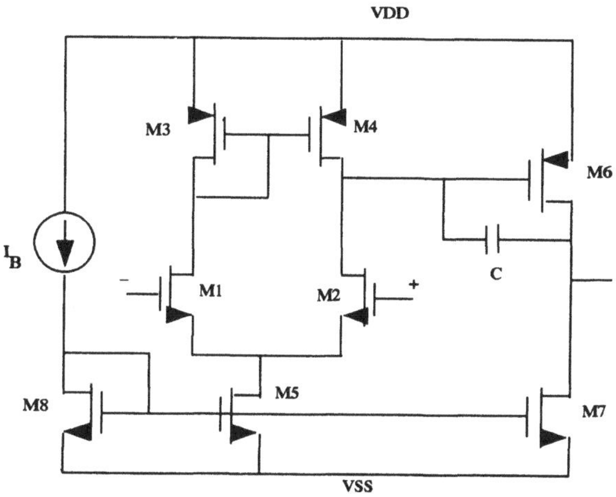

Fig. 5. Two-stage CMOS operational amplifier.

Settling time	$< ST_{max}$
PSRR	$> PSRR_{min}$
Phase margin	$> \Phi_{min}$

We can rewrite these specifications in the following form:

$$F_i(x) < spec_i \quad i = 1 \ldots m$$

where m is the number of specifications and x is the vector of circuit parameters.

Table 2. Design example, specifications and simulated results.

Parameter	Specification	p	HSPICE	μ
gain	≥70 dB	10 dB	84.3 dB	1.0
gain-bandwidth	≥1 MHz	0.2 MHz	0.9 MHz	0.5
power dissipation	≤ 10 mW	2 mW	1.7 mW	1.0
phase margin	≥ 60°	5°	58°	0.6
slew rate	≥ 2.0 V/μs	0.2 V/μs	1.9 V/μs	0.5
settling time	≤ 1 μs	0.1 μs	0.5 μs	1.0
PSRR	≥ 60 dB	5 dB	89 dB	1.0

The source of fuzziness in these specifications is the meaning of "<," because when we expect a specification $F_i(x) < spec_i$ we may accept values around $spec_i$ with different degrees of acceptability. In this case "<" may be replaced by its fuzzy version "$\lesssim$" which means essentially smaller than [13]. Each specification is now characterized by a membership function $\mu_i(x)$, Figure 6.

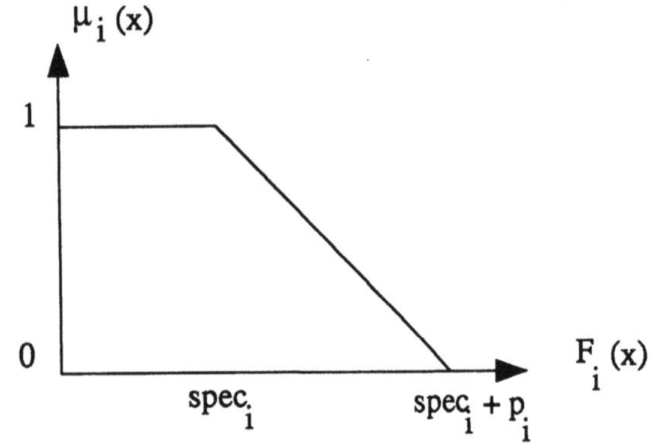

Fig. 6. Membership function for the fuzzy specification.

where p_i represents admissible tolerances of the ith specification.

We need to compose the m membership functions in a single synthesis membership function $\psi(x)$, which can be interpreted as a *functionality measure* of the circuit. Here, we use the following composition form [7]:

$$\psi(x) = \sum_{i=1}^{m} \omega_i \mu_i(x)$$

where ω_i is a weight reflecting the importance of each specification.

The term *good circuit* has a fuzzy meaning and can be described by a fuzzy set where $\psi(x)$ may be interpreted as its membership function.

The membership function $\psi(x)$ takes its value in the interval (0, 1). A circuit with $\psi(x) = 0$ is a bad one and therefore is not acceptable. When $\psi(x) = 1$ we have a circuit that fully satisfies all the specifications. An intermediate value, between 0 and 1, indicates the circuit's grade of membership in the set of good circuits. Table 2 provides results for a sample op amp.

We obtained the parameter measurement with the HSPICE simulator [14]. Here the specifications are assumed to be equally important ($\omega_i = 1/7$). The value of ψ for the op amp is $\psi(x) = 0.8$. This means that our circuit belongs to the set of good circuits with a grade of membership equal to 0.8. Note that $\psi(x)$ is less than 1, because not all the specifications are fully satisfied. However, the value 0.8 shows that we have a relatively good circuit.

8. Conclusion

In this article, we have proposed fuzzy optimization models to help in the formulation of analog test problems. The basic concepts of our approach make it applicable to various decision problems, in both the test space and the design space. It has been shown that it is possible to use the approach for aggregating test performances into a single test measure that evaluates the global performances of the circuit. The fuzzy membership functions can be derived easily from the cost functions or penalty-credit functions usually used in design optimization. The grade of membership of a circuit can be interpreted as a score with respect to the various constraint specifications. The global grade of membership is a composite score over all objectives.

References

1. T.W. Williams and P.K. Parker, "Design for Testability—A Survey," *IEEE Trans. Computers*, vol. C-31, pp. 2–15, Jan. 1982.
2. M. Abramovici, M.A. Breuer, and A.D. Friedman, *Digital Systems Testing and Testable Design*, Computer Science Press, New York, 1990.
3. M.S. Abadir and M.A. Breuer, "A Knowledge-Based System for Designing Testable VLSI Chips," *IEEE Design & Test of Computers*, vol. 2, pp. 56–68, Aug. 1985.
4. X.A. Zhu and M.A. Breuer, "A Knowledge-Based System for Selecting Test Methodologies," *IEEE Design & Test of Computers*, vol. 5, pp. 41–59, Oct. 1988.
5. X.A. Zhu and M.A. Breuer, "Analysis of Testable PLA Designs," *IEEE Design & Test of Computers*, vol. 5, pp. 14–28, Aug. 1988.
6. M. Fares and B. Kaminska, "A Fuzzy Decision-Making Approach for Test Space Exploration," *Proceedings of the Third European Test Conference*, IEEE Computer Society Press, Los Alamitos, 1993.
7. R.E. Bellman and L.A. Zadeh, "Decision-Making in a Fuzzy Environment," *Management Science*, vol. 17, pp. 141–164, Dec. 1970.
8. R.R. Yager, "Multiple Objective Decision-Making Using Fuzzy Sets," *Int. J. Man-Machine Studies*, vol. 9. pp. 375–382, July 1977.
9. L.A. Zadeh, "Outline of a New Approach to the Analysis of Complex Systems and Decision Processes," *IEEE Trans. Systems, Man and Cybernetics*, vol SMC-3, pp. 28–44, Jan. 1973.
10. D. Dubois and H. Prade, *Fuzzy Sets and Systems: Theory and Applications*, Academic Press, New York, 1980.
11. S. Bhawmik, V.K. Narang, and P.P. Chaudhuri, "Selecting Test Methodologies for PLAs and Random Logic Modules in VLSI Circuits—An Expert System Approach," *INTEGRATION, The VLSI Journal*, vol. 7, pp. 267–281, 7, 1989.
12. A. Walker and W.E. Alexander, "Fault Diagnosis in Analog Circuits Using Element Modulation," *IEEE Design & Test of Computers*, vol. 9, pp. 19–29, March 1992.
13. H.J. Zimmermann, "Description and Optimization of Fuzzy Systems," *Int. J. General Systems*, vol. 2, pp. 209–215, 1976.
14. HSPICE User's Manual: H9007, Meta-Software, Inc., 1991.

Bozena Kaminska is an associate professor of Electrical Engineering Department at Ecole Polytechnique, University of Montreal, Canada. Previously, she was with the Institute of Electron Technology in Warsaw, Poland. Her research interests include analog and digital testing, DFT, BIST, and high-level synthesis. Kaminska received the MS and PhD degrees in electronic engineering from the Technical University of Warsaw, Poland. She is a member of the IEEE Computer Society.

Mounir Fares is a research assistant in the Electrical Engineering Department of the Ecole Polytechnique, University of Montreal, Canada. His research interests include design for test, multichip modules, and system-level testing. He received the BS degree in electrical engineering from the Ecole nationale Polytechnique d'Alger, Algeria, and the MS degree from the Ecole Polytechnique de Montreal, Canada.

JOURNAL OF ELECTRONIC TESTING: Theory and Applications, 5, 307–312 (1994)

Self-Test of Sequential Circuits with Deterministic Test Pattern Sequences*

ARNO KUNZMANN AND FRANK BOEHLAND
Forschungszentrum Informatik (FZI), Haid-und-Neu-Strasse 10-14, 76131 Karlsruhe

Abstract. This article describes a new approach for synthesizing a cost-efficient self-test hardware for a given set of deterministic test pattern sequences. To minimize the test hardware effort instead of all the test sequences, only a very small subset will be selected such that a simple generation of all necessary test sequences will be ensured. This procedure drastically decreases the storage requirements (about 80%) and therefore distinctly reduces the necessary test hardware overhead. Experimental results on the ISCAS-S-benchmarks emphasize the efficiency of our approach.

Keywords: Automatic Test Pattern Generation (ATPG), Design-for-Testability, Deterministic Test Pattern Sequences, Field-Programmable Gate-Arrays (FPGAs), Self-Test, Sequential Circuits

1. Introduction

In the field of VLSI testing there exist several reasons for a minimization of test sets. The main objectives are a shortening of the test application time especially for complex circuits with complete or partial scan design, and a reduction of the storage requirements. The latter is of special interest for this article in order to synthesize a cost-efficient self-test hardware for deterministic test pattern sequences. This solution can be used for both an external and an internal self-test.

Up to now the proposed test pattern generating hardware solutions are mainly restricted to random and deterministic test pattern sets or to specific circuit structures (e.g., pipelines, combinational circuits). Since the generation of deterministic test patterns by different types of (non-)-linear shift registers has not been very promising and in general requires complex hardware structures (e.g., [1]), conventional ROM-based approaches are still of much importance. Obviously, the test hardware effort is dominated by the ROM size. This dominance is of much more importance if test sequences are also taken into account.

In this article a new approach will be presented that significantly reduces the storage requirements, since only a small subset of a given deterministic test sequence has to be stored. Based on this subset all applicable test sequences can easily be derived and realized with very little effort by a specific test chip or additional hardware on the same chip.

*This work is supported by the ESPRIT project 6855 (LINK).

2. Related Work

In order to discuss the important work in the field of test set minimization, two different approaches can be distinguished:

1. The test pattern generation algorithm itself is modified.
2. The minimization procedure is independent of the test generation algorithm and can be used in an additional post-processing phase.

The following references described procedures for combinational circuits, but all these proposals can easily be adapted, if test sequences for sequential circuits have to be generated.

The proposals for the modification and extension of test generation algorithms are mainly based on an analysis of the fault sets [2]. This includes the dynamic test compaction where tests for additional faults are generated as long as sufficient unspecified inputs are available [3]. In [4] an innovative approach is presented to further reduce the number of test patterns. The basic idea of this procedure consists of unspecifying those primary inputs that are not essential for the fault detection. In this procedure the average number of faults detected by a single test pattern can be enlarged. Experimental results, based on the PODEM algorithm, showed the effectiveness of this improved dynamic compaction. In [4] a procedure for the generation of independent fault sets is presented. Based on these fault sets an efficient test set reduction can be performed, but this approach is restricted to small and medium sized circuits.

In a post-processing phase after the completion of the test pattern generation, static compaction algorithms can be applied [6] typically by a resimulation of the test patterns in reverse order [7]. But since the computation of a minimal test pattern set is an NP-complete problem, in general heuristics are used to find optimized solutions.

In this article we propose a new procedure that combines the two ideas of (1) minimizing the number of deterministic test pattern sequences to be stored and (2) enabling the synthesis of an efficient self-test hardware structure. Furthermore, this procedure can optionally be applied in a post-processing phase based only on a set of deterministic test pattern sequences. Therefore, no modifications or extensions of the test pattern generation algorithm are required.

3. Selection of Basic Test Sequences

The main idea of our approach can be explained by a simple observation if deterministic test pattern sequences are compared: Very often pairs of test sequences can be found that differ in only one bit position. Starting with stuck-at faults at primary inputs, the sensitized paths for fault propagation can remain unchanged, and only the corresponding bit position has to be inverted.

Similar observations can also be made for internal faults for given fault propagating paths. Of course, it is not ensured that only one-bit position has to be changed to detect additional faults, but the experimental results will show that this is often the case. A further important observation is the sensitizing of operation units. Take, for instance, a multiplication unit: Here, by any input combination, paths are sensitized and faults will be propagated. This also holds if sequential realizations are considered.

These observations led to the approach for deriving all distinct test sequences from each given deterministic sequence that differ in only a single bit from the original.

Given a test sequence of length k and of width w, $k*w$ additional test sequences can be derived. This resulting set is the so-called toggle sequence. The originally given test sequence will be denoted as the "basic test sequence." In Figure 1 this procedure is illustrated for the test sequence of length 2.

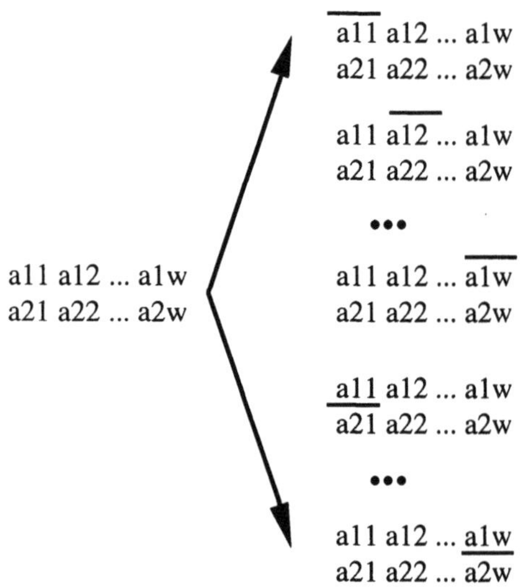

Fig. 1. Test sequence and its toggle sequence.

It should be noted that test sequences may contain several unspecified bit positions ("don't cares"). In this case, the overall number of different derivable test sequences will be reduced.

To give a first exampe of the efficiency of this approach we will analyze the sequential benchmark circuit s382 assuming a partial scan. To obtain full fault coverage, 53 test pattern sequences are generated with the SPROUT9v algorithm [8]. To maintain full fault coverage, it is sufficient to select only 10 test sequences of length 2 and of width 18. Out of this basic set 360 additional test sequences have to be derived for test application. Obviously the storage requirement is significantly reduced from 53 to 10, at the same time offering a simple generation of the overall test pattern set by test hardware (see next section).

Now the question will be discussed how, for a given set of deterministic test sequences, a minimized basic test sequence can be determined. The proposed selection procedure consists of the following two phases:

1. computation and minimization of a basic test sequence
2. computation of the bit positions to be inverted.

Each task can be solved by an additional fault simulation phase. The main difference is the order the test sequences have to be simulated. In phase (1) after each deterministic test sequence all the derivable test sequences are grouped and fault simulated. Only if a group detects at least one additional fault the corresponding original deterministic test sequence will be part of the final basic test sequences. In [9], where the proposed method is restricted to combinational circuits, an optimized minimization procedure is described.

Up to now we have assumed that each bit position has to be inverted. But in practice only a small subset has to be taken into account. This subset will be selected by phase (2) as a direct result of the first fault simulation phase with a different grouping of test sequences. In the following example w is the product of test sequence length and test pattern width, uniquely indicating the bit position and the corresponding test pattern number. In this fault simulation phase, the test sequences are divided into $w + 1$ different groups:

1. All basic test sequences
2. All basic test sequences with inverted bit position 1
3. All basic test sequences with inverted bit position 2

$w + 1$. All basic test sequences with inverted bit position w.

Bit position j has to be inverted if and only if group $j + 1$ detects at least one additional fault that was not yet detected by one of the previous groups.

4. Synthesis of the Self-Test Hardware

The proposed selection strategy is well suited for self-test and/or external test generation by special test chips. Figure 2 shows the principal structure of the test generation hardware. The most area consuming module of the test chip is the ROM containing the deterministic basic test sequences. Each row contains a test sequence of length k and pattern width w. To reduce the ROM size the test sequences can be compacted by conventional procedures (e.g., use of compatible test patterns). This compaction is independent of the selected bit positions that have to be inverted. For the row-selection (ith sequence) a simple shift register can be used with the shown initialization. The selection of the bit positions is enabled by a cyclic shift register, also initialized by having all zeros except a single "1."

The inversion of selected bit positions can be performed by a third feedback shift register of length $k * w + 1$. Starting with the shown initialization, first the selected test sequence will be shifted out unchanged, and thereafter the first bit position will contain the value

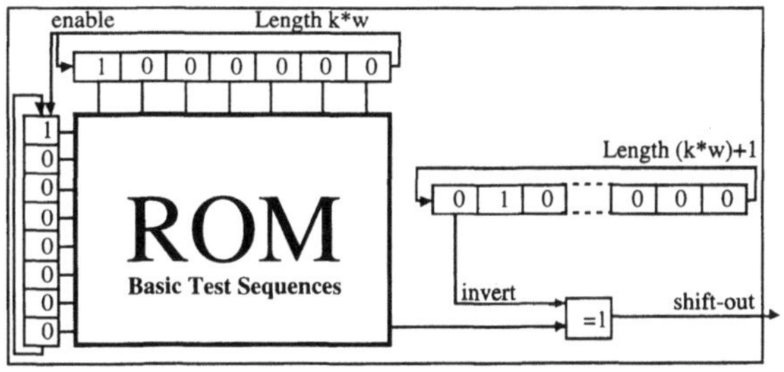

Fig. 2. Structure of test generation hardware.

1. With each following shift-out-phase all the test sequences will be changed by one single bit position.

If the test application time is critical and the overall number of test patterns has to be reduced by inverting only a few selected bit positions, the feedback function has to be modified or, if possible, the order of the bit positions must be rearranged such that only the first few bits are inverted.

5. Experimental Results

The described procedure was applied on the ISCAS-89 sequential benchmark circuits]10]. The first table shows the results for the seven largest S-benchmarks assuming a complete scan design, i.e., the application of the proposed approach on combinational circuits. Table 1 gives an overview on the original, on the minimized test pattern sets and on the size of the basic test pattern. The overall number of test patterns is the product of column 3 and column 5 incremented by 1.

Compared with the original test pattern number the average relative savings is about 83%. Table 1 also shows that these savings are rather independent of the circuit size and also of the test pattern width. More detailed results for combinational circuits including the C-benchmark circuits can be found in [9]. Here a direct comparison with [4] is given that requires over 70% more test patterns than the proposed basic test pattern approach. A direct comparison with [5] is not possible due to the fact that this approach is restricted to small and medium circuits.

The second table shows the resuts of some benchmarks. Since without scan design only a very low fault coverage can be achieved, at least a partial scan is necessary. It should be noted that the application of the basic test sequence approach is totally independent of the scan/non-scan decision. In this article the partial scan selection strategy is tuned on generating acyclic circuit structures as described in [8]. For the resulting modified sequential circuits deterministic test sequences were determined with SPROUT9v [8].

Table 1. Analysis of ISCAS-benchmark circuits with complete scan.

Circuit	Original Number of Det. Test Patterns	Minimized Basic Test Set	Relative Reduction vs. Original Patterns	Bit Positions to be Inverted (absolute)	Bit Positions to be Inverted (relative)	Overall Number of Test Patterns
s5378	315	62	80.3%	97	45.1%	6.076
s9234	481	78	83.8%	146	58.9%	11.466
s13207	578	91	84.3%	200	28.5%	18.291
s15850	500	76	84.8%	259	42.3%	19.760
s35932	74	14	81.1%	168	9.5%	2.336
s38417	1.107	166	85.0%	536	32.2%	89.142
s38584	883	146	83.5%	474	32.4%	69.350
Total	**3.938**	**633**	**83.9%**			

Table 2 contains the necessary number of test patterns, the length and width of each pattern, and the achieved fault coverage relative to the number of detectable faults. The results of the minimization procedure are depicted in Table 3 containing the absolute and relative number of basic test sequences. To determine the relative number of bit positions to be inverted, the overall number of bit positions must be taken into account: the product of test pattern width and the test sequence length. To calculate the overall number of test patterns we have to multiply the number of bit positions to be inverted (incremented by 1 due to the unchanged test sequence) the test sequence length and the primary inputs. This number also includes the pseudo-primary inputs of the scan elements and therefore is equal to the test pattern width.

In summary, an average reduction of about 80% of the test sequence number could be reached. This distinctly reduces the storage requirements and enables cost-efficient self-test hardware solutions.

The results also show an increased number of test patterns to be applied that obviously could be kept within a reasonable upper limit and have, as a side effect and similar to random test patterns, the possible advantage of covering additional faults not addressed by the original deterministic test sequence.

Based on the achieved results a more detailed cost analysis was performed. In order to avoid cost-intensive test equipment, a realization with field-programmable gate-arrays (FPGAs) will be discussed. This special purpose chip(s) must generate all test pattern sequences and analyze the test responses. The latter task can be realized by adding a conventional linear feedback shift register (LFSR) of sufficient length (e.g., 32) to minimize the error mask probability. While this part of the test chip may be the same for several designs, the pattern generating hardware will be based on a single design. Since subsequently each design needs its own test chip for test pattern genration and test response analysis, a FPGA realization with XILINX was selected.

Table 4 shows the number of combinational logic blocks (CLBs) necessary to realize the modules of the test chip. The number of the 3 shift registers is based on the assumption that two flip-flops can be realized within one CLB. Since each CLB has the ability to store up to 32 bit, the product of test pattern number and pattern width can be divided by 32 to compute the CLB count. For the necessary address decoding a 20% overhead is assumed. The depicted comparison shows that the theoretical results on the test lengths also result in similar reduction in the test hardware.

Based on the CLB counts for the test chip realization the final table gives a direct comparison of the appropriate XILINX devices. In each case the device size could be reduced. The relative cost savings are based on the suggested retail prices for the corresponding devices. However, the given number should be interpreted as rough estimation values.

6. Conclusion

In this article a new strategy was presented to select a minimal set of deterministic test pattern sequences as a basis for specific cost-efficient self-test hardware. The described procedure can be realized independently of the used test pattern generation algorithm. Furthermore, based on the basic deterministic test pattern sequence the test generation hardware can be synthesized

Table 2. Characteristics of ISCAS-benchmark circuits with partial scan.

Circuit	Primary Inputs	Flipflops	Deterministic Test Pattern Sequences (SPROUT9v)	Test Sequence Length	Test Pattern Width	Fault Coverage [%]
s382	18	6	53	2	18	100.0
s444	18	6	46	2	18	100.0
s641	50	4	71	2	50	100.0
s713	50	4	85	2	50	100.0
s953	22	23	134	2	22	100.0
s1196	15	17	205	4	15	100.0
s1238	15	17	217	4	15	99.4
s1423	88	3	89	2	88	100.0
s5378	88	126	291	15	88	97.3
s9234	171	76	455	5	171	98.7

Table 3. Analysis of ISCAS-benchmark circuits with partial scan.

Circuit	Number of Basic Test Sequences	Relative Reduction vs. Original Patterns	Bit Positions to be Inverted (absolute)	Bit Positions to be Inverted (relative)	Overall Number of Test Patterns to be Applied
s382	10	81.1%	19	52.8%	720
s444	12	73.9%	20	55.6%	756
s641	17	76.1%	31	31.0%	3.200
s713	17	80.0%	33	33.0%	3.400
s953	28	79.1%	34	77.3%	1.540
s1196	46	77.6%	36	60.0%	2.220
s1238	47	78.3%	32	53.3%	1.980
s1423	25	71.9%	51	29.0%	9.152
s5378	59	79.7%	135	10.2%	179.520
s9234	102	77.6%	201	23.5%	172.710

Table 4. XILINX realization of the test chips.

	Original Deterministic Patterns			Basic Test Patterns			
Circuit	RAM Size	Registers + LFSR	Total Number of CLBs	RAM Size	Registers + LFSR	Total Number of CLBs	Relative Reduction [%]
s381	72	78	150	14	45	59	60.7
s444	62	71	133	16	49	65	51.1
s641	266	112	378	64	75	139	63.2
s713	319	126	445	64	75	139	68.8
s953	221	161	382	46	83	129	66.2
s1196	461	434	895	104	208	312	65.1
s1238	488	458	946	106	212	318	66.4
s1423	587	149	736	165	110	275	62.4
s5378	14.405	2.243	16.648	2.921	945	3.866	76.8
s9234	14.588	1.239	15.827	3.270	612	3.882	75.5

Table 5. Comparison of necessary XILINX devices.

Circuit	XILINX Device (convent. approach)	XILINX Device (new approach)	Relative Cost Savings [%]
s382	XC4005	SC4002	49.5
s444	XC4004	XC4003	10.8
s641	XC4010	XC4004	81.9
s713	XC4010	XC4004	81.9
s953	XC4010	XC4004	81.9
s1196	*	XC4008	can
s1238	*	XC4008	be
s1423	*	XC4008	realized
s5378	*	*	cannot be
s9234	*	*	realized

very easily. It can be shown that the storage requirements could be drastically reduced, on average by about 80%, compared with the original number of deterministic test sequences. Similar cost savings could be obtained if FPGA realizations are compared. The resulting overall number of test patterns is increased, but as the results show the test application time can be kept within a reasonable limit (under 1 second assuming a MHz clock frequence).

Besides the shown technique of using a ROM module for storing the basic test sequences, also other alternative approaches can also be applied to generate these sequences. Since most of these solutions are limited to a small number of test patterns, the achieved reduction enhances the application range.

Further reductions in the test hardware can be achieved if the described approach is combined with random test pattern sequences realizable by an additional linear feedback shift register.

Acknowledgments

The author would like to thank Ulrich Weinmann and Wolfgang Rosenstiel for their valuable contributions to this work. I also want to express my appreciation to André Hergenhan for his implementation of a first synthesis system that generates xnf-descriptions of the resulting test chip.

References

1. W. Daehn, *Deterministische Testmustergeneratoren für den Selbsttest von integrierten Schaltungen*, Ph.D. Thesis, University of Hannover, 1983.
2. S.B. Akers et al., "On the Role of Independent Fault Sets in the Generation of Minimal Test Sets," *Proc. International Test Conference 1987*, pp. 1100–1107, Sept. 1987.
3. P. Goel and B.C. Rosales, "Test Generation and Dynamic Compaction of Tests," *Test Conference 1979, Digest of Papers*, pp. 189–192, 1979.
4. I. Pomeranz, L.N. Reddy, and S.M. Reddy, "COMPACTEST: A Method to Generate Compact Test Sets for Combinational Circuits, *Proc. Int. Test Conference 1991*, pp. 194–203, Oct. 1991.
5. G. Tromp, "Minimal Test Sets for Combinational Circuits," *Proc. International Test Conference 1991*, pp. 204–209, Oct. 1991.
6. M. Abramovic et al., *Digital Systems Testing and Testable Designs*, Computer Science Press, New York, 1990.
7. J.A. Waicukauski et al, "ATPG for ULTRA-Large Structured Designs," *Proc. International Test Conference 1990*, pp. 44–51, Sept. 1990.
8. A. Kunzmann and H.-J. Wunderlich, "An Analytical Approach to the Partial Scan Problem," *JETTA*, vol. 1, pp. 163–174, 1990.
9. A. Kunzmann, "Generation of Deterministic Test Patterns by Minimal Basic Test SEts," *EuroDAC*, Sept. 1992.
10. F. Brglez, D. Bryan, and K. Kozminski, "Combinational Profiles of Sequential Benchmark Circuits," *Int. Symposium on Circuits and Systems*, 1989, pp. 1929–1934, May 1989.
11. A. Kunzmann, "FPGAs for Deterministic Test Pattern Generation," Workshop on Programmable Logic & Applications, Vienna, Aug. 1992.

Arno B. Kunzmann received the diploma degree and the Dr. rer. nat. (Ph.D.) degree in computer science from the University of Karlsruhe, in 1982 and 1989, respectively. Since 1990 he joins the Computer Science Research Center karlsruhe (FZI). Here he is head of the department, Microelectronic System Design. His research activities include design for testability methods, especially self-test strategies, and algorithms for deterministic test pattern generation.

Frank Boehland received the diploma degree in computer science from the University of Karlsruhe in 1994. His current research activities are concentrated on ATPG algorithms for both stuck-at and delay faults.

Subject Index*

*Page numbers in this index are the ones appearing at the bottom of pages.

MIX
Papier aus verantwortungsvollen Quellen
Paper from responsible sources
FSC® C105338

If you have any concerns about our products,
you can contact us on
ProductSafety@springernature.com

In case Publisher is established outside the EU,
the EU authorized representative is:
Springer Nature Customer Service Center GmbH
Europaplatz 3, 69115 Heidelberg, Germany

Printed by Libri Plureos GmbH
in Hamburg, Germany